2

Springer Series on
Chemical Sensors and Biosensors

Methods and Applications

Series Editor:
O.S. Wolfbeis

Vladimir M. Mirsky

Ultrathin Electrochemical Chemo- and Biosensors

Technology and Performance

 Springer

Chemical sensors and chemical biosensors are becoming more and more indispensable tools in life science, medicine, chemistry and biotechnology. The series covers exciting sensor-related aspects of chemistry, biochemistry, thin film and interface techniques, physics, including opto-electronics, measurement sciences and signal processing. The single volumes of the series focus on selected topics and will be edited by selected volume editors. The Springer Series on Chemical Sensors and Biosensors aims publish state-of-the-art articles that can serve as invaluable tools for both practitioners and researchers active in this highly interdisciplinary field. The carefully edited collection of papers in each volume will give continuous inspiration for new research and will point to existing new trends and brand new applications.

ISSN 1612-7617
ISBN 3-540-21285-X
Springer-Verlag Berlin Heidelberg New York

Library of Congress Control Number 2004109052

Springer-Verlag is a part of Springer Science+Business Media

springeronline.com

© Springer-Verlag Berlin Heidelberg 2004
Printed in Germany

The use of general descriptive names, registered names, trademarks, etc. in this publication does not imply, even in the absence of a specific statement, that such names are exempt from the relevant protective laws and regulations and therefore free for general use.

Typesetting: medio Technologies AG, Berlin
Cover: design & production, Heidelberg

Printed on acid-free paper   52/3020xv – 5 4 3 2 1 0

Series Editor

Professor Dr. Otto S. Wolfbeis
University of Regensburg
Institute of Analytical Chemistry, Chemo- and Biosensors
Universitätsstraße 31
93053 Regensburg
Germany
e-mail: wolfbeis@chemie.uni-regensburg.de

Volume Editor

Dr. Vladimir M. Mirsky
University of Regensburg
Institute of Analytical Chemistry, Chemo- and Biosensors
Universitätsstraße 31
93053 Regensburg
Germany
e-mail: vladimir.mirsky@chemie.uni-regensburg.de

Preface

The traditional concept of biological sensors, based on enzymatic receptors and potentiometric or amperometric transducers has undergone several generations of development. Such types of biosensors have been extensively reviewed, described in many textbooks and commercialized. This book is focused on alternative types of chemical and biological sensors or sensor-like structures and approaches, exploring electrical or electrochemical signal detection. Special attention is paid to applications of linear and nonlinear impedance. Some basic ideas in this field are very old – first described, for example, in the classical work by Warburg at the end of the 19th century. Later impedance spectroscopy became a popular approach for studying adsorption of organic molecules on polarizable metal electrodes. However, analytical applications of this approach have only been developed over the last decade, after the establishment of the technology of self-assembled monolayers. In that time, when many scientists were disappointed with attempts to use the Langmuir-Blodgett technique for manufacturing electrochemical devices, the self-assembled monolayers became a viable technology for immobilization of organic molecules on electrodes and for the formation of covalently stabilized receptor layers and even more sophisticated organic nano- and microstructures. This resulted in the development of numerous analytical applications of impedometric methods which are the main topic of the present book.

The book consists of four parts. The first is focused on the analysis of two alternative types of receptors for affinity sensors – antibodies and molecularly imprinted polymers ('plastibodies'); the first approaches are reviewed by Kramer and Hock, the second by Haupt. At least for small analytes, competition between these receptor types is probably near its critical point: inexpensive and highly stable plastibodies have become more and more selective and sensitive but still have not reached the selectivity and sensitivity of antibodies, especially of the recombinant antibodies optimized by genetic engineering. Notably, during the last several years, the first impedometric chemical sensors based on ultrathin layers of molecularly imprinted polymers have been developed.

The main topic of the second part of the book is the application of impedometric transducers in chemical and biological sensors. Berney reviews direct capacitive detection with metal and semiconductor electrodes. Katz and Willner present a detailed analysis of different detection and amplification schemes as well as numerous applications in immunosensors and in DNA sensors of various types. Schöning describes a new principle of analysis of heavy metals based on scattering of conductive electrons on atoms of the analyte. This detection principle has been known for 30 years to work for mercury vapor; its combination with voltage sweep led to a general technique for analysis of heavy metal ions and their mixtures. Lisdat analyzed electrochemical approaches for development of sensors for two biologically important radicals – nitric oxide and superoxide.

The third part of the book is focused on application of thin layer sensing structures for non-invasive monitoring of living cells. Brischwein et al. describe the design and application of miniaturized chip-based sensing systems for multiparameter investigations of single cells growing directly on the chip surface. Farace and Vadgama report on the combination of conducting polymers and impedance measurements. Janshoff, Steinem and Wegener analyze the application of impedance spectroscopy, piezoelectric approaches, as well as a combination of these techniques for analysis of cell-cell and cell-substrate contacts.

The fourth part of the book comprises applications of model lipid membranes for biosensing. Knoll et al. describe design and analytical applications of covalently stabilized tethered lipid membranes. Sokolov and Mirsky analyze experimental approaches for the analysis of electrostatic potentials of lipid membranes and applications of these techniques to sensors, drug discovery, and in investigation of drug-membrane interactions; essential attention is paid to the application of non-linear impedance. The review by Hianik extends this topic to application of non-linear impedance to the analysis of mechanical properties of supported lipid membranes as well as to analytical applications of this technique. Fendler et al. report applications of model lipid systems to the electrical investigation of membrane transport proteins and describe a device for functional screening of these proteins in drug discovery.

This volume covers a fast developing area of science. This inevitably leads to variations in opinions, evaluations and outlooks in presentations of different experts. Therefore, some overlap between the different contributions of this book is complementary rather than being repetitive. The combination of the comprehensive analysis and straightforward description of main concepts and approaches, allows one to hope that this book will be interesting not only for scientists working in- and near the described fields, but also for students studying analytical chemistry, biophysics, biochemistry, electrochemistry, material science, and micro- and nanotechnology.

July 2004, Vladimir M. Mirsky

Contents

Part II: Impedometric and amperometric chemical and biological sensors

Chapter 3
Capacitance Affinity Biosensors
HELEN BERNEY

Chapter 4
Immunosensors and DNA Sensors Based on Impedance Spectroscopy
EUGENII KATZ, ITAMAR WILLNER

Chapter 5
"Voltohmmetry" – a New Transducer Principle for Electrochemical Sensors 117

MICHAEL J. SCHÖNING

Chapter 6
Electrochemical Sensors for the Detection of Superoxide and Nitric Oxide –
Two Biologically Important Radicals 141

F. LISDAT

Part III: Non-invasive electrical monitoring of living cells

Chapter 7
Living Cells on Chip: Bioanalytical Applications

MARTIN BRISCHWEIN, HELMUT GROTHE, ANGELA M. OTTO,
CHRISTOPH STEPPER, THOMAS WEYH, BERNHARD WOLF

Chapter 8
Bioanalytical Application of Impedance Analysis: Transducing in Polymer-Based Biosensors and Probes for Living Tissues

G. FARACE, P. VADGAMA

Chapter 9
Noninvasive Electrical Sensor Devices to Monitor Living Cells Online

ANDREAS JANSHOFF, CLAUDIA STEINEM, JOACHIM WEGENER

Part IV: Lipid membranes as biosensors

Chapter 10
Functional Tethered Bilayer Lipid Membranes

WOLFGANG KNOLL, KENICHI MORIGAKI, RENATE NAUMANN,
BARBARA SACCÀ, STEFAN SCHILLER, EVA-KATHRIN SINNER

Contributors

Ernst Bamberg
Max Planck Institute for Biophysics
Marie-Curie-Straße 15
60439 Frankfurt/Main
Germany
e-mail: ernst.bamberg@mpibp-frank-furt.mpg.de

Helen Berney
INEX
Herschel Building Annex
University of Newcastle
Newcastle upon Tyne
NE1 7RU
United Kingdom
e-mail: helen.berney@ncl.ac.uk

Martin Brischwein
Heinz Nixdorf Chair of Medical
Electronic
Technical University of Munich
Theresienstraße 90
80333 Munich
Germany
e-mail: martin.brischwein@ei.tum.de

Jan Joep H.H.M. De Pont
Department of Biochemistry
Nijmegen Center for Molecular Life
Sciences
Nijmegen
The Netherlands

Wolfgang Dörner
IonGate Biosciences GmbH
Industriepark Höchst D528
65926 Frankfurt/Main
Germany
e-mail: doerner@iongate.de

Giosi Farace
IRC in Biomaterials
Queen Mary University of London
Mile End Rd.
London, E1 4NS
United Kingdom

Klaus Fendler
Max Planck Institute for Biophysics
Marie-Curie-Straße 15
60439 Frankfurt/Main, Germany
e-mail: klaus.fendler@mpibp-frank-furt.mpg.de

Helmut Grothe
Heinz Nixdorf Chair of Medical
Electronic
Technical University of Munich
Theresienstraße 90
80333 Munich
Germany

Karsten Haupt
University of Technology of
Compiègne
B.P. 20529
60205 Compiégne cedex
France
e-mail: karsten.haupt@utc.fr

TIBOR HIANIK
Department of Biophysics and
Chemical Physics
Comenius University
Mlynská Dolina F1
84248 Bratislava,
Slovak Republic
e-mail: hianik@fmph.uniba.sk

B. HOCK
Chair of Cell Biology
Center of Life Sciences
Technical University of Munich
Alte Akademie 12
85350 Freising
Germany

ANDREAS JANSHOFF
Institute for Physical Chemistry
Johannes-Gutenberg-University
Jakob-Welder-Weg 11
55128 Mainz
Germany

EUGENII KATZ
Institute of Chemistry
The Hebrew University of Jerusalem
Jerusalem 91904
Israel
e-mail: ekatz@vms.huji.ac.il

BELA KELETY
IonGate Biosciences GmbH
Industriepark Höchst D528
65926 Frankfurt/Main
Germany
e-mail: kelety@iongate.de

WOLFGANG KNOLL
Max-Planck-Institute of Science for
Polymer
Ackermannweg 10
55128 Mainz
Germany
e-mail: knoll@mpip-mainz.mpg.de

M. KLINGENBERG
Institute for Physical Biochemistry
University of Munich
80336 Munich
Germany

KARL KRAMER
Chair of Cell Biology
Center of Life Sciences
Technical University of Munich
Alte Akademie 12
85350 Freising
Germany
e-mail: kramer@wzw.tum.de

GERARD LEBLANC
Laboratory Jean Maetz
CEA LRC16V
University of Nice Sophia-Antipolis
CNRS ERS1253, BP 68
06238 Villefranche-sur-mer cedex
France

FRED LISDAT
Institute of Biochemistry and Biology
University of Potsdam
Karl-Liebknecht-Straße 24–25
14476 Golm
Germany
e-mail: flisdat@rz.uni-potsdam.de

VLADIMIR M. MIRSKY
University of Regensburg
Institute of Analytical Chemistry,
Chemo- and Biosensors
Universitätsstraße 31
93053 Regensburg
Germany
*e-mail: vladimir.mirsky@chemie.uni-
regensburg.de*

KENICHI MORIGAKI
Special Division for Human Life
Technology
National Institute of Advanced
Industrial Science and Technology
(AIST)
Midorigaoka, Ikeda 563-8177
Japan

RENATE NAUMANN
Max-Plank-Institute of Science
for Polymer
Ackermannweg 10
55128 Mainz
Germany

ANGELA M. OTTO
Heinz Nixdorf Chair of Medical
Electronic
Technical University of Munich
Theresienstraße 90
80333 Munich
Germany

BARBARA SACCA
Max-Plank-Institute of Science
for Polymer
Ackermannweg 10
55128 Mainz
Germany

STEFAN SCHILLER
Max-Plank-Institute of Science
for Polymer
Ackermannweg 10
55128 Mainz
Germany

MICHAEL. J. SCHÖNING
University of Applied Sciences
Aachen (Div. Jülich) and
Research Centre Jülich
Institute of Thin Films and Interfaces
Ginsterweg 1
52428 Jülich
Germany
e-mail: m.j.schoening@fz-juelich.de

EVA-KATHRIN SINNER
Department Membran-Biochemical
Max-Planck-Institute for Biochemical
Am Klopferspitz 18A
82152 Martinsried
Germany

VALERI S. SOKOLOV
Frumkin Institute of
Electrochemistry
Russian Academy of Sciences
117071 Moscow
Russia
e-mail: sokolov@elchem.ac.ru

CLAUDIA STEINEM
University of Regensburg
Institute of Analytical Chemistry,
Chemo- and Biosensors
Universitätsstraße 31
93053 Regensburg
Germany

CHRISTOPH STEPPER
Heinz Nixdorf Chair of Medical
Electronic
Technical University of Munich
Theresienstraße 90
80333 Munich
Germany

PANKAJ VADGAMA
IRC in Biomaterials
Queen Mary
University of London
Mile End Rd.
London, E1 4NS
United Kingdom
e-mail: p.vadgama@qmul.ac.uk

JOACHIM WEGENER
Institute for Biochemistry
Westfaelische Wilhelms-University
Wilhelm-Klemm-Str. 2
48149 Münster
Germany
e-mail: wegenej@uni-muenster.de

THOMAS WEYH
Heinz Nixdorf Chair of Medical Electronic
tronic
Technical University of Munich
Theresienstraße 90
80333 Munich
Germany

ITAMAR WILLNER
Institute of Chemistry
The Hebrew University of Jerusalem
Jerusalem 91904
Israel
e-mail: willnea@vms.huji.ac.il

BERNHARD WOLF
Heinz Nixdorf Chair of Medical
Electronic
Technical University of Munich
Theresienstraße 90
80333 Munich
Germany
e-mail: bernhard.wolf@ei.tum.de

Part I
Receptors

Antibodies for Biosensors

K. Kramer, B. Hock

Abstract. Biosensor sensitivity and selectivity essentially depend on the properties of the biorecognition elements to be used for analyte binding. Immunosensors employing antibodies as binding proteins are effective tools for the analysis of a wide variety of analytes from the group of antigens and haptens. Genetic engineering provides an elegant way not only for the generation of virtually unlimited amounts of biorecognition molecules, but also for the alteration of existing properties and supplementation with additional functions. Different strategies for the synthesis of antibody fragment libraries, followed by the selection of specific antibody variants, were examined. An antibody library was derived from a set of B cells. Chain shuffling of the antibody heavy and light chains provided the improved binders. An ELISA was achieved for the herbicide atrazine with an IC_{50} of 0.9 µg/l and a detection limit of 0.2 µg/l, which corresponds to an equilibrium dissociation constant of 7.46×10^{-10} M as determined by a surface plasmon resonance-based biosensor. The close relations between the optimization of recombinant antibodies by evolutionary strategies and genetic algorithms are considered.

Keywords: Biosensors, Immunosensors, Recombinant antibodies, Ab libraries, s-Triazines

1
Introduction

Biosensors take advantage of biomolecular interactions and have been used in the past most frequently in the field of clinical analysis and biochemistry [1–3]. More recently, biosensor applications have extended to environmental analysis [4, 5]. The biological structures required for the selective receptor unit of the sensor are usually derived from subcellular components and include enzymes, antibodies (Ab), and hormone receptors. More recently, DNA, membrane components, and even organelles or intact cells have complemented the set of components available for biorecognition. This review is focused on Ab-based biosensors, known as immunosensors.

The proper functioning essentially depends on the immobilization of the respective ligands, Ab or coating conjugates, on the sensor surface, their correct orientation and homogeneity. It is obvious that recombinant approaches for their synthesis can accomplish the demand for homogeneous preparations in virtually unlimited amounts. In addition they are also intensively used for the modification of available structures, e.g., for the attachment of anchor groups, change of binding properties, or improvement of stability.

As the ultimate goal of Ab production by the immune system is the recognition and labeling of foreign antigens, immunosensors are used to supplement chemical analysis. Their main strength is a potentially high affinity and selectivity toward the analyte(s). Consequently Ab engineering is a powerful tool for modifying Ab properties.

2
Structure and Function of Antibodies

The key reaction at the recognition part of immunosensors is based on the interaction between the Ab molecule and the corresponding antigen as ligand. The potential of immunosensors is largely determined by the selectivity and affinity of the individual Ab applied for the detection of a particular ligand. Ab molecules are members of the immunoglobulin superfamily. They consist of two identical heavy chains and two identical light chains (Fig. 1). Two antigen-binding F_{ab} fragments are assembled, each by the complete light chain combined with the corresponding part of the first two N-terminal heavy-chain domains. Two F_{ab} fragments are linked via flexible hinge regions with the crystallizable fragment F_c, which harbors the remaining constant heavy-chain domains C_H2 and C_H3. The globular structure of Ab domains is caused by the characteristic immunoglobulin fold [7]. Antiparallel β-strands form a typical double layer in each domain, which is stabilized by hydrophobic interactions and conserved intradomain disulfide bonds. The antigen binding site or paratope is localized at the N-terminal moiety of the variable regions. Due to the frequency of sequence variation between different Ab molecules, sections with hypervariable or complementarity determining regions (CDR) and conserved or framework regions can be distinguished for these domains [8, 9]. Each of the variable (V) regions contains three CDR loops that are embedded into four sections of the frame-

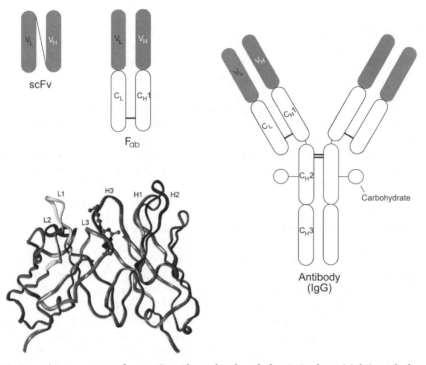

Fig. 1. Domain structure of a native Ab molecule of the IgG class (right) and the corresponding recombinant single-chain Fv (scFv) and F_{ab} fragments. The antigen-binding domains V_L and V_H are depicted in gray. A ribbon model of the Fv domains is shown with the analyte complexed in the Ab binding site. The colored CDR loops are designated as L1–3 for the light-chain and H1–3 for the heavy-chain variable region. The ribbon model was kindly provided by S. Hörsch [6]

work region. In addition to their elevated sequence variability, most of the CDR loops are characterized by significant differences in length [10].

The binding of the antigen is crucially determined by the three CDRs of each V domain. The composition and conformation of the six CDRs (three from the light-chain and three from the heavy-chain V domain) in each paratope determine the topography of the antigen binding site and, hence, the "recognition" of the ligand. Additional interactions by residues of the framework region are reported for discrete ligands [11]. The binding energy between the paratope and the antigenic epitope involves hydrogen bond, hydrophobic, van der Waals, and coulombic interactions as well as calcium bridges [12]. Since these multiple forces are exclusively effective on short molecular distances, they require a complementary topography at the Ab–antigen interface. The affinity and selectivity of an Ab is thereby defined by the total of the intermolecular forces and by the degree of complementarity.

3
Production of Antibodies

Abs are components of the immune system, which are capable of binding antigens with high selectivity. Virtually any component can serve as an antigen if it is able to challenge the immune system. Abs to large molecules are generated by injecting vertebrates with the corresponding macromolecule. Abs to small and therefore nonimmunogenic molecules (haptens) such as pesticides are generated by conjugation to large immunogenic carriers for the immunization. The elicited immune response in both cases is a cascade of lymphocyte cell activation and differentiation, resulting finally in the production of a multitude of Abs that are able to bind the antigen.

3.1
Antibodies Obtained from Sera and Hybridoma Cell Cultures

Abs obtained from sera are traditionally designated as polyclonal Abs. Polyclonal Abs consist of a wide variety of Ab molecules comprising different affinities and selectivities. The antigenicity of the immunogen determines whether or not a significant portion of the serum Abs will be directed against the target molecule. Major advantages of polyclonal Abs are considered in the convenient, fast, and economic production of immunoreagents. However, polyclonal Abs suffer from their heterogenicity and from their limited supply that results in difficulties in the standardization required in immunosensor fabrication. For instance, even if the identical immunogen, immunization route, and donor species are used for the production of polyclonal Abs, the composition of the sera will differ from donor to donor as well as from bleed to bleed. Because polyclonal Abs are produced in vertebrates the amount of serum is limited to the lifespan and to the blood capacity of individual donors.

The technical invention of Köhler and Milstein [13] provided an excellent alternative facilitating the production of monoclonal Abs. The technique enables the propagation of Ab-producing B cells in culture by fusion with a transformed cell line that allows the immortality of the fused cells. The fusion products of Ab-producing B cells and tumor cells are termed hybridomas. They generally continue to secrete Abs. Due to the clonal expression ability, individual hybridoma lines can be selected and cultured after screening for a particular target molecule. Hence, immortal cell lines can be produced that secrete exclusively one type of Ab of defined specificity and affinity. These features of the hybridoma technology permit an unlimited supply of conveniently standardized Abs.

3.2
Recombinant Antibodies

Many applications of Abs have been reported in the area of immunosensor development and application (for latest reviews see [14–17]). But a huge gap exists between the vast number of priority analytes and the limited number of available immunosensors. This can be attributed to the effort which is necessary for the generation of new Abs. Independent of the Ab type, polyclonal Ab (pAb), or monoclonal Ab (mAb), time-consuming immunizations are required for each individual Ab.

For this reason, the generation of recombinant Abs has become part of modern biotechnology. First concepts emerged in the late 1980s [18–21]. Again, progress was originally driven by medical applications and pushed at human Abs. The efforts to generate Abs by recombinant technologies were eventually extended to applications in various areas such as food safety and environmental monitoring [22–25].

Ab engineering is a powerful tool for modifying Ab properties [26–28]. *In vivo*, Ab variable (V) genes encoding the antigen-binding molecular domains are modified during the secondary immune response in the microenvironment of lymphoid germinal centers by somatic hypermutation. Appropriate variants are subsequently selected from this pool of mutant immunoglobulins upon their improved affinity to the antigen [29]. A strategy for mimicking Ab maturation in vitro is based on the evolutionary concept of variation and selection. Similar to the *in vivo* process, the Ab gene is diversified in an initial step by mutational procedures such as nucleotide variation in mutator strains or PCR-based sequence randomization (Fig. 2). In the latter case, the mutated Ab gene repertoire is subsequently cloned into an expression vector and displayed at the protein level, for instance on the surface of bacteria [30], yeast cells [31], filamentous phage [32], or alternatively as stable protein–ribosome–mRNA complexes termed polysomes [33]. The surface presentation of the Ab phenotype enables the selection of improved variants, for example by their enhanced affinity to the antigen. Similar to the natural process, the in vitro evolution is an iterative process. Hence, improved Ab variants can be utilized as template structures for subsequent evolutionary cycles, until they meet the requirements of their designated application.

The following sections deal with two major aspects of recombinant Ab synthesis: generation of an Ab gene library comprising a pool of group-selective Ab sequences, followed by the selection of new variants, and alteration of the affinity profile of an existing Ab by molecular evolution. For this purpose, s-triazine-selective Abs were chosen. The chemical structures of some of the priority target analytes are shown in Fig. 3. In the context of Ab engineering, biosensors provide a convenient tool for determining the kinetic constants of analyte binding.

3.3
Antibody Library Derived from a Set of B Cells

A strategy was developed to generate a group-selective library representing a large Ab gene pool. This library was designed to include a range of s-triazine-specific Abs in order to facilitate the selection of Abs against defined members of the s-triazine family.

Ab-secreting B cells were derived from Balb/c mice, which had previously been immunized with different s-triazine immunogens including derivatives of atrazine, ametryn, terbuthylazine, deethylatrazine, and simazine. B cells secreting s-triazine-selective Ab were separated from unspecific B cells by means of

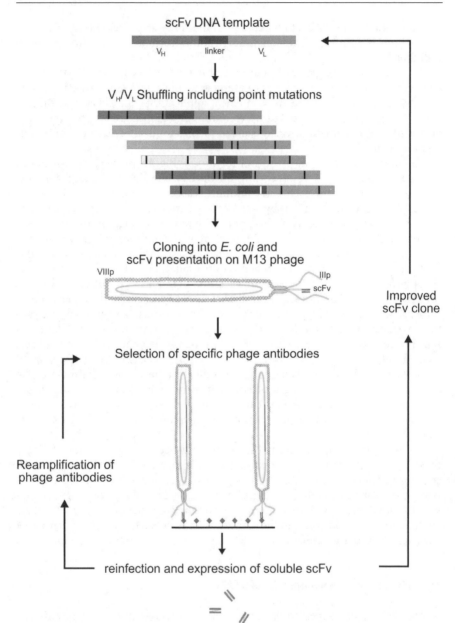

Fig. 2. Principle of molecular Ab evolution. The scFv Ab gene template is randomized by chain shuffling as indicated by V genes in different gray scales. Additional point mutations introduced by error prone PCR are depicted as vertical lines

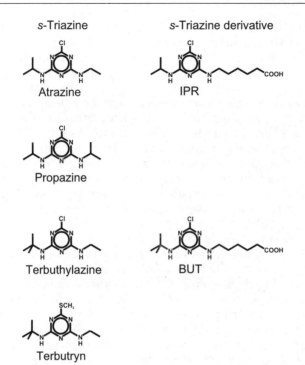

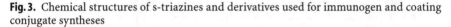

Fig. 3. Chemical structures of s-triazines and derivatives used for immunogen and coating conjugate syntheses

immunomagnetic separation. This method takes advantage of the membrane-bound receptor molecules on the B-cell surface, i.e., transmembrane protein complexes that share their ligand-binding characteristics with the secreted Ab. These surface receptors can be tagged by target molecules covalently linked to paramagnetic particles. After magnetizing the particles in a magnetic field, specific hybridomas are removed from bulk cultures by magnetic force [34].

The magnet-bound B cells derived from several mice were used for mRNA extraction and subsequent cloning of Ab encoding genes into the phage display system (see Fig. 2). The resulting library comprises all target-relevant sequences that have been included in the starting B cell repertoire.

This group-selective Ab library was then used to screen for Ab variants selective for s-triazines containing a tert-butyl group (BUT), i.e., terbutryn and terbuthylazine, as well as those s-triazines bearing an isopropylamino residue (IPR), i.e., atrazine and propazine (see Fig. 3). Specific phages were enriched by three repetitive cycles of selection using immunoaffinity chromatography with BUT derivative-coated and IPR derivative-coated columns. An aliquot of 95 individual clones was removed from the final, third selection cycle and employed for the expression of soluble Ab fragments, which were tested by ELISA for specific displacement by terbutryn and atrazine. Several clones were identified and characterized.

Binding studies for the characterization of Abs can be performed by different means. Enzyme-linked immunosorbent assay (ELISA), fluorescence quenching, and optical sensor designs are most frequently used for this purpose. Changes in the mass bound to the sensor surface are commonly monitored using surface plasmon resonance detection [35]. The antigen or, in the case of haptens, coating conjugate with an exposed analyte derivative is covalently bound to the sensor chip surface, e.g., through carboxymethylated dextran mediated by carbodiimide. Then a stream carrying the binding Ab is directed over the surface. In the case of inhibition binding assays, varying analyte concentrations and the Ab at a concentration roughly equaling the affinity constant is included in the sample. In the absence of analyte the Abs are maximally bound to the coating conjugate. With increasing analyte concentrations decreasing amounts of the binding protein become attached. In the case of direct kinetic analysis, varying Ab concentrations are directed over the sensing surface.

The reaction kinetics of the three best binding clones for triazine herbicides containing tert-butylamino and isopropylamino groups were determined with the BIAcore 2000 sensor. Since the affinity calculation is based on known Ab concentrations, the percentage of functional Abs in the samples was determined according to Kazemier et al. [36]. For this purpose the affinity-purified Ab samples were incubated overnight with BUT and IPR derivatives coupled to sepharose. Then the affinity-bound protein representing the active Ab was removed by centrifugation. The protein concentration of the supernatant and the original affinity-purified Abs was determined spectroscopically. The supernatant contains inactive Abs which are not correctly folded, although soluble, or inactivated by low-pH elution during affinity purification. The percentage of functional Abs ranged between 66% for clone BUT-56 and 81% of total protein for clone BUT-8.

Figure 4 shows representative sensorgrams for the binding of a recombinant Ab, derived from clone BUT-4, to a tert-butyl derivative–ovalbumin conjugate or ovalbumin as a control. The binding constants (Table 1) are in the range of affinity-matured Abs that are obtained during the secondary immune response.

The mutants bearing the highest affinity to terbutryn are located within a very close area of the sequence space, as revealed by the few amino acid deviations within the group of terbutryn-selective variants (six positions with amino acid differences, data not shown). This sequence area probably occupies a unique peak position for terbutryn selectivity within the Ab repertoire enclosed in the library.

Isopropylamino-selective clones were isolated from the group-specific Ab library in a similar way to that described above for the terbutryn-selective clones. Following three repetitive cycles of selection with deethylatrazine columns, an aliquot of 95 individual clones derived from the third selection cycle was employed for the expression of soluble Ab fragments and tested by ELISA for deethylatrazine selectivity. Several clones were identified revealing specific binding to the IPR–ovalbumin conjugate. The clones IPR-7 and IPR-32 could be selectively displaced by atrazine as indicated by the corresponding calibration curves (Fig. 5).

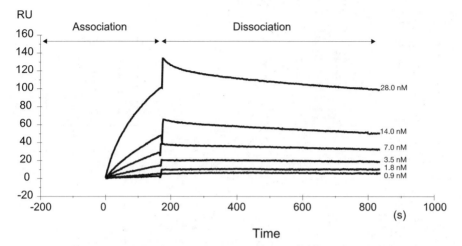

Fig. 4. Sensorgrams of scFv binding (clone 4). CM5 chips were coated with 4,000 RU terbuthylazine-ovalbumin conjugate (channel 2) or ovalbumin (channel 1) as a control using amine coupling. Affinity measurements were performed at 25 °C with 1–28 nM of active scFv (diluted in PBS containing 0.005% Tween 20) for 3 min using a flow rate of 30 µl/min

Table 1. Association rate k_a, dissociation rate k_d, and equilibrium dissociation constant K_D of Ab for terbuthylazine obtained from the clones BUT-4, BUT-8, and BUT-56

Clone	k_a (M^{-1}s^{-1})	k_d (s^{-1})	K_D (M)
BUT-4	8.49×10^3	2.87×10^{-4}	3.38×10^{-8}
BUT-8	3.51×10^3	2.49×10^{-4}	7.09×10^{-8}
BUT-56	4.09×10^3	3.28×10^{-4}	8.01×10^{-8}

4
Diversification by Chain Shuffling

The alteration of the affinity profile of existing Abs by genetic engineering [37] is considered a powerful instrument to circumvent classical Ab production schemes, which require extended immunization periods. The main purpose of generating the group-selective recombinant Ab library presented above was to utilize its gene repertoire as a source for subsequent fine-tuning of Ab molecules by directed evolution, rather than to provide Ab variants from the cloned V_H–V_L pairings. It was expected that this library would substantially facilitate the engineering of desired Ab specificities and affinities to any member of the triazine group without the need for new immunizations.

The highest affinity obtained from the group-selective library was reached by the Ab clone IPR-7 with an IC$_{50}$ of 14 µg/l for the detection of atrazine (see

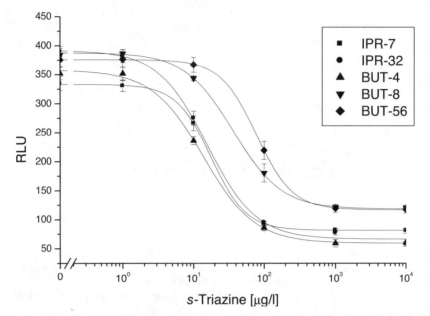

Fig. 5. Calibration curves of IPR and BUT clones for the analysis of atrazine and terbutryn, respectively, by direct, heterogeneous, competitive ELISA. IPR and BUT clones were selected by phage display technology employing s-triazine derivatives, which contain isopropylamino (IPR) and butylamino (BUT) residues. RLU: relative luminescence units. Experimental: Polystyrene microtiter plates (Greiner, Germany) were precoated overnight with 150 µl of anti-E-tag antibody (Pharmacia; 5 µg/ml in carbonate buffer: $NaHCO_3$/ Na_2CO_3, 50 mM, pH 9.6) at 4 °C. All further reactions were carried out at room temperature. Following three wash steps with PBS–Tween 20 (4 mM phosphate-buffered saline, pH 7.2, supplemented with 0.05% Tween 20), 150 µl affinity-purified scFv was filled into each well. Unbound scFv was removed by rinsing three times with PBS–Tween 20 after 2 h incubation. Triazine standard or PBS (100 µl) were added together with 50 µl enzyme tracer (BUT– and IPR–horseradish peroxidase conjugate) and incubated for 1 h. After a final washing step, 150 µl of Super Signal ELISA Pico chemiluminescent substrate (Pierce, USA) was pipetted into each well. The luminescence was measured after 5 min shaking incubation at 430 nm with a Tecan reader

Fig. 5). However, cross-reactivity estimates for a selection of 12 s-triazines indicated that this clone preferentially binds to sebuthylazine and propazine, whereas atrazine is recognized only to a lower extent (78% cross-reactivity to atrazine, if sebuthylazine is arbitrarily set to 100%).

Starting with this clone, an evolutionary strategy was applied to searching for a variant with superior affinity to atrazine (100% cross-reactivity) and improving the sensitivity in the ELISA performance by at least one order of magnitude. For this purpose chain shuffling (see Fig. 2) was applied as a combinatorial approach. This was carried out with the aid of splicing by overlap extension. For V_H-shuffling the light chain of the template Ab IPR-7 and the heavy-chain repertoire of the group-specific library (see above) were amplified in two separate PCRs. Then the light-chain amplification product was assembled with the amplified heavy-chain repertoire in a subsequent PCR, resulting in combinations of the template light chain with all heavy-chain variants enclosed in the group-specific library. This library was subsequently selected by phage display and screened for the best binder to atrazine. Then, the heavy chain of the best binder was shuffled against the library light-chain repertoire.

The kinetic characteristics of the shuffled IPR clones were determined by surface plasmon resonance (BIAcore 2000) as already described above. The kinetic data of the scFv variants are detailed in Table 2. The equilibrium dissociation constants of the scFv variants are approaching the typical K_D level of affinity-matured Abs *in vivo* [38]. The 1.4-fold improvement in K_D from IPR-7 to IPR-26 is in line with the 1.5-fold diminished IC_{50} and the 1.6-fold improved detection limit in the corresponding atrazine ELISAs (see Fig. 6). The molecular evolution from IPR-26 to IPR-83 resulted in a 12.3-fold reduction of the K_D, which is mirrored by a 10.1-fold decrease of the IC_{50} atrazine concentration and a 16.0-fold decline of the detection limit (see Fig. 4). The overall improvement in the K_D from the template Ab IPR-7 to the optimized variant IPR-83 by a factor of 17 is consistent with the 15-fold lowered atrazine concentration for the IC_{50}. The detection limit was even diminished by a factor of 25. The optimized mutant Ab IPR-83 provided a preferred recognition of atrazine and an IC_{50} of 0.9 µg/l in the corresponding atrazine ELISA (see Fig. 6). In this way, a detection limit of 0.2 µg/l atrazine was reached.

Table 2. Association rate constant k_a, dissociation rate constant k_d, and equilibrium dissociation constant K_D for IPR obtained from the scFv clones IPR-7, IPR-26, and IPR-83. The values for k_a, k_d, and K_D were measured utilizing the BIAcore 2000 system

Clone	k_a ($M^{-1}s^{-1}$)	k_d (s^{-1})	$K_D = k_d/k_a$ (M)
IPR-7	1.38×10^5	1.75×10^{-3}	1.27×10^{-8}
IPR-26	2.10×10^5	1.93×10^{-3}	9.20×10^{-9}
IPR-83	6.73×10^5	5.02×10^{-4}	7.46×10^{-10}

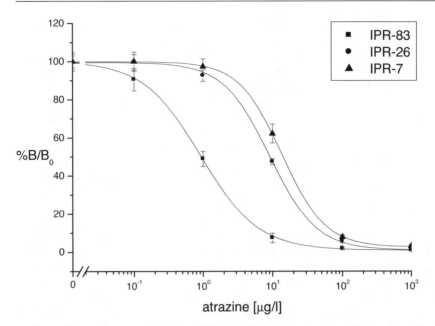

Fig. 6. Calibration curves of IPR-7 and $V_{H/L}$-shuffled clones for the analysis of atrazine by direct, heterogeneous, competitive ELISA. The original curves, which are based on relative luminescence units (RLU), were normalized according to $\%B/B_0=(L-L_{excess})/(L_0-L_{excess})$, where L is the luminescent signal, L_{excess} the luminescent signal at excess of atrazine, and L_0 the luminescent signal at zero dose of atrazine. L_0 corresponds to the maximum achieved signal of the calibration curve and was determined as 362 RLU for IPR-7, 647 RLU for IPR-26, and 407 RLU for IPR-83. L_{excess} corresponds to the minimum achieved signal of the calibration curve and was determined as 61 RLU for IPR-7, 48 RLU for IPR-26, and 63 RLU for IPR-83. Experimental according to Fig. 5

5
Antibody Optimization as a Special Case of Genetic Algorithms

Evolutionary strategies for Ab optimization can be comprehended as a special case of genetic algorithms. This optimization theory was introduced in 1975 by Holland [39] in order to provide a general framework for viewing natural as well as artificial adaptive systems, and show how the evolutionary process can be applied to artificial systems. The particular efficiency and robustness of this approach relates to the fact that it is a highly parallel algorithm that introduces diversification at the coding level (i.e., DNA) rather than dealing directly with functions (protein affinities, cross-reactivities) [40]. This means that a population of character strings patterned after DNA sequences, each with an associated fitness value, is transformed into a new population of character strings that is subjected to screening at the functional level (i.e., at the protein level).

This strategy is mirrored in Ab optimization with the aid of phage display. The key steps in the case of Ab optimization are (1) generation of an Ab library representing a large repertoire of Ab genes, (2) selection of the highest affinity Ab by phage display (this step links the selection of the highest affinity Ab to the recovery of the gene encoding this protein), and (3) an affinity maturation strategy for improving and manipulating the selected Ab [41]. It is evident that this approach mimics the highly successful natural process of Ab generation and diversification, which includes reassortment and splicing of DNA fragments from Ab genes as well as somatic mutations and selection.

Considering *in vivo* Ab generation, it is obvious that the reach of molecular modeling, which is confined to the protein level, must be limited. The same holds true for Ab engineering restricted to error-prone PCR as the sole means of diversification. It is not necessarily a larger Ab repertoire of the immune system which provides an advantage over recombinant libraries in terms of efficiency and speed of Ab optimization. It is rather the fast screening cycle performed in parallel by a huge number of B cells that are simultaneously subjected to somatic mutation and recombination as basic events of Ab diversification and maturation. Furthermore, natural libraries are continuously supplemented under the selection pressure whereas recombinant libraries are depleted by affinity screening.

6
Biosensors as a Tool in Evolutionary Antibody Synthesis

Biosensors are now considered as standard tools in Ab engineering. The univocal and solid identification of improved variants is one of the key steps in molecular Ab evolution. As demonstrated above, it is advantageous to characterize Ab variants for improved binding in parallel by competitive enzyme-linked immunosorbent assay (ELISA) and by use of a surface plasmon resonance-based optical sensor BIAcore 2000. ELISA is primarily designed to deliver analytical information such as the IC_{50} and detection limit of a calibration for a defined analyte, whereas the biosensor measurements provide kinetic rankings of the Ab variants. These biosensors are proving invaluable tools in protein engineering, particularly in research aimed at the isolation and improvement of protein binders and catalysts from macromolecular repertoires containing billions of individual members [42].

6.1
Antibodies as Receptors in Biosensor Designs

Beyond their use as a tool for Ab characterization, biosensors with Abs or Ab fragments are found in a variety of applications. Ab-based biosensors are termed immunosensors. The broad panel of applications of immunosensors include clinical medicine, food safety, bioprocess monitoring, military applications, environmental monitoring, and more. Since none of the reactants needs to be labeled, artifactual changes in binding properties that often result from labeled molecules are avoided.

Biosensor instruments are well-suited for rapid mapping of viral epitopes and for identifying optimal combinations of capturing and detector Abs in sandwich assays. Biosensor binding data are also useful for selecting peptides to be used in diagnostic solid-phase immunoassays. Thus, very small changes in binding affinity can be measured with considerable precision, which is a prerequisite for analyzing the functional effect and thermodynamic implications of limited structural changes in interacting molecules. On-rate (k_a) and off-rate (k_d) kinetic constants of the interaction between, for instance, a virus and a therapeutic Ab can be readily measured and the equilibrium affinity constant K can be calculated from the ratio $k_a/k_d=K$ [43].

Numerous biosensor techniques have been reported that allow improved studies of the kinetics, structure, and (solid/liquid) interface phenomena associated with protein–ligand binding interactions. In addition, promising application areas for affinity-based biosensor techniques include clinical/diagnostics, food processing, military, and antiterrorism as well as environmental monitoring [17]. Research and development applications of optical biosensors are found in many areas of drug discovery, including target identification, ligand fishing, assay development, lead selection, early ADME, and manufacturing quality control [14].

Biosensors are suitable for real-time monitoring, e.g., in bioreactors, and for the determination of various physiological and pharmacological parameters. They may be employed in home testing (glucose, lactate), hospitals (bedside testing, emergency, surgery, dialysis monitoring etc.), clinical laboratory analyses (immunoassays, DNA analysis), and research centers. The biosensor may be miniaturized for single use or for implementation in sensor arrays. Applications to microenvironments (in vivo, single cell) or discrete one-shot decentralized tests may also be considered [15].

In order to meet the specific requirements of this multitude of application areas, immunosensor formats have been developed for the last two decades. Depending on the method of signal transduction, biosensors may be divided into

Table 3. Categories of immunosensors according to different physical principale

Category	Sensor type
Optical	Surface plasmon resonance
	Grating couplers
	Differential interferometry
	Reflectometric interference spectroscopy
	Double- beam waveguide interferometers
	Resonant mirror (mode coupling)
Electrochemical	Potentiometric
	Immunofield- effect transistors
	Amperometric
	Voltammetric
Mass sensitive	Piezoelectric
	Accoustic surface waves technique
Thermo electric	Thermo electric

four different groups: optical, mass, electrochemical, and thermal sensors (see Table 3). In addition, biosensors can be classified into sensors for direct and indirect (labeled) detection. Analyte binding is directly determined in real time in the first case by measuring the physical changes induced by biospecific immune complex formation. In contrast, a preliminary reaction takes place in indirect sensor formats, where the products are subsequently detected by the sensor [44].

6.2
Limitations of Antibody-Based Biosensor Designs

In contrast to classical sensors, where reversibility is one of the key issues, immunosensors circumvent this property by quasi-reversible approaches. Since high-affinity binding is a fundamental requirement for immunosensors, high sensitivity (meaning low detection limits) and fast reversibility exclude each other. This becomes obvious if the relation between the affinity constant K and the middle of a binding assay is considered. The middle of an assay indicates the range where the assay operates and consequently the sensitivity. It also corresponds to the region where the most accurate measurements are possible.

In the equilibrium reaction between the free analyte H (here a hapten) and the monovalent Ab towards the analyte–Ab complex Hab (=bound analyte)

$$H + Ab = HAb \tag{1}$$

the equilibrium affinity constant K determines the concentration ratio between the Ab-bound hapten and the free reaction partners

$$K = k_a / k_d = [HAb]/[H][Ab]\left(1\,mol^{-1}\right). \tag{2}$$

At the middle of the test, half of the Ab binding sites are occupied. Therefore

$$[HAb] = [Ab]. \tag{3}$$

In this case Eq. 2 becomes

$$K = 1/\left[H_{0.5}\right] \tag{4}$$

with $[H_{0.5}]$=free hapten concentration at 50% occupation of the Ab binding sites (half saturation). It is inversely proportional to the affinity constant. This means that at high K values only low analyte concentrations are required for half saturation.

Therefore it is of practical interest to relate the analyte concentration required for half saturation to K and the Ab concentration. For this purpose, the total hapten concentration $[H_{tot}]$, which is equal to the analyte concentration in the sample, is related to K and the total Ab concentration $[Ab_{tot}]$. Since

$$\left[H_{tot}\right] = \left[H\right] + \left[HAb\right] \tag{5}$$

it can be derived for half saturation (Eq. 3) from Eqs. 4 and 5 that

$$\left[H_{0.5,\,tot}\right]=1/K+\left[Ab\right] \tag{6}$$

$$=1/K+\left[Ab_{tot}/2\right] \tag{7}$$

with [$H_{0.5,\,tot}$]=total hapten concentration (=analyte concentration in the sample) resulting in half saturation. This means that the analyte concentration at half saturation is inversely related to the equilibrium affinity constant K and directly related to the Ab concentration. Optimizing assay sensitivity therefore must consider high affinity Abs and low Ab concentrations.

Several strategies have been used to approach reversible operation in the immunochemical field. The most simple and effective way is a quasi-reversible strategy. It was employed for the first time in the field of immunosensing by the group of Rolf D. Schmid with a sensing device, the flow-injection immunoanalyzer (FIIA [45]). This technology is a further development of FIA (flow-injection analysis), introduced in 1975 by Ruzicka and Hansen [46]. FIA is now widely used as a method for the continuous, automated performance of chemical and biochemical detection methods. It is based on the injection of a liquid sample into a continuous flow of a liquid carrier. The sample is introduced through an injection valve with a loop providing an accurately known volume and is transported to the biosensor.

In the example cited above [45], FIA was combined with a competitive immunoassay for the detection of atrazine. An atrazine–peroxidase conjugate was used as tracer. The central part is a membrane reactor with an immobilized atrazine Ab. The following components are consecutively pumped through the membrane reactor: sample (or standard), enzyme tracer peroxide (H_2O_2), and the second substrate hydroxyphenylpropionic acid. The enzyme activity of the Ab-bound tracer is determined fluorimetrically. Sensitivities were obtained comparable to best immunoassays (down to 1 ng/l atrazine). But even more important, the membrane reactor specifically developed for FIIA enables the automated change of the Ab membrane after the completion of the assay. Therefore this sensing device could in principle be used for the continuous measurement of real samples.

A similar approach termed flow-injection immunoaffinity analysis was developed by Krämer [47]. The central component consists of a column reactor harboring protein A, which is immobilized on glass beads. Anti-pesticide Abs are injected that bind to protein A. This is followed by the consecutive injection of pesticide-containing sample, enzyme tracer, and substrate. The product of the enzymatic reaction is monitored by a flow-through fluorimeter connected with the reactor. Multiple measurements are conveniently feasible by regenerating the protein A affinity. Thus, up to 1,600 repetitive measurements could be achieved without any compression of the analytical quality. A single measurement is performed within 50 min. Applying different Abs enabled the determination of various pesticides at concentrations as low as 20 ng/l for atrazine, 10 ng/l for terbutryn, and 10 ng/l for diuron. This sensor design moves online monitoring within reach.

Immunosensors could also be operated in a quasi-continuous mode. However, the same result is usually achieved by other means, namely the regeneration of free Ab binding sites, which is due to a change of the Ab affinity in the presence of the regeneration medium. Experiences with optical immunosensors (e.g., BIAcore) have shown that up to 100 regeneration cycles can be carried out with a single preparation of immobilized Abs. But it seems to be advantageous to apply this method to immobilized coating conjugates and to use a new aliquot of Abs after each regeneration cycle. This holds particularly true in the case of nonprotein analytes, which are usually less sensitive to the regeneration cycle than Ab binding sites.

An early example was provided by the group of Mascini [48]. An atrazine derivative was immobilized at the surface of a plasmon resonance sensor. Then a stream containing a mixture of the Ab and the sample was directed over the surface. A secondary Ab binding to the Fc moiety of the primary Ab was added in order to enhance the optical signal. The corresponding calibration curve revealed a detection limit of 50 ng/l for the determination of the herbicide atrazine. A single measurement cycle including the regeneration step was completed within 15 min. This example underpins that immunosensors are basically suited for measurements combining both velocity and sensitivity.

Whereas the examples cited above are designed for single analyte analysis, different multianalyte designs utilize Ab patterns. The first multianalyte approach designated as microspot immunoassay was described by Ekins et al. [49]. They suggested that a multitude of different fluorescence-labeled Abs are deposited at defined microspots onto a solid sensor surface. Each spot contains a minimum amount of a distinct Ab. After incubation with the analyte in the sample, a fluorescence-labeled tracer competes with the analyte for the limited number of Ab binding sites. Since Ab and tracer are labeled with different fluorescence markers, the fractional occupancy of the Ab binding sites, and thus the analyte concentration in the sample, can be derived from the fluorescence signal ratio. The signal ratio is independent of the Ab concentration if the microspots contain a very low amount of Ab almost approaching the zero dose. Applying laser scanning confocal fluorescence microscopy, up to 10,000 microspots per cm^2 could be optically screened, for instance on the surface of a silicon chip. Table 4

Table 4. Examples forof Ab-based multi-analyte strategies

Assay	Detection mode	Analytes	Citation
Microspot-Immunoassay	Confocale fluorescence microscopy		Ekins et al. 1990 [49]
RIANA-Sensor	Evanescent field fluorescence	Herbicides (atrazine, 2,4-D, isoproturon)	Brecht et al. 1998 [50]
Biochip	Chemiluminescence; CCD detection	Herbicides (s-triazine, 2,4-D), TNT	Weller et al. 1999 [51]
Immuno-CD	IR fluoreszcence	Pesticides (hydroxyatrazine, carbaryl, molinate)	Kido et al. 2000 [52]

summarizes some multianalyte immunosensors. These include the RIANA sensor, the biochip, and the Immuno-CD, which utilize optical detection modes.

7
Conclusions and Outlook

The advantages of Abs as biorecognition elements in immunosensors are clearly seen in their high affinity toward the analyte and their selectivity, which can conveniently be assayed by immunosensors. In addition, the accessibility of proteins to genetic engineering facilitates their supplementation with additional functions and the alteration of existing properties. Changes of binding properties are particularly attractive in the case of biosensor applications. The generation of group-selective recombinant Ab libraries combined with additional methods of diversification such as chain shuffling provides a huge in vitro Ab repertoire for selecting new binding properties. The examples given in this paper dealt with the optimization of recombinant Abs directed against herbicidal haptens. In this chapter efforts toward high affinity Abs were described. For this purpose a group-selective Ab library was generated with B cells from immunized mice as a source of Ab genes. This library allowed the subsequent generation of Ab fragments with selectivities not contained in the basic library.

Currently the potential of alternative Ab molecules is exploited in order to supplement approaches utilizing the classical Ab scaffold. For instance, single-domain Abs are naturally occurring in camelids and sharks. The large CDR loop of camelids, stabilized by an intraloop disulfide bond, is considered a critical component in providing high affinity for the corresponding single-domain Abs [53]. The extracellular domain of the T cell activation marker, CTLA4, represents a domain that shares common features of Ab variable regions. This protein has been successfully adopted as a scaffold to engineer large CDR loops that can penetrate clefts in potential target antigens such as receptors, enzymes, and viruses [54]. Smaller proteins with Ab-like properties such as anticalins [55] are reported to offer a better access to engineering. An entirely different approach is the application of molecular imprinting technologies (MIPs [56]). Synthetic hexapeptide ligands derived from combinatorial peptide libraries are suitable for large-scale synthesis. These hexapeptides are synthesized on the surface of beads and subsequently screened for target molecule selectivity [57]. They are expected to offer better performance in organic solvents and other extreme conditions. Alternative affinity sensors are based on aptamers, artificial nucleic acid ligands that can be generated against amino acids, drugs, proteins, and other molecules. They are isolated from combinatorial libraries of synthetic nucleic acids by an iterative process of adsorption, recovery, and reamplification. Aptamers, first reported in 1990, are attracting interest in the areas of therapeutics and diagnostics and offer themselves as ideal candidates for use as biocomponents in biosensors (aptasensors), possessing many advantages over state-of-the-art affinity sensors [58].

It remains to be seen which kind of binding molecules are best suited for biotechnological applications. One of the vital goals for the future is the simple and cheap access to evolutionary technologies for tailored binders with predefined properties such as selectivity, affinity, stability, and more.

Acknowledgements: Financial support was obtained from the EC (DG XII Environment and Climate 1994-8, project no. ENV4-CT96-0333, Envirosense project).

References

1. Scheller F, Schubert F (eds) (1989) Biosensoren. Akademie, Berlin
2. Canh TM (1993) Biosensors. Chapman & Hall, London
3. Spichiger-Keller UE (1998) Chemical sensors and biosensors for medical and biological applications. Wiley-VCH, Weinheim
4. Hock B, Barceló D, Cammann K, Hansen P-D, Turner APF (eds) (1998) Biosensors for environmental diagnostics. Teubner, Stuttgart
5. Bilitewski U, Turner APF (eds) (2000) Biosensors for environmental monitoring. Harwood, Amsterdam
6. Hörsch S (1996) Diploma thesis, University of Stuttgart
7. Poljak RJ, Amzel LM, Avey HP, Chen BL, Phizackerley RP, Saul F (1973) Proc Natl Acad Sci USA 70:3305
8. Wu TT, Kabat EA (1970) J Exp Med 132:211
9. Kabat EA, Wu TT, Perry HM, Gottesman KS, Foeller C (1991) Sequences of proteins of immunological interest, 5th edn. US Department of Health and Human Services, Public Health Service, National Institutes of Health (NIH Publication No 91-3242), Washington
10. Padlan EA (1994) Mol Immunol 31:169
11. Tulip WR, Varghese JN, Laver WG, Webster RG, Colman PM (1992) J Mol Biol 227:122
12. Van Oss CJ (1992) Antibody–antigen intermolecular forces. In: Roitt IM, Daves PJ (eds) Encyclopedia of immunology, vol 1. Academic, London, p 97
13. Köhler G, Milstein C (1975) Nature 256:495
14. Cooper MA (2002) Nat Rev Drug Discov 1:515
15. Kauffmann JM (2002) Ann Pharm Fr 60:28
16. Raman SC, Raje M, Varshney GC (2002) Crit Rev Biotechnol 22:15
17. Rogers KR (2000) Mol Biotechnol 14:109
18. Better M, Chang CP, Robinson RR, Horwitz AH (1988) Science 240:1041
19. Skerra A, Plückthun A (1988) Science 240:1038
20. Huse WD, Sashy L, Iverson SA, Kang AS, Alting-Mees H, Burton DR, Benkovic JS, Lerner RA (1989) Science 246:1275
21. McCafferty J, Griffiths AD, Winter G, Chiswell DJ (1990) Nature 348:552
22. Ward VK, Schneider PG, Kreissig SB, Hammock BD, Choudary PV (1993) Protein Eng 6:981
23. Kramer K, Hock B (1996) Food Agric Immunol 8:97
24. Lee N, Holtzapple CK, Stanker LH (1998) J Agric Food Chem 46:3381
25. Li Y, Cockburn W, Kilpatrick J, Whitelam GC (1999) Food Agric Immunol 11:5
26. Boder ET, Midelfort KS, Wittrup KD (2000) Proc Natl Acad Sci USA 97:10701
27. Daugherty PS, Chen G, Iverson BL, Georgiou G (2000) Proc Natl Acad Sci USA 97:2029
28. Hanes J, Schaffitzel C, Knappik A, Plückthun A (2000) Nature Biotechnol 18:1287
29. Rajewsky K (1996) Nature 381:751
30. Georgiou G, Stathopoulos C, Daugherty PS, Nayak AR, Iverson BL, Curtiss L (1997) Nature Biotechnol 15:29
31. Boder ET, Wittrup KD (1997) Nature Biotechnol 15:553
32. Smith GP, Scott JK (1993) Meth Enzymol 217:228
33. Hanes J, Plückthun A (1997) Proc Natl Acad Sci USA 94:4937
34. Kramer K, Giersch T, Hock B (1994) Food Agric Immunol 6:5
35. Johnson B, Löfas S, Lindquist G (1991) Anal Biochem 198:268
36. Kazemier B, de Haard H, Boender P, van Gemen B, Hoogenboom H (1996) J Immunol Meth 194:201

37. Kipriyanov SM, Little M (1999) Mol Biotechnol 12:173
38. Winter G, Griffiths AD, Hawkins RE, Hoogenboom HR (1994) Ann Rev Immunol 12:433
39. Holland J (1975) Adaption in natural and artificial systems. University of Michigan Press, Ann Arbor
40. Goldberg DE (1989) Genetic algorithms in search, optimization, and learning. Addison-Wesley, Boston
41. Irving R, Hudson P (1995). Superseding hybridoma technology with phage display libraries. In: Zola H (ed) Monoclonal antibodies. The second generation. Bios, Oxford, p 119
42. Huber A, Demartis S, Neri D (1999) J Mol Recognit 12:198
43. Van Regenmortel MH, Altschuh D, Chatelier J, Rauffer Bruyere N, Richalet Secordel P, Saunal H (1997) Immunol Invest 26:67
44. Luppa PB, Sokoll LJ, Chan DW (2001) Clin Chim Acta 314:1
45. Wittmann C, Schmid RD (1994) J Agric Food Chem 42:1041
46. Ruzicka J, Hansen EH (1975) Chim Acta 78:145
47. Krämer PM (1998) Food Technol Biotechnol 36:111
48. Minnuni M, Mascini M (1993) Anal Lett 26:1441
49. Ekins RP, Chu R, Biggart E (1990) Ann Biol Clin 48:655
50. Brecht A, Klotz A, Barzen C, Gauglitz G, Harris RD, Quigley GR, Wilkinson JS, Sztajnbok P, Abuknesha R, Gascon J, Oubina A, Barcelo D (1998) Anal Chim Acta 362:69
51. Weller MG, Schuetz AJ, Winklmair M, Niessner R (1999) Anal Chim Acta 393:29
52. Kido H, Maquieira A, Hammock BD (2000) Anal Chim Acta 411:1
53. Decanniere K, Desmyter A, Lauwereys M, Ghahroudi MA, Muyldermans S, Wyns L (1999) Structure 7:361
54. Nutall SD, Rousch MJM, Irving RA, Hufton SE, Hoogenboom HR, Hudson PJ (1999) Proteins 36:217
55. Beste G, Schmidt FS, Stibora T, Skerra A (1999) Proc Natl Acad Sci USA 96:1898
56. Chen B, Bestetti G, Day RM, Turner APF (1998) Biosens Bioelectron 13:779
57. Haupt K, Mosbach K (2000) Chem Rev 100:2495
58. O'Sullivan CK (2002) Anal Bioanal Chem 372:44

Chapter 2

Molecularly Imprinted Polymers as Recognition Elements in Sensors

Karsten Haupt

Abstract. The technique of molecular imprinting allows for the formation of specific recognition sites in synthetic polymers through the use of templates or imprint molecules. These recognition sites mimic the binding sites of biological receptor molecules such as antibodies and enzymes. Molecularly imprinted polymers can therefore be used in applications based on specific molecular binding events, such as affinity separation, assays and sensors, and in organic synthesis and catalysis. The stability, ease of preparation, and low cost of these materials make them particularly attractive. This review focuses on recent developments and advances in the molecular imprinting technique with special emphasis on applications in sensors.

Keywords: Molecularly imprinted polymer, Template polymerization, Artificial receptor, Biomimetics, Sensor

Abbreviations

MIP	Molecularly imprinted polymer
SAW	Surface acoustic wave
QCM	Quartz crystal microbalance
2,4-D	2,4-Dichlorophenoxyacetic acid
FTIR	Fourier-transform infrared

1.
Molecularly Imprinted Polymers

1.1
General Principle of Molecular Imprinting

The design and synthesis of biomimetic receptor systems capable of binding a target molecule with similar affinity and specificity to antibodies has been a long-term goal of bioorganic chemistry. One technique that is being increasingly adopted for the generation of artificial macromolecular receptors is molecular imprinting of synthetic polymers [1]. This is a process where functional and cross-linking monomers are copolymerized in the presence of a target analyte (the imprint molecule), which acts as a molecular template. The functional monomers initially form a complex with the imprint molecule, and following polymerization, their functional groups are held in position by the highly cross-linked polymeric structure. Subsequent removal of the imprint molecule reveals binding sites that are complementary in size and shape to the analyte. In that way, a molecular memory is introduced into the polymer, which is now capable of selectively rebinding the analyte (Fig. 1).

The complex between monomers and imprint molecule can be formed via reversible covalent bonds or via noncovalent interactions such as hydrogen bonds, ionic bonds, hydrophobic interactions, van der Waals forces, etc. A combination of the two can also be used. In order to compare the covalent and noncovalent imprinting approaches, different aspects have to be taken into account. The noncovalent imprinting approach, which has been pioneered by Mosbach and

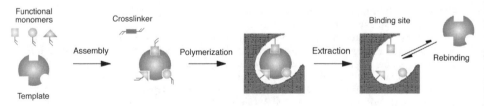

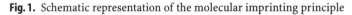

Fig. 1. Schematic representation of the molecular imprinting principle

coworkers [2], is more flexible concerning the choice of functional monomers, possible target molecules and the use of the imprinted materials. After polymerization, the imprinted molecule can be removed from the polymer by simple solvent extraction. However, the prepolymerization complex is an equilibrium system, the stability of which depends on the affinity constants between imprint molecule and functional monomers. This may yield a certain heterogeneity of the binding sites. For covalent imprinting, a polymerizable derivative of the imprint molecule has to be synthesized, and after synthesis of the polymer, the imprint molecule has to be removed by chemical cleavage. If upon use of the polymer the covalent bonds have to be re-formed, association kinetics may be slow. On the other hand, owing to the greater stability of covalent bonds, covalent imprinting protocols should yield a more homogeneous population of binding sites. Moreover, the yield in binding sites relative to the amount of imprint molecule used (imprinting efficiency) should be higher than with noncovalent protocols. This approach has been developed primarily by Wulff and coworkers [3].

Protocols have also been suggested that combine the advantages of both covalent and noncovalent imprinting, that is, the target molecule is imprinted as a stable complex with the functional monomers formed via covalent interactions, whereas upon later use of the molecularly imprinted polymer (MIP), only noncovalent interactions come to play. As an example, Whitcombe and coworkers have reported the imprinting of a tripeptide (Lys-Trp-Asp) using a sacrificial spacer (o-hydroxybenzamide) between imprint molecule and monomer. In addition to these covalent bonds, noncovalent interactions have also been used. After polymerization, the covalent bonds between the imprint molecule and the monomers are hydrolyzed leaving precisely positioned carboxyl groups (Fig. 2). During rebinding the peptide interacts with the polymer only via noncovalent interactions [4].

A simple demonstration by our group [5] of the molecular imprinting effect is shown in Fig. 3. A nonimprinted copolymer of trifluoromethylacrylic acid and divinylbenzene was synthesized. To the same monomer mixture were also added increasing amounts of a template molecule (theophylline) before polymerization in order to create imprinted sites. When the resultant polymers were

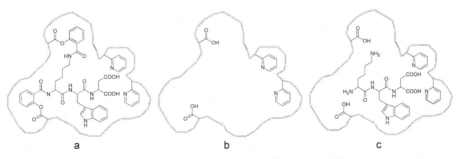

a   b   c

Fig. 2a–c. Molecular imprinting of the tripeptide Lys-Trp-Asp using both covalent and noncovalent interactions. **a** Binding site with covalently bound imprint molecule; **b** binding site after chemical cleavage and extraction of the imprint molecule; **c** rebinding of the imprint molecule via only noncovalent interactions [4]

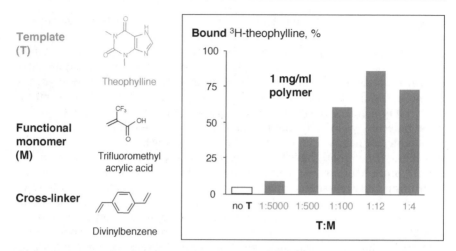

Fig. 3. The imprinting effect: ingredients of a theophylline-imprinted polymer (left), and the binding capacity for radiolabeled theophylline of a series of polymers synthesized with different amounts of template (right) [5]

checked for binding of radiolabeled theophylline, it turned out that a very small quantity of template (a molar ratio of 1:5,000 relative to the amount of the functional monomer trifluoromethylacrylic acid) had doubled the binding capacity of the polymer with respect to the nonimprinted material. The more template was present during polymerization, the higher the capacity of the polymer. At a ratio of 1:12 the maximum capacity was reached.

1.2
The Imprinting Matrix

1.2.1
Acrylic and Vinyl Polymers

At the present time, the majority of reports on MIPs describe organic polymers synthesized by radical polymerization of functional and cross-linking monomers having vinyl or acrylic groups, and using noncovalent interactions between monomers and template. This can be attributed to the rather straightforward synthesis of these materials, and to the vast choice of available monomers. These can be basic (e.g., vinylpyridine) or acidic (e.g., methacrylic acid), permanently charged (e.g., 3-acrylamidopropyl trimethylammonium chloride), hydrogen bonding (e.g., acrylamide), hydrophobic (e.g., styrene), and others. These rather "simple" monomers normally have association constants with the template that are too low to form a stable complex. Thus, they have to be used in excess to shift the equilibrium toward complex formation. Somewhat more sophisticated monomers are also starting to appear that form stable interactions with the template molecule or substructures thereof, and that can sometimes be used in a stoichiometric ratio [6–10]. Two examples of monomers rec-

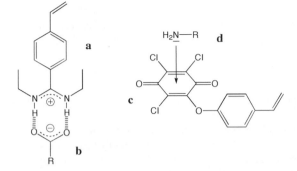

Fig. 4. Amidine functional monomer a binding to a carboxyl group b [8]; tetrachloroquinone monomer c complexing an amino group d [10]

ognizing amino and carboxyl groups are depicted in Fig. 4. Another possibility to obtain stronger interactions in the prepolymerization complex, in particular in polar solvents like water, is by using coordination bonds with metal chelate monomers [11–13].

In order to obtain an optimized polymer for a given target analyte, combinatorial approaches to MIP synthesis have been used [5, 14, 15] where the ingredients of the imprinting recipe, in particular the kind and molar ratio of the functional monomers, are varied. This is ideally done using automated procedures [14]. As an example, a MIP selective for the triazine herbicide terbuthylazine was optimized using a combinatorial approach where a number of different MIPs were synthesized on a small scale (ca. 55 mg) [15]. The functional monomer was selected from a library composed of six different molecules (methacrylic acid, methyl methacrylate, hydroxyethyl methacrylate, trifluoromethylacrylic acid, 4-vinylpyridine, and N-vinyl-α-pyrrolidone). An initial screening was performed for the type of functional monomer that retained the template most strongly. Among the six monomers tested, methyl methacrylate, 4-vinylpyridine, and N-vinyl-α-pyrrolidone led to polymers from which the imprint molecule was rapidly and quantitatively extracted, whereas methacrylic acid and trifluoromethylacrylic acid led to polymers that retained the template more strongly. Based on these two monomers, secondary screening for selectivity was performed. For that purpose, nonimprinted control polymers were also prepared and analyte binding to the MIPs and control polymers evaluated in batch mode. The polymer showing the highest selectivity was found to be the one based on methacrylic acid.

1.2.2
Other Organic Polymers

During recent years other polymers have started to appear that are either better suited for a specific application or easier to synthesize in the desired form. For example, polymers such as polyphenols [16], poly(aminophenyl boronate) [17], poly(phenylenediamine) [18], poly(phenylenediamine-co-aniline) [19], polyurethanes [20], overoxidized polypyrrole [21], and others have been used. Compared to polymers based on acrylic and vinyl monomers, the use of the above-mentioned polymers seems to be somewhat restricted due to the limited choice of functional monomers.

1.2.3
Other Imprinting Matrices

Sol–gels such as silica and titanium dioxide are now gaining in importance as imprinting matrices, although they were introduced years ago. Silica has been used as the imprinting matrix for imprinting with inorganic ions [22] and organic molecules [23–27]. Thereby, either the bulk material can be imprinted by the sol–gel method, thus creating microporous materials with specifically arranged functional groups [22, 24, 27, 28], or an imprinted polysiloxane layer is deposited at the silica surface [23, 26, 29–31]. Recently, Katz and Davis have reported the molecular imprinting of bulk amorphous silica with single aromatic molecules using a covalent monomer template complex, creating shape-selective catalysts [28]. They have also been able to directly observe molecular-imprint-generated microporosity (additional porosity was created in the silica upon template removal) using physical adsorption experiments. Another material that has been imprinted using the sol–gel technique is titanium oxide [32–34].

1.3
Target Molecules

One of the many attractive features of the molecular imprinting technique is that it can be applied to a wide range of target molecules. The imprinting of small, organic molecules (e.g., pharmaceuticals, pesticides, amino acids and peptides, nucleotide bases, steroids, and sugars) is now well established and considered almost routine. Metal and other ions have also been used as templates to induce the specific arrangement of functional groups in the imprinting matrix [22, 35–37]. Somewhat larger organic compounds (e.g., peptides) can also be imprinted via similar approaches, whereas the imprinting of much larger structures is still a challenge. Specially adapted protocols have been proposed to create imprints of proteins in a thin layer of acrylic polymer on a silica surface [13], of cells using a lithographic technique [38], and even of the surface structure of mineral crystals [39].

Figure 5 shows an interesting approach to create imprints of proteins in a surface [40]. The protein of interest is first adsorbed onto an atomically flat mica surface. It is then spin-coated with a disaccharide solution that upon drying forms a thin layer (1–5 nm) attached via multiple hydrogen bonds to the protein. This protective disaccharide shell is then covered with a fluoropolymer layer via glow-discharge plasma deposition, which covalently incorporates the sugar molecules. Finally, the polymer layer is attached to a glass substrate using an epoxy glue. After peeling off the mica, the protein is removed by treatment with aqueous NaOH/NaClO, leaving nanocavities as revealed by tapping-mode atom-

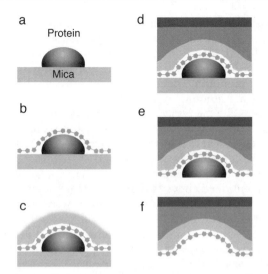

Fig. 5a–f. Generation of a protein imprint. The protein is adsorbed onto mica (**a**), coated with disaccharide (**b**), a fluoropolymer layer is overlaid by plasma deposition (**c**), the polymer is glued to a glass substrate (**d**), the mica is peeled off (**e**), and the protein is removed leaving a binding site (**f**) [40]

ic force microscopy. The authors reported that the cavities were complementary in size and, it seems, to some extent also in functionality, to the template protein. For example, they could show that a surface imprinted with bovine serum albumin preferentially adsorbed the template protein from a binary mixture with immunoglobulin G. Moreover, an imprint preferentially adsorbed the template protein over lysozyme that is similar in size and isoelectric point, and vice versa. This protocol, although somewhat complex, might be shown to be more or less generally applicable to create imprints of proteins.

1.4
Physical Forms and Preparation Methods of MIPs

1.4.1
Imprinted Particles: Making MIPs Smaller

Traditionally MIPs have been prepared as bulk polymer monoliths followed by mechanical grinding to obtain small micrometer-sized particles. Whereas the materials obtained through this somewhat inelegant method still seem to be useful for many applications, other applications require MIPs in defined physical forms for which specially adapted synthesis methods are needed. During the past few years, three aspects have mainly been addressed: the synthesis of small, spherical particles of below-micrometer size, the synthesis of thin layers, and the creation of surface imprints. MIP nanobeads can be synthesized by different methods such as precipitation polymerization and emulsion polymerization. Precipitation polymerization can be performed with similar prepolymerization mixtures as for bulk polymers, except that the relative amount of solvent present in the mixture is much higher. When polymerization progresses, imprinted nano- or microspheres precipitate instead of polymerizing together to form a polymer monolith. The method has the drawback that because of the dilution factor, higher amounts of imprint molecule are needed, although this may be compensated by the typically higher yields. This method has been successfully used by Ye at al. to prepare imprinted particles for binding assays [41, 42], and it has been shown that in some applications these particles performed better than particles obtained by grinding [43].

Wulff's group has used an approach somewhat similar to the precipitation polymerization mentioned above [44]. However, they adjusted the polymerization conditions so that soluble polymer microgels were produced. These had a molecular weight in the range of 10^6 g/mol, which places them close to proteins with respect to molecular size. Although microgels could be synthesized using a monomer mixture adapted to the imprinting process, to obtain selective, imprinted materials proved to be less straightforward with this technique and more optimization work needs to be done. Ishi-i et al. created "imprints" at the surface of fullerenes. They introduced two boronic acid groups into [60]fullerene using saccharides as template molecules, and observed regioselective rebinding of the saccharide [45]. Very recently, the group of Zimmermann published a report on molecular imprinting inside dendrimers [46]. Their method involved the covalent attachment of dendrons to a porphyrin core (the template), crosslinking the end-groups of the dendrons, and removal of the porphyrin template by hydrolysis. This technique seems to yield homogeneous binding sites, to allow quantitative template removal, produces only one binding site per polymer molecule, and the materials are soluble in common organic solvents.

1.4.2
Imprinting at Surfaces

Imprinted materials with binding sites situated at or close to the surface of the imprinting matrix have many advantages: the sites are more accessible, mass transfer is and binding kinetics may be faster, target molecules conjugated with bulky labels can still bind, etc. That such materials are not universally used is due to their preparation being less straightforward than that of a bulk polymer and requiring specially adapted protocols. Whitcombe and coworkers have developed a technique based on emulsion polymerization, i.e., small beads are created in an oil-in-water biphasic system stabilized by a surfactant. The imprint molecule (here cholesterol) is part of the surfactant (pyridinium 12-(cholesteryloxycarbonyloxy)dodecane sulfate) [47]. As a result, all binding sites are situated at the particle surface, which was demonstrated by flocculation experiments using polyethylene glycol-bis-cholesterol. Another protocol for the creation of surface binding sites has been introduced by our group. The imprint molecule is immobilized onto a solid support, such as porous silica beads, prior to polymerization [48]. The pores are then filled with the monomer mixture, and the polymerization is initiated. The silica is removed by chemical dissolution, which

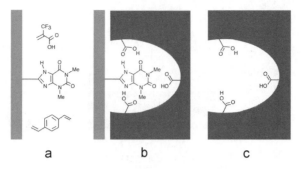

Fig. 6a–c. Molecular imprinting of theophylline immobilized on a solid support: immobilized template with monomers (**a**), composite material after polymerization (**b**), imprinted polymer after dissolution of the support (**c**) [48]

leaves behind a porous polymeric structure which is the negative image of the original bead. The binding sites are now all situated at the surface of the polymer, and are uniformly oriented (Fig. 6).

1.4.3
Thin Imprinted Polymer Films

These have been reported on several occasions and are necessary in many applications of MIPs. For example, they can be synthesized *in situ* at an electrode surface by electropolymerization [16, 18, 19, 21], or at a nonconducting surface by chemical grafting [17, 49]. To grow the imprinted polymer from a surface by using polymerization initiators chemically bound [50] or physically adsorbed [51] to the surface is another recent development. An elegant way would be to apply the soft lithography technique [52] to create thin MIP layers and surface patterns. There has been a first report on the use of this technique in combination with molecular imprinting, although no details are given concerning the binding performance of the obtained MIP microstructures [53]. It seems also that the current imprinting recipes are not always compatible with the poly(dimethylsiloxane) stamps used for soft lithography, and thus more development efforts are needed.

2
Applications of Imprinted Polymers in Sensors

2.1
General Considerations

MIPs are synthetic macromolecular receptors and can be used in applications where specific molecular binding events are of interest. Currently the main application area for MIPs is analytical chemistry [54], and many reports have been published on the use of MIPs for separation, in particular enantioseparation and solid-phase extraction, but also in chemical synthesis and catalysis. An-

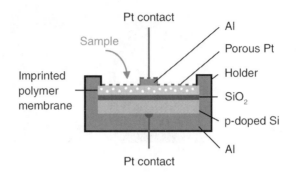

Fig. 7. Capacitance sensor employing a field-effect capacitor as the transducer [58]

other application area is as antibody mimics in immunoassays and biosensors. In chemical sensors and biosensors, a chemical or physical signal is generated upon binding of the analyte to the recognition element. A transducer then translates this signal into a quantifiable output signal. The same general principle applies if a MIP is used as the recognition element instead of a biomolecule. MIPs have been used in combination with optical, acoustic, and electrochemical transducers. Certain changes in one or more physicochemical parameters of the system (such as mass accumulation or adsorption heat) that occur upon analyte binding are used for detection. This principle is widely applicable and more or less independent of the nature of the analyte. Alternatively, reporter groups may be incorporated into the polymer to generate or enhance the sensor response. In other cases, the analyte may possess a specific property (such as fluorescence or electrochemical activity) that can be used for detection.

Early attempts to utilize the recognition properties of MIPs for chemical sensing were, for example, ellipsometric measurements on thin vitamin K_1-imprinted polymer layers [55], the measurement of changes in the electrical streaming potential over a HPLC column packed with a MIP [56], or permeability studies of imprinted polymer membranes [57]. The first reported integrated sensor based on a MIP was a capacitance sensor. The device consisted of a field-effect capacitor containing a thin phenylalanine anilide-imprinted polymer membrane. Binding of this model analyte resulted in a change in capacitance of the device, thus allowing for the detection of the analyte in a qualitative manner (Fig. 7) [58].

2.2
General Transducer Types

During the last few years, there has been a big boost in the use of mass-sensitive acoustic transducers for the design of MIP-based sensors. These are either piezoelectric crystals in which the resonance frequency of the bulk material is measured (for frequencies in the lower MHz range), or surface acoustic wave devices (SAWs) that comprise a separate waveguide and piezoelectric transmitter and receiver (for frequencies up to 2.5 GHz) [59]. SAWs are limited to gas-phase applications due to heavy damping of the acoustic wave in liquids. A variant of SAWs, so-called shear transverse wave resonators (STW, for frequencies in the higher MHz range) show excellent performance even in aqueous solutions [60, 61]. In these devices the oscillation frequency changes in response to mass changes at the transducer surface upon analyte binding to the recognition element.

QCM sensors, which are bulk acoustic wave devices, have been particularly popular because of their relatively low price, their robustness, and ease of use. They consist of a thin quartz disk with electrode layers on both sides, which can be put into oscillation using the piezoelectric effect. A thin imprinted layer is deposited on one side of the disk. Analyte accumulation in the MIP results in a mass change, which in turn causes a decrease in oscillation frequency that can easily be quantified by frequency counting. In addition, it is relatively easy to interface the MIP with the sensor. Measurements with QCMs can be performed both in solution [18, 20, 62, 63] and in the gas phase [64, 65]. A problem associated with this transducer type, in particular if used in the liquid phase, is that they are sensitive not only to mass changes but also to changes in viscoelasticity close to the surface. Thus, the sample matrix can cause artifacts, and the use of a reference sensor coated with a nonselective polymer of the same type as the MIP has been proposed to eliminate these. Such reference sensors can be made of the same quartz crystal as the selective sensor [66], which eliminates temperature effects at the same time. For example, a QCM has been used to construct an imprinted polymer-based sensor for glucose [18]. The polymer, poly(o-phenylenediamine), was electrosynthesized directly at the sensor surface in the presence of 20 mM glucose. In that way, a very thin (10 nm) polymer layer was obtained that could rebind glucose with certain selectivity over other compounds such as ascorbic acid, paracetamol, cysteine, and to some extent fructose.

Thin TiO_2 sol–gels have been used for imprinting of azobenzene carboxylic acid [33]. In a recent application for the detection of cells [66], imprints of whole yeast cells in polyurethane layers and in sol–gel layers have been produced at the surface of a QCM crystal using a stamping method. The sensor could be used to quantify yeast cells in suspension at concentrations between 1×10^4 and 1×10^9 cells per ml under flow conditions. Others have relied on common acrylic polymers for the design of MIP-based QCM sensors [62, 63, 65, 67]. One reason for that was probably the abundance of know-how available on such polymers, and their adaptability to many different template molecules due to the plethora of available functional monomers. With such polymers, it has been demonstrated that the sensor selectivities are similar to those obtained in other applications of acrylic MIPs. For example, a QCM sensor coated with a polymer imprinted with S-propranolol (a β-blocker) was able to discriminate between the R- and S-enantiomers of the drug with a selectivity coefficient of 5 [62].

Other sensors have been designed based on conductometric transducers. These measure the change in conductivity of a selective layer in contact with two electrodes upon its interaction with the analyte. Conductometric sensors are often based on field-effect devices. For example, capacitance sensors such as the above-mentioned field-effect capacitor [58] belong to this group. Capacitive detection was also employed in conjunction with imprinted electropolymerized polyphenol layers on gold electrodes [16]. The sensitive layer was prepared by electropolymerization of phenol on the electrode in the presence of the template phenylalanine. The insulating properties of the polymer layer were studied by electrochemical impedance spectroscopy. Electrical leakages through the polymer layer were suppressed by deposition of a self-assembled monolayer of mercaptophenol before polymerization and of alkanethiol after polymerization. After that, the template was removed. The multilayer system obtained displayed a decrease in electrical capacitance on addition of phenylalanine. Only a low response was observed toward other amino acids and phenol. The same authors also reported a capacitive creatinine sensor based on a photografted molecularly imprinted polymer [68]. Another example of this sensor type is a sensing device for the herbicide atrazine, which is based on a freestanding atrazine-imprinted acrylic polymer membrane and conductometric measurements [69]. The authors carefully optimized the polymer recipe, in particular with respect to the kind and molar ratio of cross-linking monomers used, and the relative amount of porogenic solvent in the imprinting mixture. This turned out to be an important factor not only in obtaining flexible and stable membranes, but also because the conductometric response seemed to depend on the ability of the MIP to change its conformation upon analyte binding. Attractive features of this sensor were the comparatively short time required for one measurement (6–10 min), its rather low detection limit of 5 nM, and its good selectivity for atrazine over structurally related triazine herbicides.

A very general means of transducing the analyte binding to the recognition element is by measuring the adsorption heat that is produced or taken up upon any adsorption or desorption process. This heat, although small, can be measured using a sensitive calorimetric device. MIP sensors with calorimetric transducers have not yet been described, but the feasibility of quantifying analyte binding to a MIP by isothermal titration microcalorimetry has recently been demonstrated [70]. The use, for example, of thermistor-based devices [71] should allow for the construction of integrated sensors.

2.3
The Analyte Generates the Signal

If the target analyte exhibits a special property such as fluorescence or electrochemical activity, this can be exploited for the design of MIP-based sensors. Optical sensors for the detection of fluorescent analytes belong to this group [20, 72]. A potential problem that can arise when this detection principle is used is that traces of the imprint molecule can remain entrapped in the polymer, which may cause a high background signal resulting in decreased sensitivity. A remedy could be to imprint the polymer with a structurally related molecule similar to the analyte [73] but not having that special property, providing that the MIP is still able to bind the target analyte. If the analyte lacks a specific property useful for detection, a competitive or displacement sensor format may be used. A labeled analyte derivative or a nonrelated probe is allowed to compete with the analyte for the binding sites in the MIP [74–78]. In one application, a voltammetric sensor for the herbicide 2,4-D was reported [79], where the electroac-

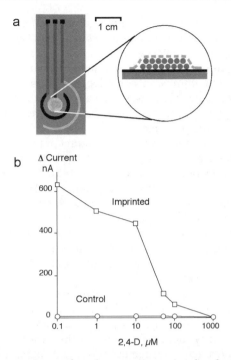

Fig. 8a. Disposable sensor element based on a screen-printed carbon electrode. The MIP was coated onto the carbon working electrode (middle) which is surrounded by a carbon counter electrode (small arc) and a Ag/AgCl reference electrode (large arc). **b** Calibration curve for 2,4-D, with the imprinted polymer (squares) and the control polymer (circles) [79]

tive compound 2,5-dihydroxyphenylacetic acid was used as a probe instead of the labeled analyte. MIP particles were coated as a thin layer onto a disposable screen-printed carbon electrode. The electrode was then incubated with the sample to which the probe was added. In the presence of the analyte, some of the probe was competed out of the imprinted sites, whereas the remaining probe was directly quantified by differential pulse voltammetric measurements (Fig. 8).

2.4
The Polymer Generates the Signal

An attractive design of the recognition element/transducer couple is to have the signal generated by the polymer itself. This approach appears promising since it should facilitate the construction of simple integrated sensing devices. One example for such a format is an optical sensing system where fluorescent reporter groups are incorporated into the MIP, the properties of which are altered upon analyte binding [80, 81]. For example, a fluorescent functional monomer, *trans*-4-[*p*-(*N*,*N*-dimethylamino)styryl]-*N*-vinylbenzylpyridinium chloride, has been used together with a conventional functional monomer to prepare a polymer imprinted with cyclic adenosine monophosphate [80]. Upon binding to the imprinted sites, the analyte interacts with the fluorescent groups and their fluorescence is quenched, thus allowing the analyte to be quantified. Others have used a similar system with a fluorescent metalloporphyrin as the reporter group, of which a polymerizable derivative was used as one of the functional monomers (Fig. 9) [82]. Binding of the analyte 9-ethyladenine then resulted in quenching of the fluorescence of the polymer.

A very sensitive sensor for a hydrolysis product of the chemical warfare agent soman has been described based on a polymer-coated fiber optic probe and a luminescent europium complex for detection [83]. The complex of europium ligated by divinylmethyl benzoate (ligating monomer) and by the analyte pinacolyl methylphosphonate was copolymerized with styrene, whereafter the analyte molecule was removed by washing. Rebinding of the analyte was quantified from laser-excited luminescence spectra. Although it is not clear whether imprinting has contributed to the selectivity of the sensor, this detection principle appears attractive as very low detection limits can be obtained (7 parts per trillion in this particular case).

The signals generated by most of the above-mentioned transducer types are two-dimensional and provide only limited information about the composition of the sample. Although this is normally compensated by the high selectivity of MIPs, a different strategy could conceivably be the use of "intelligent" transducer mechanisms, which generate signals with higher inherent information content. One way to achieve that is to exploit the high molecular specificity of absorption spectra in the mid-infrared spectral region (3500–500 cm^{-1}). The combination of MIPs and FTIR spectrometry might allow analytical problems to be addressed where the selectivity of the MIP alone is not sufficient, e.g., when samples with complex matrices are to be investigated, or when structurally very similar analytes are present in the sample. A recent report described an approach toward a chemical sensor based on an imprinted polymer and infrared evanescent-wave spectroscopy [84]. A polymer molecularly imprinted with 2,4-

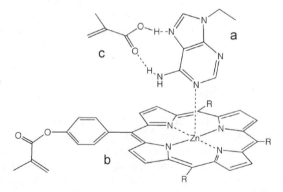

Fig. 9. Prepolymerization complex of 9-ethyladenine (imprint molecule, a), a Zn-porphyrin signalling monomer b, and methacrylic acid c as a comonomer [82]

D was coated in the form of a thin film onto a ZnSe attenuated total reflection element, which was mounted in a flow cell. Accumulation of 2,4-D in the MIP layer could be followed online and in real time by FTIR spectrophotometric measurements. Analyte binding was concentration dependent and could be quantified by integrating characteristic analyte bands.

3
Outlook

Even though MIPs have already found some niche applications that are close to commercialization, such as solid-phase extraction, more work needs to be done to make them a real alternative or complement to biomolecules. In particular, what one hopes to achieve is the development of MIPs that contain a more homogeneous binding site population, have a higher affinity for the target analyte, and that can be routinely used in aqueous solvents. A considerable part of the current research effort on MIPs already deals with these problems. In fact, some of the above-mentioned detection methods are, apart from their use in sensors, equally well suited for investigating the recognition of analytes by MIPs at the molecular level. On the other hand, the outstanding stability of MIPs, their low price, and the fact that they can be tailor-made for analytes for which a biological receptor cannot be found, are among the properties that make them especially suitable for sensor applications. In terms of sensitivity, MIP-based biomimetic sensors are, with some exceptions, still somewhat inferior to biosensors. MIP sensors measure only analyte binding; until now, no MIP sensor has been described that uses an enzyme reaction for signal amplification, as do certain biosensors. However, the situation will certainly improve through further optimization of the MIPs and the transducers. It appears that the development of imprinted polymer-based sensors is just about to leave the proof-of-principle stage, and researchers are starting to address specific analytical problems and to measure real-world samples.

References

1. Sellergren B (2001) Molecularly imprinted polymers – Man-made mimics of antibodies and their applications in analytical chemistry. Elsevier, Amsterdam
2. Mosbach K, Ramström O (1996) Biotechnology (NY) 14:163
3. Wulff G (1995) Angew Chem Int Ed 34:1812
4. Klein JU, Whitcombe MJ, Mulholland F, Vulfson EN (1999) Angew Chem Int Ed 38:2057
5. Yilmaz E, Mosbach K, Haupt K (1999) Anal Commun 36:167
6. Steinke JHG, Dunkin IR, Sherrington DC (1999) Trends Anal Chem 18:159
7. Spivak D, Shea KJ (1999) J Org Chem 64:4627
8. Wulff G, Gross T, Schönfeld R (1997) Angew Chem Int Ed 36:1962
9. Yano K, Tanabe K, Takeuchi T, Matsui J, Ikebukuro K, Karube I (1998) Anal Chim Acta 363:111
10. Lübke C, Lübke M, Whitcombe MJ, Vulfson EN (2000) Macromolecules 33:5098
11. Mallik S, Johnson RD, Arnold FH (1994) J Am Chem Soc 116:8902
12. Hart BR, Shea KJ (2001) J Am Chem Soc 123:2072
13. Kempe M, Glad M, Mosbach K (1995) J Mol Recognit 8:35

14. Takeuchi T, Fukuma D, Matsui J (1999) Anal Chem 71:285
15. Lanza F, Sellergren B (1999) Anal Chem 71:2092
16. Panasyuk TL, Mirsky VM, Piletsky SA, Wolfbeis OS (1999) Anal Chem 71:4609
17. Piletsky SA, Piletska EV, Chen B, Karim K, Weston D, Barrett G, Lowe P, Turner APF (2000) Anal Chem 72:4381
18. Malitesta C, Losito I, Zambonin PG (1999) Anal Chem 71:1366
19. Peng H, Liang C, Zhou A, Zhang Y, Xie Q, Yao S (2000) Anal Chim Acta 423:221–228
20. Dickert FL, Tortschanoff M, Bulst WE, Fischerauer G (1999) Anal Chem 71:4559
21. Deore B, Chen Z, Nagaoka T (2000) Anal Chem 72:3989
22. Dai S, Shin Y, Barnes CE, Toth LM (1997) Chem Mater 9:2521
23. Glad M, Norrlöw O, Sellergren B, Siegbahn N, Mosbach K (1985) J Chromatogr 347:11
24. Makote R, Collinson MM (1998) Chem Commun 3:425
25. Sasaki DY, Rush DJ, Daitch CE, Alam TM, Assink RA, Ashley CS, Brinker CJ, Shea KJ (1998) ACS Symp Ser 703:314
26. Markowitz MA, Kust PR, Deng G, Schoen PE, Dordick JS, Clark DS, Gaber BP (2000) Langmuir 16:1759
27. Sasaki DY, Alam TM (2000) Chem Mater 12:1400
28. Katz A, Davis ME (2000) Nature 403:286
29. Lulka MF, Chambers JP, Valdes ER, Thompson RG, Valdes JJ (1997) Anal Lett 30:2301
30. Lulka MF, Iqbal SS, Chambers JP, Valdes ER, Thompson RG, Goode MT, Valdes JJ (2000) Mat Sci Eng C-Bio S 11:101–105
31. Iqbal SS, Lulka MF, Chambers JP, Thompson RG, Valdes JJ (2000) Mat Sci Eng C-Bio S 7:77–81
32. Lahav M, Kharitonov AB, Katz O, Kunitake T, Willner I (2001) Anal Chem 73:720
33. Lee SW, Ichinose I, Kunitake T (1998) Langmuir 14:2857
34. Lee SW, Ichinose I, Kunitake T (1998) Chem Lett 12:1193
35. Kato M, Nishide H, Tsuchida E, Sasaki T (1981) J Polym Sci Polym Chem 19:1803
36. Kido H, Miyama T, Tsukagoshi K, Maeda M, Takagi M (1992) Anal Sci 8:749
37. Chen H, Olmstead MM, Albright RL, Devenyi J, Fish RH (1997) Angew Chem Int Ed 36:642
38. Aherne A, Alexander C, Payne MJ, Perez N, Vulfson EN (1996) J Am Chem Soc 118:8771
39. D'Souza SM, Alexander C, Carr SW, Waller AM, Whitcombe MJ, Vulfson EN (1999) Nature 398:312
40. Shi HQ, Tsai WB, Garrison MD, Ferrari S, Ratner BD (1999) Nature 398:593
41. Ye L, Weiss R, Mosbach K (2000) Macromolecules 33:8239
42. Ye L, Cormack PAG, Mosbach K (1999) Anal Commun 36:35
43. Surugiu I, Ye L, Yilmaz E, Dzgoev A, Danielsson B, Mosbach K, Haupt K (2000) Analyst 125:13
44. Biffis A, Graham NB, Siedlaczek G, Stalberg S, Wulff G (2001) Macromol Chem Phys 202:163–171
45. Ishi-i T, Nakashima K, Shinkai S (1998) Chem Commun 9:1047
46. Zimmerman SC, Wendland MS, Rakow NA, Zharov I, Suslick KS (2002) Nature 418:399
47. Pérez N, Whitcombe MJ, Vulfson EN (2001) Macromolecules 34:830
48. Yilmaz E, Haupt K, Mosbach K (2000) Angew Chem Int Ed 39:2115
49. Piletsky SA, Matuschewski H, Schedler U, Wilpert A, Piletska EV, Thiele TA, Ulbricht M (2000) Macromolecules 33:3092
50. Sulitzky C, Ruckert B, Hall AJ, Lanza F, Unger K, Sellergren B (2002) Macromolecules 35:79
51. Panasyuk-Delaney T, Mirsky VM, Ulbricht M, Wolfbeis OS (2001) Anal Chim Acta 435:157
52. Xia Y, Whitesides GM (1998) Angew Chem Int Ed 37:550
53. Yan M, Kapua A (2000) Polymer Prepr 41:264–265
54. Haupt K (2001) Analyst 126:747–756

55. Andersson L, Mandenius CF, Mosbach K (1988) Tetrahedron Lett 29:5437
56. Andersson LI, Miyabayashi A, O'Shannessy DJ, Mosbach K (1990) J Chromatogr 516:323
57. Piletsky SA, Parhometz YP, Lavryk NV, Panasyuk TL, El'skaya AV (1994) Sens Actuators B Chem 18–19:629
58. Hedborg E, Winquist F, Lundström I, Andersson LI, Mosbach K (1993) Sens Actuators A Phys 36–38:796
59. Benes E, Groschl M, Burger W, Schmidt M (1995) Sens Actuators A Phys 48:1
60. Tom-Moy M, Baer RL, Spira-Solomon D, Doherty TP (1995) Anal Chem 67:1510
61. Dickert FL, Hayden O, Halikias KP (2001) Analyst 126:766–771
62. Haupt K, Noworyta K, Kutner W (1999) Anal Commun 36:391
63. Liang C, Peng H, Bao X, Nie L, Yao S (1999) Analyst 124:1781
64. Dickert FL, Forth P, Lieberzeit P, Tortschanoff M (1998) Fresenius J Anal Chem 360:759
65. Ji H-S, McNiven S, Ikebukuro K, Karube I (1999) Anal Chim Acta 390:93
66. Dickert FL, Hayden O (2002) Anal Chem 74:1302
67. Kugimiya A, Takeuchi T (1999) Electroanal 11:1158
68. Panasyuk-Delaney T, Mirsky VM, Wolfbeis OS (2002) Electroanal 14:221
69. Sergeyeva TA, Piletsky SA, Brovko AA, Slinchenko EA, Sergeeva LM, Panasyuk TL, Elskaya AV (1999) Analyst 124:331
70. Weber A, Dettling M, Brunner H, Tovar GEM (2002) Macromol Rapid Commun 23:824
71. Xie B, Ramanathan K, Danielsson B (1999) Adv Biochem Eng Biotechnol 64:1
72. Kriz D, Ramström O, Svensson A, Mosbach K (1995) Anal Chem 67:2142
73. Andersson LI, Paprica A, Arvidsson T (1997) Chromatographia 46:57
74. Levi R, McNiven S, Piletsky SA, Cheong SH, Yano K, Karube I (1997) Anal Chem 69:2017
75. Haupt K, Mayes AG, Mosbach K (1998) Anal Chem 70:3936
76. Haupt K (1999) React Funct Polym 41:125
77. Piletsky SA, Terpetschnik E, Andersson HS, Nicholls IA, Wolfbeis OS (1999) Fresenius J Anal Chem 364:512
78. Kriz D, Mosbach K (1995) Anal Chim Acta 300:71
79. Kröger S, Turner APF, Mosbach K, Haupt K (1999) Anal Chem 71:3698
80. Turkewitsch P, Wandelt B, Darling GD, Powell WS (1998) Anal Chem 70:2025
81. Liao Y, Wang W, Wang B (1999) Bioorg Chem 27:463
82. Matsui J, Higashi M, Takeuchi T (2000) J Am Chem Soc 122:5218
83. Jenkins AL, Uy OM, Murray GM (1999) Anal Chem 71:373
84. Jakusch M, Janotta M, Mizaikoff B, Mosbach K, Haupt K (1999) Anal Chem 71:4786

Part II

Impedometric and Amperometric Chemical and Biological Sensors

Capacitance Affinity Biosensors

Helen Berney

Abstract. In the 1980s work began on the development of a new type of biosensor, where changes in the dielectric properties, between electrodes or at an electrode surface, were monitored. In this way a biorecognition event, where a biomolecule (analyte) in solution binding to a capture layer immobilised on an electrode surface or between electrodes, could be detected directly without the need for labels or indicators. Initial work in this field tended to focus on enzymatic or antibody antigen sensing as a replacement for immunoassay formats. However, with the current interest in elucidating the human genome there has been a rise in DNA-based sensing systems. In this chapter an introduction to the principle of capacitance-based transduction will be given. The electrode/electrolyte interface for metal electrodes and semiconductors will be described and a simple model of these interfaces outlined. An overview of capacitance-based sensors using metal and semiconductor electrodes will be given.

Keywords: Capacitance, Impedance, Biosensor, Biorecognition, Affinity

Abbreviations

Ab	Antibody
AC	Alternating current
Ag	Antigen
C	Capacitance
C_V	Capacitance–voltage
DC	Direct current
DNA	Deoxyribonucelic acid
EDC	1-(3-Dimethylaminopropyl)-3-ethylcarbodiimide hydrochloride
EIS	Electrolyte–insulator–semiconductor
GOx	Glucose oxidase
I	Current
IC	Integrated circuit
IDE	Interdigitated electrode
IHP	Inner Helmholtz plane
MOS	Metal oxide semiconductor
OHP	Outer Helmholtz plane
PEGDGE	Polyethylene glycol diglycidyl ether
R	Resistance
V	Voltage
X	Reactance
Z	Impedance

1
Introduction

In recent times it has become apparent that more sophisticated rapid measurement devices are necessary to collect real-time or in-line information from a variety of environments, from bioprocessing to healthcare. It is hoped that pressure to meet these demands will be answered by the development of chemical and biological sensors. Modern biosensors have evolved from the marriage of two disciplines: engineering and molecular biology. The former provides a transducer, the latter, biomolecules forming a sensing layer that can be used to detect specific analytes [1]. This biotransducer combination exploits the molecular resolving power of a biological reagent directly.

Biological sensors are relatively new measurement devices. The path towards modern-day sensor technology began in the mid-1950s, with Leland C. Clark Jr. He invented an electrode to measure dissolved oxygen in the blood of patients undergoing surgery. A modification of this system led to the first blood glucose sensing device, an area of diagnostics now dominated by sensors. The first glucose sensor was a centimetre in diameter. During the past decade, manufacturing techniques, initially developed for integrated circuits, have made possible the production of sensor electrodes hundreds or thousands of times smaller. This chapter will concentrate on the development of capacitance-based affinity sensors.

2
Capacitance-Based Transduction

The idea that changes in the electrical conductivity or impedance (Z) of a biological system may be used to sense biological activity goes back to the previous century [2]. Impedance has been used to determine the molecular thickness of biological membranes and the voltage-gated conductivity of nerve axons [3]. Impedance is a term used to describe how much effort is required to produce a response. Electrically the effort is voltage (V) and the response is current (I). Resistance is a measure of how much a system resists the flow of a DC signal. Impedance is a measure of how much a system resists the flow of an AC signal at a given frequency. In DC systems only resistors impede current flow. However, in AC systems resistors, capacitors and inductors impede current flow[1].

In DC theory resistance is defined by Ohm's Law:

$$V = IR \tag{1}$$

This applies as a special case in AC theory where the frequency is 0 Hz. In AC theory, where the frequency is non-zero, the analogous equation is:

$$V = IZ \tag{2}$$

In an electrochemical cell, slow electrode kinetics, slow chemical reactions and diffusion can all impede electron flow, and can be considered analogous to the resistors, capacitors and inductors that impede the flow of electrons in an AC circuit. The total impedance in a system is the combined opposition of all the resistors, capacitors and inductors to the flow of electrons. Reactance (X) is the name given to the opposition of capacitors and inductors.

The applied AC signal has an amplitude (magnitude) and a time-dependent characteristic (phase). An AC voltage applied across a resistor is in phase with the current, so only the magnitude of the signal is affected. Capacitors and inductors affect not only the magnitude of the AC current but also its time-dependent characteristic. An AC voltage across a capacitor results in a current that leads the applied voltage by $\pi/2$. The voltage is 90° out of phase. An AC voltage across an inductor results in a current that lags the applied voltage by $\pi/2$. This is also a 90° phase shift. For real systems, which at a given frequency behave as a resistor and capacitor in series (or parallel), the phase angle takes a value between zero and 90°.

Because an AC waveform has both a magnitude and a phase it is useful to view it as a vector (Fig. 1). The in-phase signal (resistance) is plotted on the x axis and the out-of-phase signal (capacitance plus inductance, which is the reactance) is plotted on the y axis. For numerical analysis it is convenient to define the axes as real (in-phase) and imaginary (out-of-phase). Impedance is the

---

[1] EG & G Princeton Applied Research, Application note AC-1, Basics of Electrochemical Impedance Spectroscopy.

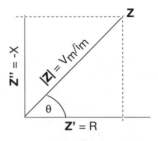

Fig. 1. Impedance is a complex quantity, containing both real and imaginary parts. The modulus of the impedance and the phase angle are used to obtain the resistance and reactance according to the relations $R=|Z| \cos \theta$ and $X=-|Z| \sin \theta$, where $|Z|$ is the modulus of impedance (ohms) and θ is the phase angle (degrees)

complex quantity given by $Z=R+jX$, where R is the resistance (ohms), X the reactance which is equal to $-1/\omega C$ (ohms) and $j=(-1)^{1/2}$.

The impedance technique used to evaluate the electrical behaviour of biosensors consists of applying a sinusoidal voltage of small amplitude and fixed frequency to the system and measuring the current. From this response signal the in-phase and out-of-phase components of the total impedance are measured. This information can be used to determine the degree of phase shift between the output and the input waveforms and thus the magnitude of the impedance can be calculated. This response is monitored over time.

A conducting electrode immersed in electrolyte solution can generally be described as resembling a capacitor in its ability to store charge. In most cases, the capacitance is measured at the metal/solution interface in the system. Ions and dipoles are ordered outside a metal electrode in such a way that charges in the metal are balanced, thereby forming the electrical double layer. Since capacitive measurements give information about the metal/solution interface, a chemical modification of this structure will lead to a change in capacitance. The electric capacitance between two parallel plates separated a distance, d (m), is given by the equation

$$C = \varepsilon \varepsilon_0 A / d \tag{3}$$

where ε_0 is the permittivity of free space (Fm^{-1}), ε is the dielectric constant (dimensionless) of the material between the plates and A is the surface area (m^2). Reducing the distance between the two electrodes results in increasing the capacitance. The limiting factor in forming constant and reproducible distances in the micrometre range between two plates is mechanical. In the case of electrolytic capacitors, where one of the plates is covered with an insulator and the second plate, or counter electrode, is represented by an electrolyte, this limitation does not apply. The capacitance is determined by the thickness and dielectric properties of the insulating layer and the solid/solution interface, both of which constitute the dielectric. An electrolytic capacitor allows the detection of material bound to a plate, enabling the detection of any analyte binding to specific ligands immobilised on the insulating layer of that plate.

The sensitivity of the sensor increases with decreasing thickness of the insulating layer. Consider the initial capacitance, C_i, formed by the dielectric and the primary biomolecular layer immobilised on the dielectric surface, and the capacitance, C_a, due to a biorecognition event where a second biomolecule binds to the immobilised biolayer. The total capacitance, C_t, measured through the sensor terminals, is given by the relation for two capacitors connected in series:

$$1/C_t = 1/C_i + 1/C_a \quad \text{or} \quad C_t = C_i C_a / C_i \tag{4}$$

The sensitivity is given by the ratio of change of the total capacitance C_t to the change in capacitance C_a due to the biorecognition event.

$$\left(\delta C_t / \delta C_a\right)_{C_i} = C_i^2 / \left(C_i + C_a\right)^2 \tag{5}$$

The sensitivity tends to 1 when C_i is large compared to C_a. This means that a change due to the biorecognition event will produce a maximum change in the sensor output when the initial capacitance, formed by the insulating layer and the immobilised biolayers, is very large. The capacitance of the biolayer depends on properties that cannot be used as design parameters. The only parameter left to be optimised, from a design point of view, is the thickness of the insulating layer. In order to have a maximum capacitance this layer should have a minimum thickness and be non-porous and non-soluble [4].

If a biomolecule is immobilised on the surface of an electrode pair, it introduces a capacitive component in the impedimetric response of the sensor. This originates from the low dielectric permittivity of the immobilised biomolecule. Further, when the pair of modified electrodes is exposed to a complementary binding partner, a complex is formed. The dielectric layer covering the electrodes will thicken, resulting in a corresponding decrease of the sensor series capacitance [2]. An alternative mode of operation exists where the biomolecule is immobilised between pairs of electrodes in an interdigitated array format. Binding of complementary partner molecules perturbs the electric field above the electrodes, causing a change in the measured impedance [5].

2.1
The Electrode/Electrolyte Interface

2.1.1
Metal Electrodes

Helmholtz realised, almost a century ago, that the interface between a metal electrode and an electrolyte behaves like a capacitor in that it is capable of storing electric charge [6]. The model proposed to account for this phenomenon assumed excess charge on the metal located at its surface and a corresponding layer of oppositely charged ions in a plane parallel to the surface of the electrode and very close to it (Fig. 2, Helmholtz). However, accurate measurements of the numerical value of the double layer capacitance show that the interface can nev-

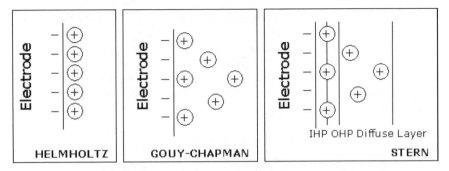

Fig. 2. Schematic illustration of the stages of development of a model for the electrode/electrolyte interface: Helmholtz, where the ions in solution are at a minimum distance from the electrode (only valid for very high concentration electrolyte solutions); Gouy–Chapman, which becomes inadequate if the Debye length or surface potential is large; and Stern, which combines the Helmholtz and Gouy–Chapman models, and takes into account the finite size of counterions and their binding properties at the surface

er be represented simply by a parallel plate capacitor. There is a dependence on potential and electrolyte concentration that must be taken into account.

The diffuse double layer model, as proposed independently by Gouy and Chapman, predicted a dependence on potential and electrolyte concentration. The theory describes a region in the solution near the electrode in which there is a space charge due to an excess of free ions of one sign. The ions attracted to the outer Helmholtz plane are not sufficient to compensate all the charges on the electrode, and the residual electric field directed normal to the surface results in a charged Gouy–Chapman layer (Fig. 2, Gouy–Chapman). There is always an electric field in the interface associated with an excess charge on the surface of the metal. A positive charge on the electrode will attract the negative ions in the solution and repel the positive ions. In the view of Gouy and Chapman, a Helmholtz-type double layer does not form because the electrostatic interaction between the field and the charges on the ions is counteracted by random thermal motion, which tends to level all concentration differences in the solution. This model represents a substantial improvement over the Helmholtz model, in that a dependence of the differential capacitance on both potential and concentration is predicted. However, wide discrepancies between theoretical predictions and experimental results still exist. The measured capacitance is much lower than that calculated except in very dilute solutions. Its variation with potential does not follow the predicted model except in very dilute solutions and at potentials near the potential of zero charge.

The next step in the development of the double layer theory was a simple one. A combination of the two previous models, with some of the ions adhering to the electrode as suggested by Helmholtz and some forming a Gouy–Chapman-type diffuse layer, was suggested by Stern as a more realistic way of describing the physical situation at the interface. In the Stern model, two capacitors effectively determine the total differential capacitance of the interface. One is due to the Helmholtz layer and one due to the diffuse layer (Fig. 2, Stern). These capacitors are in series and will give a total capacitance according to the equation

$$1/C_{\text{Total}} = 1/C_{\text{Helmholtz}} + 1/C_{\text{Gouy-Chapman}} \tag{6}$$

The total capacitance is dominated by the smaller of the two capacitances. In very dilute solutions $C_{\text{Gouy-Chapman}}$ will be predominant.

Stern also recognised the effect of specific interactions and suggested a distinction between ions which were adsorbed onto the electrode and those which merely approached it to the distance of closest approach. This view was used by Grahame to develop a model of the interface which consists of three regions, although it is still referred to as the double layer. The first region extends from the electrode to a plane passing through the centres of the specifically adsorbed ions. This is the inner Helmholtz plane (IHP). Next is the outer Helmholtz plane (OHP), which passes through the centres of the hydrated ions at their distance of closest approach to the electrode. Beyond the OHP lies the diffuse double layer [6].

2.1.2
Semiconductor Electrodes

A semiconductor is a crystalline material having a conductivity between that of a conductor and an insulator. Many types of semiconducting materials exist, but silicon is one of the most commonly used. The crystalline silicon can be doped with other atoms to enhance conductivity. A number of groups have done work using some form of silicon-based sensor format for capacitance sensing [7–10].

Using a semiconductor electrode there is a third capacitance, the semiconductor space-charge region, which must be included in a description of the electrode/electrolyte interface. The amount of charge in the space-charge region of the semiconductor depends on several processes. A semiconductor in vacuum will almost inevitably develop a space charge near the surface simply because some of the bulk carriers near the surface will be captured by surface states. When the semiconductor is dipped into a solution, the charge in the space-charge region may change further because the surface states will be affected by their interaction with the solution and the charge in the surface states will therefore change. Ions in solution with the appropriate energy levels may inject or extract electrons from the semiconductor. This will occur under conditions of no applied bias. Further variation in the space-charge region of the semiconductor will occur if a second electrode is dipped into the electrolyte and an external voltage applied between the two electrodes [11].

According to the generally accepted view, the interfacial capacitance of a semiconductor electrode is the combination of four capacitors. These capacitors are associated with the conduction band of the semiconductor (C_{SC}), the surface states (C_{SS}), the Helmholtz layer and the Gouy–Chapman layer. The total surface capacitance is considered to be

$$1/C_{\text{Total}} = 1/(C_{\text{SC}} + C_{\text{SS}}) + 1/C_{\text{Helmholtz}} + 1/C_{\text{Gouy-Chapman}} \tag{7}$$

The total potential drop across the interface is the sum of the potential drops across the capacitors in series

$$\Delta\varphi_{\text{Total}} = \Delta\varphi_{\text{SC*}} + \Delta\varphi_{\text{Helmholtz}} + \Delta\varphi_{\text{Gouy–Chapman}} \tag{8}$$

If the electrolyte has a suitably high ion content, then the Gouy–Chapman layer can be neglected [11, 12]. In practice the interface is rarely more than 100 Å thick, with a critical very thin layer extending 4–6 Å out into the solution from the electrode surface [6].

2.2
Modelling the Electrode/Electrolyte Interface

An electrode interface undergoing an electrochemical reaction is analogous to an electronic circuit consisting of a specific combination of resistors, inductors and capacitors. The equivalent circuit for a system consisting of a silicon–silicon dioxide–silicon nitride working electrode and platinum counter electrode is outlined in Fig. 3, where the symbols have the meanings given in Table 1. If the sensor was fabricated using a metal working electrode, the capacitances C_{nit}, C_{ox} and C_D are not present.

If the surface area of the platinum counter electrode is large compared to the working electrode area, the impedance of the interface between the electrode and the electrolyte can be neglected, eliminating C_{dl} (ce), R_{ct} (ce) and Z_W (ce). If the system is operated at a suitably high frequency the leakage resistance, R_{le}, and semiconductor/electrolyte interface Warburg impedance, Z_W, can also be neglected. If the semiconductor is operated in accumulation mode, the capacitance of the depletion layer is very high compared to the oxide and nitride capacitance. The model can be reduced to the equivalent circuit shown in Scheme 1. Combining the resistors and capacitors in series gives Scheme 2. This is called a Randles cell and it models the electrochemical impedance of an interface and fits many chemical systems. The capacitance measured at a certain frequency can be considered as resulting from a series connection of the capaci-

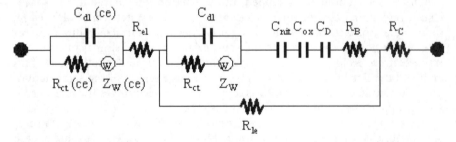

Fig. 3. Schematic representation of the equivalent circuit for a system consisting of a silicon–silicon dioxide–silicon nitride working electrode and platinum counter electrode. The symbols and what they represent are given in Table 1

The potential drop across C_{SC} is the same as across C_{SS}.

Table 1. Representation of symbols in Fig. 3

Symbol	Representation
ce	Counter electrode
w	Working electrode
re	Reference electrode
C_{dl} (ce)	Capacitance of the double layer at the interface between the platinum electrode and electrolyte
R_{ct} (ce)	Charge transfer resistance[a] of platinum electrode/ electrolyte interface
Z_W (ce)	Warburg impedance[b] of platinum electrode/ electrolyte interface
R_{el}	Resistance of the electrolyte
C_{dl}	Capacitance of the double layer at the interface between the semiconductor electrode surface and electrolyte
R_{ct}	Charge transfer resistance of semiconductor/ electrolyte interface
Z_W	Warburg impedance of semiconductor/ electrolyte interface, a value which accounts for the work done in bringing species to and from the interface
R_{le}	Leakage resistance
C_{nit}	Capacitance of the nitride dielectric layer
C_{ox}	Capacitance of the oxide dielectric layer
C_D	Capacitance of the semiconductor depletion layer
R_B	Resistance of the bulk
R_C	Resistance of the backside contact of the working electrode

[a] This represents the resistance to current flow due to the difficulty of carrying out the Faradaic electrochemical reaction. Faradaic currents are due to the electron transfers that occur across the electrode/electrolyte interface during oxidation or reduction reactions. They are different from non-Faradaic currents which are due to charging and discharging of the double- layer capacitance.

[b] The Warburg impedance is a value which describes the influence of diffusion and mass transfer of reactant towards an electrode or of a product away from an electrode surface. If diffusion effects dominate the system then the impedance is a Warburg impedance. This is a complex quantity which that has real and imaginary parts which are equal. The diffusion process is only observed at low frequencies.

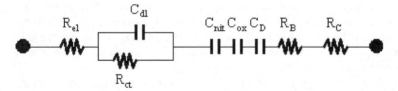

Scheme 1. Equivalent circuit model if semiconductor is operated in accumulation mode

tance of the dielectric layer and the space-charge capacitance of the silicon. The capacitance of the dielectric layers is constant, while the capacitance of the semiconductor space-charge region is dependent on the applied DC voltage.

A typical capacitance–voltage (C_V) curve for a metal oxide semiconductor (MOS) is shown in Fig. 4. There are three regions in the C_V curve: the accumulation, depletion and inversion regions. With a *p*-type silicon substrate a large

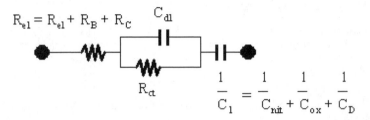

Scheme 2. Equivalent circuit model after combining the resistors and capacitors in series

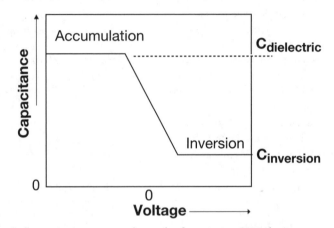

Fig. 4. A typical capacitance versus voltage plot for a p-type MOS device

negative potential applied to the electrolyte will cause an accumulation of holes at the silicon surface. The measured capacitance now approaches a value determined only by the dielectric layers, because the space charge variations due to the measuring signal no longer occur close to the interface. An increase in the DC voltage results in a complete depletion layer at the silicon surface and the capacitance will be dominated by the depletion layer capacitance. At a sufficiently high, positive voltage and low frequency an inversion layer of free electrons is formed at the surface and the measured capacitance again becomes dominated by the dielectric layer capacitance.

The position of the C_V curve along the voltage axis is influenced by several solid-state parameters, such as the silicon electron work function, the charges in the dielectric layers, the charges in the surface and interface states, electrochemical parameters, the reference electrode potential and the electrode/electrolyte interface potential [13]. If a biolayer immobilised on the surface is electrically insulating then it can be considered to add another capacitive component to the nitride, oxide and semiconductor capacitance. However, if it is not sufficiently electrically insulating, then it contributes to the resistance. In fact it is often modelled as a leaky capacitor. The semiconductor material, if heavily doped, will have a capacitance dominated by the dielectric layers and will not exhibit the characteristic C_V response. The smallest capacitance in the series capacitance measurement will dominate the overall total capacitance value.

A good insulation layer at the conducting electrode surface is important because, if an aqueous conducting solution can penetrate the layer, it will increase its capacitance due to the greater polarity of water. Furthermore, the conductivity of the water will be equivalent to a resistor in parallel with the capacitance, partially short-circuiting it [14].

3
Capacitance Sensors

There are two main approaches that have been applied to capacitance measurements. One is to monitor change in the capacitance between two metal conductors in close proximity to each other, the so-called interdigitated electrodes, where the recognition element is immobilised between the electrodes. The second is to measure the capacitance at an electrode/solution interface, where the recognition element is immobilised on the surface of the working electrode. In the case of the first approach the electrodes used are typically metal. For the second approach the electrodes can be metal or semiconductor based.

3.1
Capacitance Sensing with Interdigitated Electrodes

A normal electrical plate capacitor consists of two parallel metal plates separated by a dielectric layer of a specific thickness, d, and dielectric constant ε (Eq. 3). Changes in the dielectric properties of the layer between the metal plates result in a change in the capacitance. A number of monitoring systems have been developed based on measuring the change in impedance caused by a perturbation of the electric field above metal electrodes on binding of target biomolecules in solution to capture biomolecules immobilised on or between the electrodes. This principle has been applied by Johns Hopkins University Applied Physics group, who report measurements using planar capacitive devices. These were interdigitated copper electrodes on a glass surface. The metal was insulated by a 1 µm layer of parylene, to reduce the influence of Faradaic currents. These were then covered by a 0.3 µm layer of silicon monoxide, providing a suitable surface for immobilisation of the biorecognition element, an Ag. On addition of an Ab of concentration 1 µg ml^{-1} the capacitance decreased by 60 pF. Further addition of Ag displaced the Ab and the capacitance increased to its original value [15].

A biosensor based on acetylcholine receptors in a polymeric film on interdigitated gold electrodes has been reported to detect microgram to nanogram quantities of cholinergic ligands [16]. A change in the impedance was recorded on binding of ligands to immobilised receptors. This change was proportional to the amount of cholinergic agent present. The system had a differential set-up, with reference interdigitated electrodes and a test set. At Johns Hopkins University, Baltimore, work on a similar set of experiments, with immobilised acetylcholine receptors in a liposome layer on interdigitated gold electrodes, was performed. In this case capacitance was monitored. It increased on addition of acetylcholine but not on addition of six other neurotransmitters. An interesting aspect of this work was the investigation of different values of electrode capacitance in different solutions. The electrodes had the lowest capacitance in air (0.086 nF) and increased to 0.99 nF when in the organic solvent hexane. In phosphate-buffered saline the capacitance was measured as 4.48 nF. On immobilisation of the receptors–liposome solution the capacitance decreased to 4.43 nF [17]. In air, the total capacitance is the sum of the capacitances in series of the metal and the dielectric. Once the electrodes are introduced into a solution other capacitances in series are added due to the presence of a solid/liquid interface. In an organic solvent the absence of suitably mobile charge carriers prevents the formation of a complex series of capacitor planes. Once the electrodes are transferred to a polar solvent they form series capacitors via Helmholtz and Gouy–Chapman layers, as described in Sect. 2.1.

Nanoscaled titanium interdigitated electrodes are being developed at IMEC [18] using glucose oxidase as a model system. The electrodes, fabricated on oxidised silicon, were 400 nm wide with a spacing of 400 nm between fingers of the electrodes. In general the metal electrodes on insulator in solution can be represented by the equivalent circuit components shown in Fig. 5. The interface between the metal electrodes and the solution is comprised of the double layer capacitance of the electrodes, $C_{\text{double layer}}$, in series with the solution resistance, R_{solution}, and the dielectric capacitance of the solution, C_{solution}, in parallel. The other contributing capacitances are the insulator capacitance between the metal electrodes, $C_{\text{insulator}}$, and the capacitance formed by the stack of metal electrode/insulator/conducting substrate, $C_{\text{electrode to substrate}}$. Under specific conditions, where the insulator layer is thicker than the sum of the electrode width and spacing, the $C_{\text{electrode to substrate}}$ can be neglected. Measuring the capacitance of the electrodes in air allows the value, $C_{\text{insulator}}$, to be determined. What re-

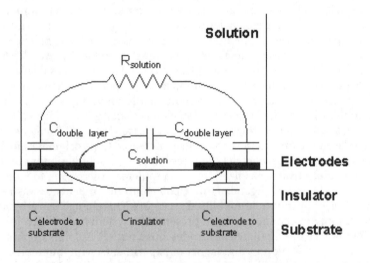

Fig. 5. An equivalent circuit model of metal electrodes on an insulator in solution (after [18])

mains are the components of interest, which yield information about the electrode/solution interface, $R_{solution}$, $C_{double\ layer}$ and $C_{solution}$.

Depending on the measurement conditions, specifically the frequency regime, the response of the system will be dominated either by the bulk solution response or by events at the metal/solution interface. This means that by measuring the impedance responses of the system, changes in conductivity and in events at the surface of the metal electrodes can be monitored. A comparison was made between Pd IDEs, Ti IDEs with a native oxide and thermally oxidised Ti IDEs. Monitoring the immobilisation of GOx, extracted values of the $R_{solution}$ and $C_{double\ layer}$ were compared. The $R_{solution}$ increased and the $C_{double\ layer}$ decreased in all cases. The solution resistance changed by a similar amount, 3–4%, in all cases. However, the double layer capacitance decreases differed for the different electrode types and were found to follow the order: Pd most sensitive, native oxide Ti next and oxidised Ti least sensitive. The electrodes with thinner native oxide were more sensitive to surface processes, as expected.

The main advantages of interdigitated electrodes over conventional macro-electrodes for biosensing are an improved sensitivity, due to the submicron widths and spacings, and the possibility of reproducible low cost device production due to the fabrication protocols. The sensitivity of the detection increases as the distance between the two conductors decreases. So, it is important that the insulating layer covering the metal electrodes be as thin as possible and has as high a dielectric constant as possible in order to reduce its importance in the overall capacitance. There have been difficulties in producing thin, pinhole-free insulation layers and there are disadvantages associated with the limitation in the production of short and reproducible distances between the two conductors [19]. However, with constant improvements in photolithography and processing technology, these can be overcome. A final challenge for this method is the sensitivity of the system to changes in the bulk solution. Adopting a differential measurement approach, where a reference sensor is used as a background signal which can be subtracted from the specific signal, may help to reduce the influence of bulk effects.

3.2
Capacitance Sensing at the Metal/Solution Interface

The application of a simple metal wire, onto which biomolecules adsorb, should be the most direct way of biosensing, comparable with many electrochemical sensors such as coated wires and metal oxide/metal electrodes. This is only possible if the frequency response of the electrode can directly be related to the concentration or activity of the adsorbed species. The specific affinity of a precoated layer of proteins should favour the desired selectivity [13]. Many protein adsorption experiments have been performed with a large variety of substrates. Under specific conditions it is possible that a metal surface can be insulated electrically by the adsorption of a layer of small organic molecules. The effect can be measured as a change in metal/electrolyte capacitance. One of the characteristics of the adsorption of species onto conducting surfaces from electrolytic solutions containing the adsorbates is its dependence on potential. Stoner and Srinivasan noted a decrease in capacitance due to adsorption of thrombin onto a platinum electrode with an applied potential of 0.2 V versus a calomel reference electrode [20]. A titanium wire electrode which had anti-human chorionic gonadotropin Ab immobilised on the surface exhibited a positive shift of 5 mV in electrode potential, versus a reference titanium wire (which had a surface that was blocked but had no Ab immobilised), on addition of human chorionic gonadotropin to a final concentration of 5.4×10^{-6} g ml^{-1} [21]. More recently, Alfonta et al., have investigated the possibility of capacitively sensing a biocatalysed precipitation of an insoluble product on the surface of a gold electrode [22]. An amplification scheme was developed by using a biotin tag bridged using avidin to a biotinylated HRP-liposome. The HRP catalyses the oxidation of 4-chloro-1-naphthol in the presence of H_2O_2 to form an insoluble product. Using this technique they could detect Ab at a concentration of 1×10^{-11} g ml^{-1} and oligonucleotide DNA interactions at 6.5×10^{-13} M.

Capacitance measurements at metal electrodes have also been used to determine enzyme activity [23]. Two platinum wires in an electrolyte solution were used to determine the activity of serine proteases. The substrate for the enzyme adsorbs onto the metal electrodes. However, the product of the enzyme substrate reaction does not adsorb. The electrode and enzyme were added to a solution of substrate at the same time. The substrate initially adsorbs, but after a while desorption occurs due to the presence of the enzyme. A trough in the measured capacitance was obtained, the position and depth of which was dependent on enzyme activity.

The response of gold electrodes to biorecognition events is a broad field of research, with groups interested in developing suitable immobilisation protocols, artificial recognition layers and monitoring affinity reactions. Self-assembled monolayers on gold surfaces introduce a method of achieving reproducible immobilisation on electrodes [24]. Thiols adsorb very strongly onto gold surfaces due to the formation of a covalent-like bond between gold and sulphur. The layers can be of crystal-like order and tend to be stable over months in different media. In order to use these monolayers for analytical purposes they have to be modified chemically with groups or molecules with appropriate recognition properties. At Kyushu University, a group at the Department of Chemical Science and Technology immobilised a monolayer of dialkyl disulphide compound carrying a terminal dinitrophenyl group on the surface of a gold disc electrode [25]. An impedimetric response was monitored in the low-frequency region, 0.1 Hz to 1 kHz, where the impedance response is dominated by the diffusion of the redox species. The measurement was performed in the presence of an electrolyte containing the redox couple $[Fe(CN)_6]^{3-/4-}$ and anti-dinitrophenyl Ab. As Ab was added and bound to the immobilised Ag, the gold electrode surface was blocked and the impedance increased. Addition of non-specific anti-IgE Ab had no effect on the impedance. The detection range was from 10 to 10^3 ng cm^{-3}.

At the University of Tübingen a similar approach was made, but chemically synthesised epitope peptides coupled to hydroxyundecanethiol were used to form the self-assembled layers [26]. The epitope peptide was for the foot-and-mouth disease virus. The advantages of using synthetic peptides include improved safety, in comparison with using a virus, and higher operational stability.

The peptide retained its immunologically active structure even after repeatedly performed denaturation steps used to cleave the Ab–Ag complex. The group also investigated the effect of Ab binding to Ag in the presence of a $[Fe(CN)_6]^{3-/4-}$ redox probe. They noted that the probe had an adverse effect on the biolayer and subsequently performed measurements in the absence of the redox probe. This resulted in a reduction of the impedance changes caused by Ab binding to the biolayer. On addition of an Ab concentration of 1.74 µg ml^{-1} the capacitance decreased by 0.02 µF min^{-1} cm^{-2}.

A very sensitive protein-based sensor has been developed capable of detecting femtomolar levels of heavy metal ions [14]. The sensor format was also based on the self-assembly of thiols on gold electrodes. Proteins with distinct binding sites for heavy metals, which were developed for this purpose, were immobilised by (i) 1-(3-dimethylaminopropyl)-3-ethylcarbodiimide hydrochloride (EDC) coupling, (ii) polyethylene glycol diglycidyl ether (PEGDGE) entrapment and (iii) glutaraldehyde coupling. Using capacitance measurements, the interface between the conducting gold electrode and the sample solution was studied. The interface consisted of the gold conductor covered by non-conducting organic molecules, outside of which ions in the aqueous conducting liquid form a space-charge region. The total capacitance is composed of a series of capacitances representing the self-assembled molecules, the protein and the space charge of ions. As the heavy metal ion binds to the protein conformational changes occur causing a change in the space-charge region, which results in a change in the measured capacitance.

A critical aspect of the development of sensitive miniaturised affinity biosensors is the optimisation of tailored biointerfaces and biorecognition layers. Work to develop alternative recognition layers, which mimic the biorecognition layer but are more robust and stable, is ongoing. Molecular imprinting is recognised as a promising technique for formation of these artificial binding sites. The polymer is prepared by cross-linking a monomer around a template molecule. The template molecule is removed after polymerisation of the monomer, and its size, shape and chemical function have formed imprint sites which allow the recognition of the template molecule or close structural molecules. Application of molecularly imprinted polymers to capacitance sensors by electropolymerisation has been investigated, using phenylalanine as a template [27]. Though the sensor proved specific and had a dynamic range of 0.5–8 mg ml^{-1}, the stability and reversibility were poor. Improvements were seen using a technique of grafting polymerisation to prepare thin-film molecularly imprinted polymers on hydrophobised gold electrodes [28]. The herbicide desmetryn was used as a template. On exposure of the sensor to the target analyte, desmetryn, the capacitance decreased. The herbicide could be detected in the range 0–180 µg ml^{-1}; this detection was reversible and the response time was of the order of minutes. The sensor did not respond significantly to additions of the triazine derivatives atrazine or terbumeton (1.5% change compared to 20% change), but addition of metribuzin resulted in capacitance changes comparable to those for desmetryn. However, this work is interesting because it combines the direct detection of a biorecognition event with the formation of a polymer-based recognition layer which is stable for 6 months in air at room temperature. The formation of the molecularly imprinted polymer recognition layer removes a level of complexity associated with the use of biological molecules, where maintaining activity, selectivity and sensitivity after immobilisation is a key area of research in the biosensors field.

Molecularly imprinted polymers and capacitance transduction have also formed the basis of a novel glucose sensor [29]. Here, o-phenylenediamine was electropolymerised on gold electrodes in the presence of the glucose template, which was removed after plugging of the polymer defects was performed.[2]

When glucose in solution binds to the imprint sites, this causes a change in the measured capacitance. A linear response was seen for a glucose concentration range of 0.1–20 mM and the limit of detection of the sensor was 0.05 mM. Addition of ascorbic acid, a common interferent, did not cause a significant change in capacitance. However, addition of fructose resulted in a 7% capacitance change. The sensor exhibited poor stability, with the capacitance signal degrading after 10 h.

Other metals which have been used for capacitive sensing include tantalum and platinum. An immunosensor for monitoring Ab–Ag reactions in real time was developed using a tantalum–tantalum oxide strip in a flow-through cell. Ab was covalently bound to the tantalum oxide surface using a silanisation step. The oxide thickness was of the order of 5 nm. The capacitance decreased by 0.2% on addition of Ag to a final concentration of 20 ng ml^{-1}. The total assay time was 20 min. The detection range of the system was from 0.2 to 20 ng ml^{-1}. It was observed that the slope of the decrease in capacitance increased with higher Ag concentrations. This is possibly an effect of reaction kinetics, but could be useful as an alternative mode of monitoring Ag concentration [4].

A differential measurement system based on impedimetric monitoring of platinum electrodes was developed for immunodetection in real time [30]. Anti-α-fetoprotein antibodies were immobilised in polysiloxane membranes via Pt-O groups on the surface of platinum electrodes for the working sensor. The reference sensor was functionalised in the same way, but the Ab immobilised was a control mouse monoclonal IgG Ab. The impedance of the sensor was determined to be composed of two main parts, the impedance of the membrane and the impedance of the electrochemical exchanges at the platinum surface. At frequencies above 10 kHz the membrane behaviour was mostly capacitive and the double layer capacitance reduced the impedance of electrochemical exchanges. However, the current crossing through the sensor was almost entirely controlled by the serial impedance of electrolyte resistance between the sensors and the counter electrode. Consequently, no significant result could be obtained above 10 kHz. Below 500 Hz the impedance was dominated by the Warburg impedance of electrochemical exchanges at the platinum surface. As a compromise, measurements were conducted at a medium frequency of 1.5 kHz. At this point the measurements were stable and mainly controlled by the phenomena occurring at the sensor surfaces. On addition of α-fetoprotein Ag there was a differential signal that was specific to the Ag–Ab binding. The response was considered to

---

2 Dodecanethiol is a chemical commonly used to fill defects in polymeric layers, generating a more dense and insulating layer on the gold electrode [14, 29].

be low and the sensitivity was reduced in solutions containing high protein concentrations. This raises questions about the usefulness of such a sensor in real samples for direct diagnostic purposes.

An alternative measurement approach for monitoring interfacial biomolecular interactions is being developed based on differential impedance spectroscopy [31]. Here, an ion displacement model was used to minimise non-specific effects and issues associated with biomolecular orientation. Ab–Ag interactions were monitored using impedance measurements of electrodes in a differential arrangement. The differential setup offers a significant advantage when compared to the single-channel system as it has the potential to amplify the signal. Background signal, due to changes in temperature, pH, electrolyte composition and non-specific binding, can be subtracted from specific signal. The group demonstrated direct observation of biomolecular interactions using covalent immobilisation on gold electrodes using APTES and a series of polymer-modified platinum electrodes based on polypyrrole without a redox probe. The surface loading using the APTES was calculated as 9×10^{16} cm^{-2} and for the polymer 9×10^{16} cm^{-2}. On addition of a specific antigen, cyanazine-BSA, the APTES electrodes exhibited a linear response region from 1 ppb to 20 ppm, and a detection limit of 0.1 ppb. The linear range of the polymer electrodes was from 0.1 ppb to 2 ppm, with a detection limit of 0.01 ppb. It is interesting that although the surface density of active binding sites was ten times higher for the APTES electrodes, the polymer electrodes were more sensitive. There may be a number of factors which account for this: steric hindrance of a closely packed monolayer, electrostatic repulsion forces to be overcome, and inherent amplification of the polypyrrole reduction process facilitated by the Ab–Ag reaction.

A biosensor comprising an anti-*Staphylococcus* enterotoxin B Ab covalently bound to an ultrathin platinum film sputtered onto a 100 nm silicon dioxide layer on a silicon chip was constructed to monitor the presence of Staphylococcus enterotoxin B [32]. The impedance drop that occurred on Ab–Ag binding was largest at a frequency of 100 Hz and a range of detection was established between 0.389 and 10.70 ng ml^{-1} *Staphylococcus* enterotoxin B. This system has the advantage that it is based on a silicon substrate and uses a metal which is compatible with standard CMOS processing.

Biomolecular binding on an ultrathin platinum film on a silicon base has been demonstrated [33], where the platinum layer is discontinuous and has an equivalent circuit of an array of resistive and capacitive elements. The dimensions of the gaps in these 25 Å films are similar to the sizes of typical macromolecules, such as Abs. The binding of small numbers of macromolecules may have a large impact on the values of individual capacitive elements of the Pt film. The sensor is fabricated by sputtering a thin layer of Ti/Pt onto a silicon dioxide on silicon surface. Frequency sweeps of the sensors before and after specific binding of alkaline phosphatase to immobilised anti-alkaline phosphatase antibody were made, where the impedance and phase angle of the sensors were measured. The impedance increase and phase angle became more negative in the frequency range 10–10,000 Hz. The sensitivity of the assay was calculated at 20 Hz. For an electrode of area 0.5 cm^2, the sensitivity for Ag binding is 5–15% at 0.04 pmol. The controls showed less than 3% change.

Although the above sensor systems have established some interesting concepts in research environments, none of them has achieved widespread use or commercialisation. An ideal sensor technology would be an integrated electrode and measurement instrument. Such sensors have many advantages over conventional ones. Arrays of sensors can be integrated for several measurands and multiplexed to a microcontroller, providing the possibility of real-time monitoring of several parameters. On-chip amplification, temperature monitoring and drift compensation, filtering and signal processing are possible. Miniaturisation increases the possibility of *in vivo* use, at least for short times, and requires smaller amounts of expensive bioactive substances. For in vitro use, the volume of analyte required decreases and in many cases higher sensitivities and better response times are obtained. Batch processing techniques as used for integrated circuits can drastically reduce costs [34]. To this end the possibility of using electrolyte–insulator–semiconductor (EIS) structures is being explored.

3.3
Capacitance Sensing at the Semiconductor/Solution Interface

The rapid development of IC technology during the last few decades has encouraged the fabrication of biosensors based on solid-state devices. Typically these sensors consist of a biorecognition layer immobilised on a signal-transducing chip, where the specific recognition of analyte modulates the space-charge region at the insulator/semiconductor interface. Originally, capacitance measurements with biologically modified EIS systems evolved from the ambition to construct an immunoFET [35, 36], but there are difficulties in monitoring a charge effect when Ab binds to Ag [7], possibly due to the formation of poorly insulating biolayers [8, 10]. It was found that C_V measurements of the structures yielded more useful information than static FET measurements. Any variation of surface potential leads to a shift of the capacitance versus voltage curve in the inversion range. The increase of the thickness of the dielectric layer induces a capacitance decrease in the accumulation range.

A study was carried out to determine if it was feasible to capacitively monitor Ab–Ag reactions directly, using EIS structures [7]. The transducer chip used was silicon with a dielectric layer of 70 nm of SiO_2. Anti-α-fetoprotein antibodies were covalently attached to the SiO_2 surface via glutaraldehyde cross-linking to 4-aminobutyldimethylmethoxy silane. Reference samples were prepared by immobilising non-specific proteins in the place of the Ab. The measurements were performed at 10 kHz. This resulted in a high-frequency C_V curve, where the capacitance in the inversion region remains low because at high frequencies the free electrons cannot follow the applied AC signal. There was a decrease in capacitance measured in the accumulation region between the bare Si–SiO_2 and the silanised sample. This indicated the formation of a new dielectric layer on the sensor chip surface. The capacitance decreased when Ab was immobilised and again when Ag bound to immobilised Ab. However, a shift in the C_V curve was only seen due to the silane layer. At the University of Genoa, the Biophysical and Electronic Engineering Department also explored the possibility of capacitive measurements of EIS structures, where the insulating layer consisted of

40 nm of Si_3N_4 and 35 nm of SiO_2 [37]. A glycoprotein called laminin, which is routinely used to promote cell adhesion to solid substrates, was used as a prototype biological material. However, here, no change in capacitance was seen on immobilisation of laminin, but a very small shift in the C_V curve towards the negative was observed.

Further work was carried out by the Laboratory of Physicochemistry of Interfaces at Lyon to characterise the EIS system. The group noted that the affinity of the Ab–Ag reaction could be characterised by the initial slope of the capacitance change on addition of Ag. The feasibility of such a system had been demonstrated for the detection of α-fetoprotein [7] and enterotoxin B from Staphylococcus aureus [10]. However, a reproducible detection of Ag-binding Ab proved more difficult to establish [38]. The problem partly stemmed from the difficulty of producing a suitable electrically blocking immobilised Ab layer on the transducer surface [39]. A study of the effects of different immobilisation procedures on the production of suitably blocking layers was undertaken. The biological molecule must also retain its recognising ability. The Ab immobilised was monoclonal anti-α-fetoprotein. The substrate was p-type silicon with 10 nm of thermal oxide on the surface. Three different coupling processes using three different spacing silanes were investigated: aminosilane, cyanosilane and heteropolysilsesquioxane. Each of the silanes provides a different hydrophobic character to the biolayer. The aminosilane yielded a hydrophilic biolayer. The C_V plot shows a decrease in capacitance for the silanisation, but only a negligible change in capacitance for the Ab immobilised layer. It is thought that inclusion of ions in the layer prevents it from having suitable electrical properties for use as a detector. The cyanosilane immobilisation results in a hydrophobic layer. The capacitance decreases on silanisation, but no further increase was seen on immobilisation of Ab. It was thought that the morphology of the layer prevented it from having an appropriate dielectric character to be used to detect Ag binding. In order to overcome such difficulties a polymeric membrane was prepared using heteropolysilsesquioxane. The formation of this layer induces a decrease in capacitance. A further decrease was seen on immobilisation of antibodies. It appeared that the Ab became incorporated into the polymeric membrane. However, on binding Ag, no decrease in capacitance was seen [40]. This work highlights the importance of developing a suitable strategy for coupling biological molecules to silicon transducers in order to carry out direct capacitance measurements of immunoreactions.

A more detailed electrical characterisation of the biolayer was needed. In an attempt to realise this, impedance measurements of grafted layers on Si–SiO_2 were investigated [38]. Initially the devices were characterised after silanisation. The impedance of the system using an electrolytic contact in an electrochemical cell was compared to a metallic contact using mercury probe impedance analysis. The measurements were performed in the accumulation regime. Silanisation was achieved using octadecyldimethyl(dimethylamino)silane, which has a long paraffinic chain and a relative permittivity close to 2. The theoretical length of the carbon chain of the silane is 23 Å. The theoretical capacitance was compared with the measured capacitance for the electrochemical and mercury probe impedance measurements. For both measurements there was a drop in capacitance for the devices with the silanised surface. The thickness of the immobilised silane was calculated as 20 Å for the electrochemical measurement and 7–11 Å for the mercury probe measurement. Interestingly, the capacitance measurement for the bare devices yielded the same calculated oxide thickness for both measurement methods, but the silane layer always had a calculated thickness two to three times lower for the mercury probe. The group postulate that the mercury compressed the silane layer. However, it is likely that in the electrolyte the silane becomes surrounded by ions, interacting with them to balance charge, and the apparent thickness increases. Nevertheless, the results showed that the silanised layer was insulating and caused a decrease in capacitance. Hysteresis experiments with the mercury probe showed that the silane layer had mobile charges associated with it, so it cannot be considered a perfect dielectric and conductance must be taken into account.

Following earlier work developing the optimum silane layer [40] the group continued to develop impedance analysis of the multi-layer structures [41]. The Si–SiO$_2$–silane–Ab stack was modelled and the impedance was analysed using a fitting program to extract the significant parameters. They found that the Ab layer coupled to aminosilane or cyanosilane did not act as a dielectric and thus, in the Ab–Ag binding, Ag could not be detected. Using polymeric membranes of different thicknesses the coupling of Ab to silane resulted in a decrease in impedance. It was thought the Ab was bound, not only to the membrane surface, but also in the membrane. This would change the electrical characteristics (conductivity, dielectric constant) of the layer. On addition of Ag the change in impedance measured was of the order of 1–2%. The dynamic range of the sensor was 10 to 150 ng ml^{-1} of specific Ag. The disadvantages of this system are that many impedance measurements need to be performed and computed before a final result can be obtained. This makes the measurement process slow and it does not allow real-time detection.

Capacitive biosensors are made of successive layers deposited on a solid support. Each immobilised layer must have suitable electrical properties and the recognition element must retain its activity. Fourier-transform infrared difference spectroscopy was used to analyse the build-up of successive layers on Si–SiO$_2$ structures [42]. The silane polymers used in this case were aminopropylmethyl–dimethylsiloxane and cyanopropylmethylsiloxane. Anti-α-fetoprotein antibodies were coupled via disuccinimidyl suberate to the aminopropylmethyl–dimethylsiloxane and coupled directly to the cyanopropylmethylsiloxane. By monitoring the successive layers capacitively it was shown that silanisation caused a decrease in capacitance. However, on immobilisation of Ab the capacitance increased and increased again on addition of Ag. FTIR analysis was performed on dried samples. Si–SiO$_2$ samples exhibited three characteristic peaks at 1074, 800 and 450 cm^{-1}. The aminopropylmethyl–dimethylsiloxane and cyanopropylmethylsiloxane had four strong peaks. At 1020 and 1080 cm^{-1} the peaks related to the Si–O–Si bond; at 800 and 1260 cm^{-1} the peaks related to the Si–CH$_3$ bond. Characteristic peaks for a protein were obtained for the Ab layer, with an amide I peak at 1660 cm^{-1} and an amide II peak at 1550 cm^{-1}. For the Ag the spectrum was similar to that for the Ab, but with a decrease of the amide I and amide II peaks. Spectral information indicated that there was some interpenetration of the Ab in the polymer, which may account for the increase in capacitance due to a change in dielectric property of the layer. This once again illustrates the importance of developing a suitable layer with which to immobilise the biorecogntion element.

An interesting approach to increasing the transducer surface area for biolayer immobilisation has been investigated by the Institute of Thin Film and Ion Technology, Jülich, Germany. The goal was to improve the stability and activity of the immobilised biolayer, in this case the enzyme penicillinase, by providing a sponge-like porous silicon surface [43–47]. Porous silicon offers a number of advantages. The pores can be adjusted by choosing appropriate etch conditions [46]. The biorecognition element can be fixed inside the pores, adding stability and preventing leaching. Due to the large surface area the transducer device can be reduced in size; high surface area means large capacitance. Heterobifunctional cross-linker molecules were employed to covalently bind enzymes to the planar substrate. The enzyme was adsorptively bound inside the porous substrate. The sensitivity of the sensors prepared by both methods was nearly identical for low penicillin concentrations, but the porous silicon-based EIS sensors exhibited a wider linear range and higher signals [43].

A surface nanostructure approach has also been investigated for use in a DNA hybridisation sensor [48]. An EIS capacitor, with a nanostructured active area formed by different approaches, was investigated. A nanoporous sensor active area was formed by electrochemical etching in hydrofluoric acid. A nanopyramidal structure was formed by anisotropic KOH etching, under specific conditions of temperature, concentration and added isopropanol. A mechanically degraded active area was formed using a diamond scribe. The responses of the devices on hybridisation of a complementary 19-mer oligonucleotide to immobilised single-stranded probe were compared. The porous silicon-based sensor showed the largest capacitance changes on hybridisation, but the coefficient of variation was high. The nanopyramidal KOH-etched devices exhibited more consistent responses. The mechanically degraded devices showed a significant improvement in sensitivity, capable of detecting 0.1 pM DNA.

To date with semiconductor-based capacitance sensors there have been two approaches: the FET and the planar EIS capacitor. A biorecognition event can be detected using an EIS capacitor either as a change in dielectric constant or a change in the thickness of a layer immobilised on the transducer. Using the FET, the biorecognition event results in a change in charge or ion concentration, which modulates the drain current of the device. The advantages of the FET sensor are those which accrue to the use of silicon processing: low cost, mass production and stable output. However, the devices are temperature sensitive, require a reference electrode and the fabrication of a different layer on the gate has yet to be perfected. The capacitive EIS is cheaper and easier to produce than the FET. However, the active area of the EIS sensor is larger than the gate region of the FET and is less suitable for miniaturisation. The choice of format will depend on the application.

4
Conclusions

Capacitance affinity biosensors, where a biorecognition element is immobilised on an electrode or between electrodes, can be used to detect specific interactions by measuring changes in the dielectric properties when an analyte binds. This recognition event can be detected directly, without the need for a label and in real time. Such sensors have been used for the detection of antigens, antibodies, proteins and oligonucleotides. However, despite potential advantages, very few practical systems are currently available. Fundamental issues still exist due to ambiguities associated with data interpretation and the complication of analyses based on equivalent circuits. In addition, compared to other biosensors based on optical or microgravimetric principles, the sensitivity of direct capacitance/impedance biosensors is relatively low, which may be due in part to nonlinear and non-stationary interfacial properties. However, as our understanding grows and advances are made in theoretical and practical means in the development of appropriate device modelling, fabrication, packaging, biomolecular immobilisation and data analysis, simple real-time measurements of analytes of interest can become a reality.

References

1. Schultz JS (1991) Sci Am 265:48
2. Varlan A (1996) PhD thesis, Katholieke Universiteit Leuven
3. Keynes RD (1983) Proc Roy Soc Lond B Biol Sci 220:1
4. Gebbert A, Alvarez-Icaza M, Stocklein W, Schmid RD (1992) Anal Chem 64:997
5. Bhunia AK, Jarad ZW, Naschansky K, Shroyer M, Morgan M, Gomez R, Bashir R, Ladisch M (2001) Impedance spectroscopy and biochip sensor for detection of Listeria monocytogenes. In: Chen Y-R, Tu S-I (eds) Photonic detection and intervention technologies for safe food. SPIE, Bellingham, p 32
6. Gileadi E, Kirowa-Eisner E, Penciner J (eds) (1975) Interfacial electrochemistry: an experimental approach. Advanced Book Program, Addison Wesley, Reading
7. Bataillard P, Gardies F, Jaffrezic-Renault N, Martelet C (1988) Anal Chem 60:2374
8. Berney H, Alderman J, Lane W, Collins JK (1997) Sens Actuators B Chem 44:578
9. Berney H, Alderman J, Lane W, Collins JK (1999) Sens Actuators B Chem 57:238
10. Billard V, Martelet C, Binder P, Therasse J (1991) Anal Chim Acta 249:367
11. Morrison SR (ed) (1980) Electrochemistry at semiconductor and oxidized metal electrodes. Plenum, New York
12. Kovacs IK, Horvai G (1994) Sens Actuators B Chem 18–19:315
13. Bergveld P (1991) Biosens Bioelectron 6:55
14. Bontidean I, Berggren C, Johansson G, Csoregi E, Mattiasson B, Lloyd JR, Jakeman KJ, Brown NL (1998) Anal Chem 70:4162
15. Newman AL, Hunter KW, Stanbro WD (1986) The capacitive affinity sensor: a new biosensor. In: Proceedings of the 2nd international meeting on chemical sensors, Bordeaux, p 596
16. Taylor RF, Marenchic IG, Cook EJ (1988) Anal Chim Acta 213:131
17. Eldefrawi NE, Sherby SM, Andreou AG, Mansour NA, Annau Z, Blum NA, Valdes JJ (1988) Anal Lett 21:1665
18. Laureyn W, Nelis D, Van Gerwen P, Baert K, Hermans L, Magnee R, Pireaux J-J, Maes G (2000) Sens Actuators B Chem 68:360

19. Berggren C, Bjarnason B, Johansson G (2001) Electroanal 13:173
20. Stoner G, Srinivasan S (1970) J Phys Chem 74:1088
21. Yamamoto N, Nagasawa Y, Sawai M, Sudo T, Tsubomura H (1978) J Immunol Methods 22:309
22. Alfonta L, Singh AK, Willner I (2001) Anal Chem 73:91
23. Arwin H, Lundstrom I, Palmqvist A (1982) Med Biol Eng Comput 20:362
24. Mirsky VM, Riepl M, Wolfbeis (1997) Biosens Bioelectron 12:977
25. Taira H, Nakano K, Naeda N, Takagi M (1993) Anal Sci 9:199
26. Rickert J, Gopel W, Beck W, Jung G, Heiduschka P (1996) Biosens Bioelectron 11:757
27. Panasyuk T, Mirsky V, Piletsky S, Wolfbeis O (1999) Anal Chem 71:4609
28. Panasyuk-Delaney T, Mirsky V, Ulbricht M, Wolfbeis O (2001) Anal Chim Acta 435:157
29. Cheng Z, Wang E, Yang X (2001) Biosens Bioelectron 16:179
30. Maupas H, Soldatkin, AP, Martelet C, Jaffrezic-Renault N, Mandrand B (1997) J Electroanal Chem 421:165
31. Sadik OA, Xu H, Gheorghiu E, Andreescu D, Balut C, Gheorghiu M, Bratu D (2002) Anal Chem 74:3142
32. De Silva MS, Zhang Y, Hesketh PG, Maclay GJ, Gendel SM, Stetter JR (1995) Biosens Bioelectron 10:675
33. Pak SC, Penrose W, Hesketh PJ (2001) Biosens Bioelectron 16:371
34. Lai R (1992) Bioelectrochem Bioenerg 27:121
35. Bergveld P, Sibbald A (1988) Analytical and biomedical applications of ion selective field effect transistors. In: Wilson and Wilson's comprehensive analytical chemistry, vol 23. Elsevier, Amsterdam
36. Schasfoort RBM, Kooyman RPH, Bergveld P, Greve J (1990) Biosens Bioelectron 5:103
37. Grattarola M, Cambiaso A, Cenderelli S, Tedesco M (1989) Sens Actuators 17:451
38. Schyberg C, Plossu C, Barbier D, Jaffrezic-Renault N, Martelet C, Maupas H, Souteyrand E, Charles M-H, Delair T, Mandrand B (1995) Sens Actuators B Chem 26–27:457
39. Gardies F, Martelet C, Colin B, Mandrand B (1989) Sens Actuators 17:461
40. Saby C, Jaffrezic Renault N, Martelet C, Colin B, Charles M-H, Delair T, Mandrand B (1993) Sens Actuators B Chem 15–16:458
41. Maupas H, Saby C, Martelet C, Jaffrezic-Renault N, Soldatkin AP, Charles M-H, Delair T, Mandrand B (1996) J Electroanal Chem 406:53
42. Sibai A, Elamri K, Barbier D, Jaffrezic-Renault N, Souteyrand E (1996) Sens Actuators B Chem 31:125
43. Thust M, Schöning MJ, Schroth P, Malkoc Ü, Dicker CI, Steffen A, Kordos P, Lüth H (1999) J Mol Catal B Enzym 7:77
44. Lüth H, Thust M, Steffen A, Kordos P, Schöning MJ (2000) Mater Sci Eng B Solid 69–70:104
45. Schöning MJ, Kurowski, Thust M, Kordos P, Schultze JW, Lüth H (2000) Sens Actuators B Chem 64:59
46. Schöning MJ, Malkoc Ü, Thust M, Steffen A, Kordos P, Lüth H (2000) Sens Actuators B Chem 65:288
47. Poghossian A, Yoshinobu T, Simonis A, Ecken H, Lüth H, Schöning MJ (2001) Sens Actuators B Chem 78:237
48. Lillis B, Hurley E, Galvin S, Mathewson A, Berney H (2002) A novel, high surface area, capacitance-based silicon sensor for DNA hybridisation detection. Eurosensors XVI, Prague, 15–18 Sept 2002

Chapter 4

Immunosensors and DNA Sensors Based on Impedance Spectroscopy

Eugenii Katz, Itamar Willner

Abstract. Impedance spectroscopy is a rapidly developing electrochemical technique for the characterization of biomaterial-functionalized electrodes and biocatalytic transformations at electrode surfaces, and specifically for the transduction of biosensing events at electrodes. The immobilization of biomaterials, e.g., antigen/antibodies or DNA on electrode surfaces alters the capacitance and interfacial electron transfer resistance of the electrodes. The impedance features of the modified electrodes can be translated into the equivalent electronic circuits consisting of capacitances and resistances. The kinetics and mechanisms of the electrochemical processes occurring at modified electrode surfaces could be derived from the analysis of the equivalent circuit elements. For example, electron transfer resistances can be found upon analysis of Faradaic impedance spectra in the form of Nyquist plots. The electron transfer rate constants can be calculated from the measured electron transfer resistances. Different immunosensors that use impedance measurements for the transduction of antigen–antibody complex formation on electronic transducers were developed. These include: (i) in-plane impedance measurements between electrodes separated by a nonconductive gap modified with antigen or antibody molecules, (ii) the amplified detection of antigen–antibody complex formation using biocatalyzed precipitation of an insoluble product as an amplification route and Faradaic impedance spectroscopy as readout signal, or (iii) the amplified detection of antigen–antibody complex by the biocatalytic dissolution of a polymer film associated with the electrode and Faradaic impedance spectroscopy as transduction method. Similarly, DNA biosensors using impedance measurements as readout signal were developed. The assembly of nucleic acid primers and their hybridization with the complementary DNA were characterized by Faradaic impedance spectroscopy using $[Fe(CN)_6]^{3-/4-}$ as a redox probe. Amplified detection of the analyte DNA using Faradaic impedance spectroscopy was accomplished by the coupling of functionalized liposomes or by the association of biocatalytic conjugates to the sensing interface, providing biocatalyzed precipitation of an insoluble product on the electrodes. The amplified detections of viral DNA and single-base mismatches in DNA were accomplished by similar methods. The theoretical background of the different methods and their practical applications in analytical procedures were outlined in the paper.

Keywords: Impedance spectroscopy, Biosensor, Immunosensor, DNA sensor

Abbreviations

AlkPh	Alkaline phosphatase
BSA	Bovine serum albumin
C_{Au}	Capacitance of a bare Au electrode
C_{dl}	Double-layer capacitance
C_{gap}	Capacitance of the gap between conductive electrodes
C_{mod}	Capacitance originating from the modified layer
CPE	Constant phase element
dATP, dTPP, dGTP, and dCTP	Deoxyribonucleoside triphosphates
DNA	Deoxyribonucleic acid
DNP	Dinitrophenyl
DNP-Ab	Dinitrophenyl antibody
ds	Double-stranded
E^0	Standard redox potential
ELISA	Enzyme-linked immunosorbent assay
f	Excitation frequency, Hz
FET	Field-effect transistor
GOx	Glucose oxidase
HIV	Human immunodeficiency virus
HRP	Horseradish peroxidase
I_C	Capacitance current
I_F	Faradaic current
$I(j\omega)$	Current phasor
IgG	Immunoglobulin G
k_{et}	Electron transfer rate constant
NADH	1,4-Dihydro-β-nicotinamide adenine dinucleotide
NAD(P)+	β-Nicotinamide adenine dinucleotide (phosphate)
PCR	Polymerase chain reaction
R_{Au}	Electron transfer resistance at a bare Au electrode
R_{et}	Electron transfer resistance
R_{gap}	Resistance of the gap between conductive electrodes
R_{mod}	Electron transfer resistance originating from the modified layer
RNA	Ribonucleic acid

R_s Ohmic resistance of the electrolyte solution
SEB   *Staphylococcus enterotoxin B*
TS   Tay–Sachs genetic disorder
$U(j\omega)$ Voltage phasor
VSV   Vesicular stomatitis virus
$|Z|$ Absolute impedance value
$Z_{im}(\omega)$ Imaginary component of complex impedance
$Z_{re}(\omega)$ Real component of complex impedance
Z_W Warburg impedance
δ_{dl} Thickness of the double-charged layer
ε_0 Dielectric constant of the vacuum
ε_{dl} Dielectric permittivity of material in the double-charged layer
ε_ϱ Effective dielectric constant
ω Excitation frequency, rad s^{-1}
ω_0 Characteristic frequency

1
Introduction

Bioelectronics is a rapidly progressing field at the junction of chemistry, bio-chemistry, physics, and material science [1]. The basic feature of a bioelectronic device is the interaction of a biomaterial with a conductive or semiconductive support, and the electronic transduction of the biological functions associated with the biological matrices. Biomaterials that can be interacted with electronic transducers include proteins, that is, enzymes [2–4], receptors [5, 6], antibodies or antigens [7–10], oligonucleotides or DNA fragments [11–13], or low molecular weight molecules exhibiting affinity interactions with other biomaterials such as cofactors, namely, $NAD(P)^+$ [14], biotin [15], or, for example, saccharides exhibiting affinity interactions with lectins [16]. Different electronic signals were employed to transduce the biological functions occurring at the electronic supports [17]. These include electrical transduction, such as current [18], potential changes [19, 20], piezoelectric transduction [21–24], field-effect transistor (FET) transduction [25, 26], photoelectrochemical transduction [27], and others [28–33]. Many of the biosensor devices involve the formation of a recognition complex between the sensing biomaterial and the analyte in a monolayer or thin film configuration on the electronic transducer [34]. Formation of the complex on a conductive or semiconductive surface alters the capacitance and the resistance at the surface/electrolyte interface. Furthermore, the buildup of the sensing biomaterial film on the conductive or semiconductive support alters the capacitance and resistance properties of the solid support/electrolyte interface. Impedance measurements provide detailed information on capacitance/resistance changes occurring on conductive or semiconductive surfaces. Thus, impedance spectroscopy [35, 36], including non-Faradaic impedance measurements resulting in capacitance sensing [37], is becoming an attractive electrochemical tool to characterize biomaterial films associated with electronic elements, thus allowing transduction of biorecognition events on the respective surfaces.

The present paper addresses recent advances in the application of impedance spectroscopy to the development of immunosensors and DNA sensors, as well as outlining different kinds of impedance spectroscopy such as non-Faradaic capacitance measurements, Faradaic impedance spectroscopy in the presence of an external redox label, and in-plane alternative voltage conductivity measurements. The paper includes a theoretical background important to understand the analytical applications of impedance measurements and an overview of the experimental results exemplifying specific biosensor systems.

2
Impedance Spectroscopy: Theoretical Background

Impedance spectroscopy is an effective method for probing the features of surface-modified electrodes [38, 39]. A small-amplitude perturbing sinusoidal voltage signal is applied to the electrochemical cell, and the resulting current response is measured. The impedance is calculated as the ratio between the system voltage phasor, $U(j\omega)$, and the current phasor, $I(j\omega)$, which are generated by a frequency response analyzer during the experiment, Eq. 1, where $j = \sqrt{-1}$ and ω and f (excitation frequency) have units of rad s^{-1} and Hz, respectively. The complex impedance $Z(j\omega)$ can be presented as the sum of the real, $Z_{re}(\omega)$, and imaginary, $Z_{im}(\omega)$ components that originate mainly from the resistance and capacitance of the cell, respectively.

$$Z(j\omega) = \frac{U(j\omega)}{I(j\omega)} = Z_{re}(\omega) + jZ_{im}(\omega); \quad \text{where } \omega = 2\pi f \quad (1)$$

Electrochemical transformations occurring at the electrode/electrolyte interface can be modeled by extracting components of the electronic equivalent circuits that correspond to the experimental impedance spectra. A general electronic equivalent circuit (Randles and Ershler model [38–41]), which is very often used to model interfacial phenomena, includes the ohmic resistance of the electrolyte solution, R_s, the Warburg impedance, Z_W, resulting from the diffusion of ions from the bulk electrolyte to the electrode interface, the double-layer capacitance, C_{dl}, and the electron transfer resistance, R_{et}, that exists if a redox probe is present in the electrolyte solution (Scheme 1A). The parallel elements (C_{dl} and Z_W+R_{et}) of the equivalent circuit are introduced since the total current through the working interface is the sum of distinct contributions from the Faradaic process, I_F, and the double-layer charging, I_C. Since all of the current must pass through the uncompensated resistance of the electrolyte solution, R_s, it is inserted as a series element in the circuit. The two components of the electronic scheme, R_s and Z_W, represent bulk properties of the electrolyte solution and diffusion features of the redox probe in solution. Therefore, these parameters are not affected by chemical transformations occurring at the electrode surface. The other two components in the scheme, C_{dl} and R_{et}, depend on the dielectric and insulating features at the electrode/electrolyte interface. The dou-

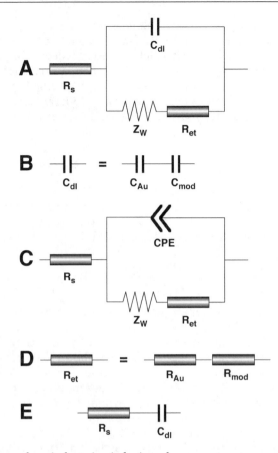

Scheme 1. A General equivalent circuit for impedance spectroscopy measurements in an electrochemical cell. **B** Equivalent circuit corresponding to the double-layer capacitance, C_{dl}, that includes the variable component, C_{mod}, controlled by the modifier layer. **C** General equivalent circuit for impedance spectroscopy with the constant phase element (CPE) depending on the roughness of the electrode surface. **D** Equivalent circuit for the electron transfer resistance, R_{et}, that includes the variable component, R_{mod}, corresponding to the different modifier states. **E** Equivalent circuit for non-Faradaic impedance spectroscopy measurements in the absence of the redox probe

ble-layer capacitance depends on the dielectric permittivity introduced into the double-charged layer molecules, ε_{dl}, and for the less polar molecules the capacitance should be smaller, Eq. 2, where $\varepsilon_0 = 8.85 \times 10^{-12}$ F m^{-1} is the dielectric constant of the vacuum, ε_ϱ is the effective dielectric constant of the layer separating the ionic charges and the electrode surface, A is the electrode area, and δ is the thickness of the separating layer.

$$C_{dl} = \frac{\varepsilon_{dl} A}{\delta}; \quad \text{where } \varepsilon_{dl} = \varepsilon_0 \cdot \varepsilon_\varrho \tag{2}$$

In the equivalent electronic circuit the double-layer capacitance, C_{dl}, can be represented as the sum of a constant capacitance of an unmodified electrode (e.g., for a polycrystalline Au electrode, $C_{Au} \approx 40$–60 μF cm^{-2}, depending on the applied potential [42]) and a variable capacitance originating from the electrode surface modifier, C_{mod}, connected as series elements (Scheme 1B). Any electrode modifier of insulating features decreases the double-layer capacitance as compared to the pure metal electrode. Thus, the double-layer capacitance could be expressed by Eq. 3.

$$\frac{1}{C_{dl}} = \frac{1}{C_{Au}} + \frac{1}{C_{mod}} \tag{3}$$

Sometimes, particularly when the electrode surface is rough [43], the electronic properties of the interface cannot be described sufficiently well with a capacitive element, and a constant phase element (CPE), Eq. 4, should be introduced instead of C_{dl} (Scheme 1C).

$$CPE = A^{-1} \cdot (j\omega)^{-n} \tag{4}$$

The constant phase element reflects nonhomogeneity of the layer, and the extent of the deviation from the Randles and Ershler model is controlled by the parameter n in Eq. 4. Introduction of a CPE element into an equivalent electronic circuit instead of a simple capacitance element was shown to be important for the modeling of primary protein layers on an electrode surface, and the parameter n was found to be ca. 0.97–0.98 [44]. The CPE has meaning of capacitance and the coefficient A becomes equal to the C_{dl} when n=1 [38, 39].

The electron transfer resistance, R_{et}, controls the electron transfer kinetics of the redox probe at the electrode interface. Thus, the insulating modifier on the electrode is expected to retard the interfacial electron transfer kinetics and to increase the electron transfer resistance. The electron transfer resistance at the electrode is given by Eq. 5, where R_{Au} and R_{mod} are the constant electron transfer resistance of the unmodified electrode and the variable electron transfer resistance introduced by the modifier, in the presence of the solubilized redox probe, respectively. These resistances are also connected as series elements in the equivalent electronic circuit (Scheme 1D).

$$R_{et} = R_{Au} + R_{mod} \tag{5}$$

A typical shape of a Faradaic impedance spectrum (presented in the form of an impedance complex plane diagram – a Nyquist plot) includes a semicircle region lying on the Z_{re}-axis followed by a straight line (Scheme 2, curve a). The semicircle portion, observed at higher frequencies, corresponds to the electron transfer-limited process, whereas the linear part is characteristic of the lower frequency range and represents the diffusionally limited electrochemical process. In the case of very fast electron transfer processes, the impedance spectrum could include only the linear part, curve b, whereas a very slow electron-transfer step results in a large semicircle region that is not accompanied by a

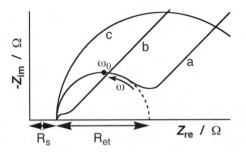

Scheme 2. Schematic Faradaic impedance spectra presented in the form of a Nyquist plot for: a) a modified electrode where the impedance is controlled by diffusion of the redox probe (low frequencies) and by the interfacial electron transfer (high frequencies); b) a modified electrode where the impedance is mainly controlled by diffusion of the redox probe; and c) a modified electrode where the impedance is controlled by the interfacial electron transfer within the entire range of the applied frequencies. The arrow shows the direction of the frequency increase

straight line, curve c. The electron transfer kinetics and diffusional characteristics can be extracted from the spectra. The semicircle diameter equals the electron transfer resistance, R_{et}. The intercept of the semicircle with the Z_{re}-axis at high frequencies ($\omega \rightarrow \infty$) is equal to the solution resistance, R_s. Extrapolation of the circle to lower frequencies yields an intercept corresponding to $R_s + R_{et}$. The characteristic frequency, ω_0, given by Eq. 6, has the meaning of the reciprocal of the time constant of the equivalent circuit. The maximum value of the imaginary impedance in the semicircle part corresponds to $Z_{im}=0.5 \cdot R_{et}$ and is achieved at the characteristic frequency, ω_0.

$$\omega_o = \left(C_{dl} \cdot R_{et} \right)^{-1} \tag{6}$$

In the absence of any redox label in the electrolyte solution, only non-Faradaic impedance is operative. The electron transfer characterizing parameters, R_{et} and Z_W, become infinite, and the equivalent circuit can be simplified, as shown in Scheme 1E. The variable component in the circuit is presented by C_{mod}, and it affects the imaginary part of the impedance, Z_{im}, Eq. 7, considering only the dense part of the double-charged layer.

$$Z_{im} = \frac{1}{\omega \left(C_{Au} + C_{mod} \right)} \tag{7}$$

The experimental results can be analyzed graphically [38, 39] (in the case of Faradaic impedance spectra usually by the use of the Nyquist coordinates: Z_{im} vs. Z_{re}) in the frame of the theoretical model [40, 41]. A computer fitting of the experimental data to a theoretical model represented by an equivalent electronic circuit is usually performed. All electronic characteristics of the equivalent circuit and the corresponding physical parameters of the real electrochemical system can be extracted from such analysis. Since the variable parameters of the system represent the functions of the modifying layer and its composition, they can be used to quantitatively characterize the layer. Analysis of the $Z_{re}(\omega)$ and $Z_{im}(\omega)$ values observed at different frequencies allows the calculation of the following important parameters: (a) the double-layer capacitance, C_{dl}, and its variable component, C_{mod}; (b) the electron transfer resistance, R_{et}, and its variable component, R_{mod}; and (c) the electron transfer rate constant, k_{et}, for the applied redox probe derived from the electron transfer resistance, R_{et}. Thus, impedance spectroscopy represents not only a suitable transduction technique to follow the interfacial interactions of biomolecules, but it also provides a very powerful method for the characterization of the structural features of the sensing interface and for explaining mechanisms of chemical processes occurring at the electrode/solution interfaces [45].

3
Immunosensors Based on Impedance Spectroscopy

Among the biomaterial-based sensing devices immunosensors are anticipated to be an important class of sensing systems in clinical diagnosis, food quality control, environmental analysis, detection of pathogens or toxins, and even the detection of explosives or drugs [46–48]. Several methodologies for the electrochemical transduction of the formation of antigen–antibody complexes on the electrode surface were reported [24, 49]. Impedance spectroscopy (including Faradaic impedance in the presence of a redox probe and non-Faradaic capacitance measurements) represents an important electronic transduction means of antigen–antibody binding interactions on electronic elements.

3.1
Immunosensors Based on In-Plane Impedance Measurements between Conductive Electrodes

In-plane impedance measurements were performed for analyzing antigen–antibody recognition processes. The detection path is based on the organization of two metallic conductive electrodes on an electrical insulator leaving a nonconductive gap between the electrodes (Scheme 3A). The gap was modified with a biomaterial capable of specific binding of a complementary unit, e.g., the gap was modified with antigen molecules for binding of the complementary antibody molecules, or an antibody was linked to the gap to bind the respective antigen molecules. Affinity binding of the respective complementary component results in a change of the electrical properties of the gap, thus affecting the impedance between the electrodes. The impedance changes can be measured as variation of conductivity, $\varrho_{gap}=R_{gap}^{-1}$, between the electrodes, that represents changes in the real part, Z_{re}, of the complex impedance [50] or variation of the gap capacitance, C_{gap}, that originates from the change of the imaginary part, Z_{im}, of the complex impedance [51]. In order to reach a high sensitivity, a short distance leaving a small dielectric gap between the conductors should separate the electrodes. For providing a large sensing surface on the thin gap, the conductive electrodes could be made as a pattern of interdigitated fingers [51] (Scheme 3B).

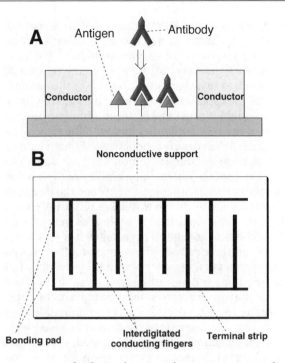

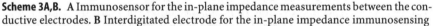

Scheme 3A,B. A Immunosensor for the in-plane impedance measurements between the conductive electrodes. B Interdigitated electrode for the in-plane impedance immunosensing

The applicability of nanoscale interdigitated electrodes for impedimetric sensing of the protein immobilization was studied and their potential use for immunosensing was addressed [52, 53].

For example, two copper conductors (25 μm high and 50 μm wide) were positioned on a surface of an insulating material, having a distance of 50 μm between them [51]. In order to observe capacitance changes for the gap, the conductors were insulated by a 1 μm layer of a nonconductive polymer that reduced the influence of Faradaic currents. A 0.3 μm layer of SiO_2 was vacuum deposited onto the polymer to provide suitable functional groups for the immobilization of a recognition element. Then antigen molecules were covalently bound to the gap surface via a silanization technique. When the functionalized sensing interface located in the gap between the electrodes was reacted with the complementary antibody units, the bulky protein molecules were bound to its surface replacing an aqueous solution. Due to the low dielectric constant of the antibody compared with water there was a change in the dielectric properties between the electrodes, causing an alteration in the capacitance. A similar approach was used for detection of neurotransmitters and toxins [54], acetylcholine [55], alkaline phosphatase [56], and IgG antibodies [57]. In order to amplify the dielectric changes resulting from the biorecognition event, bubbles of O_2 were generated in the sensing layer [58]. The sensing interface in the gap between the electrodes was functionalized with HIV antigens that were reacted with HIV antibodies. Then the sensing interface was reacted with anti-HIV antibodies conjugated to the catalase enzyme. The enzyme stimulated the biocatalytic decomposition of added H_2O_2 yielding bubbles of O_2 in the sensing layer between the electrodes. This resulted in a significant decrease of the medium dielectric constant. The extent of the capacitance change in the system was dependent on the amount of oxygen bubbles generated by the biocatalytic process and thus related to the amount of the HIV antibodies bound to the sensing interface.

Antigen molecules labeled with conductive polymer chains were applied for the competitive analysis of antigens on an antibody-functionalized gap [59]. The impedance change of the system upon binding of the labeled antigen mainly originated from the conductivity increase between the electrodes (Scheme 4A). Polyaniline (1) was covalently bound to a rabbit IgG using glutaric dialdehyde as a linker, and then the functionalized IgG molecules were added to the gap functionalized with anti-rabbit goat IgG. This resulted in a significant increase in the conductivity between the electrodes measured at a fixed frequency of 20 kHz. Competitive immunoassay was performed in the presence of different concentrations of unlabeled IgG and a constant amount of the polyaniline–IgG conjugate. A decrease of the sensor response was observed as the concentration of unlabeled IgG increased, due to a decrease in the amount of the conductive IgG associated with the gap. The decrease of the system response was proportional to the IgG analyte concentration that could be sensed, with a sensitivity limit that corresponds to 0.5 µg mL^{-1}. Another immunosensor was based on the conductance change of a phthalocyanine thin film assembled beneath the antigen–antibody sensing interface using the biocatalytic generation of iodine as a doping reagent of the phthalocyanine film [60] (Scheme 4B). Horseradish peroxidase, HRP, was covalently bound to rabbit IgG. The HRP-labeled IgG was reacted with the sensing interface functionalized with anti-rabbit IgG, and then the bound HRP biocatalyzed oxidation of I$^-$ ions to I_2 in the presence of H_2O_2. The film bridging the gap between the electrodes was composed of tetra-tert-butyl copper phthalocyanine (2) that revealed substantial decrease in the film resistance upon interaction with iodine (I_2). This allowed the direct sensing of the HRP-labeled IgG, as well as the sensing of unlabeled IgG using a competitive assay.

A discontinuous ultrathin metal film consisting of conductive metallic islands separated by nanosized gaps was placed between conductive electrodes and used for the impedimetric immunosensing of Staphylococcus enterotoxin B (SEB) [61, 62]. A Pt film was deposited on a silicon dioxide surface by electron-beam evaporation at rate of 0.1 Å s^{-1} resulting in a nominal thickness of 25 Å, as indicated by a quartz crystal microbalance analysis. Transmission electron microscopy images revealed that the film was composed of regularly distributed Pt islands of radii 30 to 60 Å, leaving uncoated silicon dioxide surface domains of 21 to 50 Å. The discontinuous metal film was further modified with a polysiloxane layer, which was used for the covalent coupling of anti-SEB antibodies. The impedance |Z| spectra obtained for the anti-SEB-functionalized sensing layer in phosphate buffer shows the highest impedance value of ca. 1.8 kΩ at the frequency of 1×10^2 Hz (Fig. 1A, curve a), that was approximately three orders of magnitude less than the impedance of the dry film. This suggests that the mech-

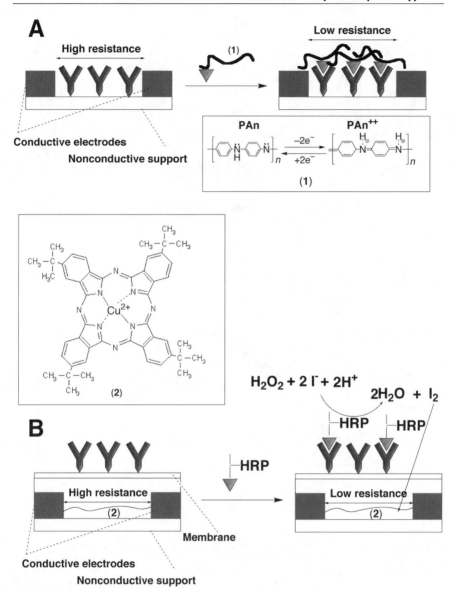

Scheme 4A,B. Various mechanisms of the in-plane impedance immunosensing. A The increase of the immunosensing film conductivity upon binding of the polyaniline-labeled IgG antigen to the anti-IgG antibody functionalized membrane. B The increase of the conductivity of the tetra-tert-butyl copper phthalocyanine film upon I_2 generation by the immunosensing membrane in the presence of HRP-labeled antigen, H_2O_2, and iodide

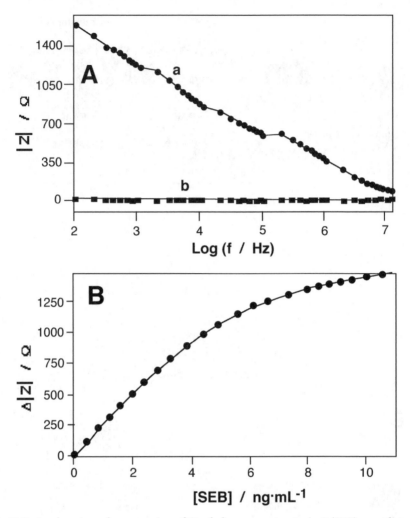

Fig. 1A,B. In-plane impedance sensing of *Staphylococcus enterotoxin B* (SEB) on a discontinuous ultrathin Pt film functionalized with anti-SEB antibodies. A Absolute impedance value, $|Z|$, as a function of the applied frequency. Data recorded: a) in phosphate buffer prior to the SEB binding; b) after the sensing film was exposed to the SEB solution, 100 ng mL^{-1}, for 1 h. **B** A calibration plot derived from the impedance measurements at a frequency of 100 Hz using different SEB concentrations. (Reproduced from Ref. [61] with permission)

anism of electrical conductance through the Pt islands was dominated by the ionic conductivity of the hydrated and buffered protein layer. The reaction of the anti-SEB-functionalized surface with the complementary SEB analyte resulted in a 98% decrease in the film impedance to the value of 25 Ω at 1×10^2 Hz (Fig. 1A, curve b). The impedance decrease originates from the increase of the amount of the hydrated proteins between the conductive Pt islands. A calibration plot for analysis of SEB was derived from the impedance spectra obtained at different concentrations of the toxin (Fig. 1B).

3.2
Immunosensors Based on Interfacial Impedance Measurements at Conductive Electrodes

Formation of antigen–antibody complexes on conductive supports yields a chemically modified film on the surface that alters the impedance features of the interface (Scheme 5). The formation of the antigen–antibody complex perturbs the double-charged layer existing at the electrode/electrolyte interface resulting in the increase of its thickness, δ_{dl}, and providing the insulation at the electrode surface with respect to redox labels added to the solution. This results in the capacitance change and electron transfer resistance change at the interface, respectively.

Various techniques were developed for the immobilization of the biorecognition components, antigens or antibodies, on electrode surfaces. For example, thin oxide layers existing on some electrode surfaces (e.g., indium–tin oxide electrode, Pt/Pt–OH electrode) provide hydroxyl groups for silanization and

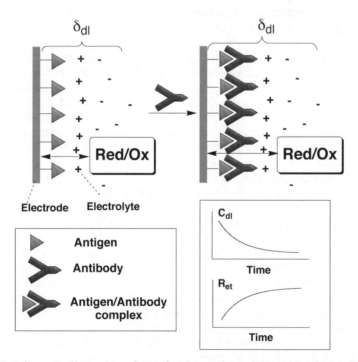

Scheme 5. Schematic illustration of interfacial impedance immunosensing. A bioaffinity interaction between an antibody and an antigen-functionalized electrode increases the double-charged layer thickness, δ_{dl}, and inhibits the interfacial electrochemical process of a redox probe. The electrode capacitance, C_{dl}, and the electron transfer resistance, R_{et}, are changed, respectively

further covalent binding of antigens or antibodies [63] (Scheme 6A). A versatile method for the organization of antigen–antibody monolayers on electrodes is based on the self-assembling of thiols on a gold electrode surface [64]. Thiol or disulfide groups were synthetically introduced into the antigen or antibody molecules in order to enable their direct tethering to Au electrodes [65–68]. For example, an epitope that consists of amino acids 135–154 of the capsid protein VP1 of the foot-and-mouth disease virus was used as a model antigen recognized by respective monoclonal and polyclonal antibodies [65, 66]. This synthetic polypeptide was covalently bound to ω-hydroxyundecanethiol, $HS(CH_2)_{11}OH$, using a succinic acid spacer, and the synthetic antigen (3) was self-assembled at a Au electrode surface (Scheme 6B). Alternatively the biorecognition molecules could be chemically bound to a thiol monolayer preassembled at a Au electrode [69], e.g., using a cystamine self-assembled monolayer as a primary functionalized interface (Scheme 6C). A large variety of chemical coupling reactions were applied to covalently link antigen or antibody molecules to functional groups provided by self-assembled thiol monolayers [70]. Noncovalent affinity binding of antibodies to an electrode surface has been exemplified with a biotinylated antibody (anti-human IgG) bound to a biotinylated polypyrrole layer through an avidin linker [71] (Scheme 6D). Direct visualization of IgG binding on different surfaces was achieved using AFM and the results were compared with impedance spectra changes [72].

Impedance spectroscopy allows the detection of capacitance changes at the interfaces that originate from biorecognition events. These capacitance changes can be derived from the imaginary part, Z_{im}, of the complex impedance spectra. The interfacial capacitance can be measured at various biasing potentials (differential capacitance), thus allowing the electrical probing of the system when different charges exist at the electrode. Since antigens and antibodies are usually charged protein molecules, formation of the bioaffinity complexes between them could be affected by the charge applied to the electrode surface. Furthermore, the orientation of the antigen–antibody complex vs. the electrode surface could be affected by the electrode charge. Therefore, capacitance changes observed upon formation of the bioaffinity complexes are potential dependent and the optimal conditions for their measurement could be found. Non-Faradaic impedance spectroscopy in the absence of a redox probe was applied to follow the biorecognition events at the functionalized electrode surfaces. For example, Fig. 2A,B shows small changes of the absolute impedance value, $|Z|$, and the impedance imaginary part, Z_{im}, upon the association of the foot-and-mouth disease virus antibody with a thiol monolayer functionalized with the synthetic polypeptide antigen [66]. Although the impedance changes are small, they allowed the analysis of the kinetics of the antigen–antibody binding (Fig. 2C). Similar capacitance measurements were applied for quantitative immunoanalysis [51]. For example, anti-rabbit IgG antibodies coupled to a self-assembled cystamine monolayer on a Au electrode were used for detection of the respective antigen in a linear range between 0.5 and 8.5 ng mL^{-1}, based on the analysis of capacitance changes by non-Faradaic impedance spectroscopy [73].

Faradaic impedance spectroscopy is usually considered to be more sensitive to the insulation of the electrode surface upon the binding of bulky antibodies

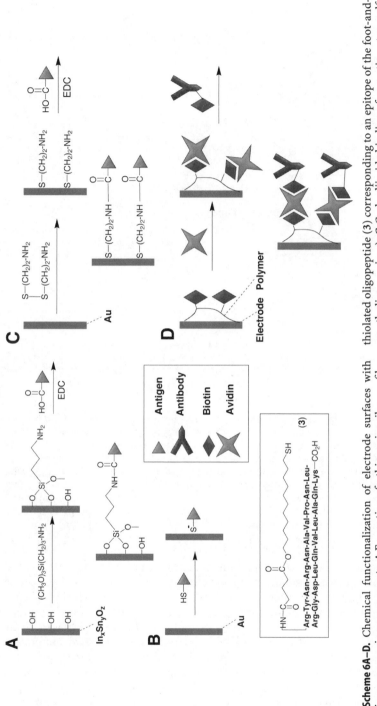

Scheme 6A–D. Chemical functionalization of electrode surfaces with immunosensing components. **A** Formation of a thin aminosiloxane film on a surface of a metal oxide electrode (e.g., indium-tin oxide) followed by carbodiimide coupling of an antigen (EDC=1-ethyl-3-(3-dimethyla minopropyl)carbodiimide). **B** Self-assembling of thiolated antigen molecules on a Au electrode surface (the exemplified antigen represents a thiolated oligopeptide (3) corresponding to an epitope of the foot-and-mouth disease virus). **C** Carbodiimide binding of an antigen to a self-assembled cystamine monolayer on a Au electrode. **D** Affinity binding of a biotinylated antibody to a biotin-functionalized polymer film mediated by avidin units

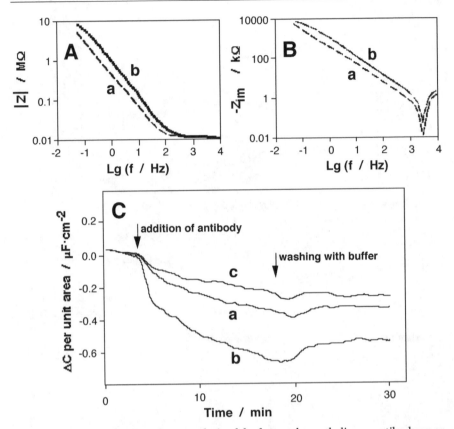

Fig. 2A–C. Non-Faradaic impedance analysis of the foot-and-mouth disease antibody on an electrode surface functionalized with the respective antigen (3). **A** The absolute impedance value, |Z|, as a function of the applied frequency: a) prior to the antibody binding; b) after the formation of the saturated antigen/antibody complex film. **B** The imaginary part of the impedance, Z_{im}, as a function of the applied frequency: a) prior to the antibody binding; b) after the formation of the saturated antigen/antibody complex film. **C** The electrode capacitance changes upon immunosensing: a) using the antibody at a concentration of 1.74 mg mL^{-1}; b) using the antibody at a concentration of 17.4 mg mL^{-1}; c) using a nonspecific antibody at a concentration of 17.5 mg mL^{-1}. (Reproduced from Ref. [66] with permission)

to the antigen-functionalized electrode surfaces. In order to measure the full electrical impedance at the interface, including its real part, Z_{re}, a redox probe (often [Fe(CN)$_6$]$^{3-/4-}$ ions are used) is added to the electrolyte solution. The redox probe provides a Faradaic current at the electrode surface, if the appropriate potential is applied to the electrode (usually the electrode is biased at the standard potential, E^0, of the redox probe). The changes in the real component of the impedance spectrum observed upon bioaffinity complex formation at the electrode surface are usually higher than the respective capacitance changes; however, they are usually measured at a fixed potential. For example, interaction of a human mammary tumor-associated glycoprotein with an electrode surface functionalized with its monoclonal antibody resulted in significant insulation of the electrode surface that was reflected by a large increase of the electron transfer resistance [74]. However, it should be noted that the changes in the Faradaic impedance spectrum upon antigen–antibody complex formation are not always large. If the sensing layer is not well organized and contains many pinholes between the biomolecules, and the redox probe is sufficiently small to freely penetrate through these pinholes, the changes observed in the Faradaic impedance spectra could be minor. For example, Fig. 3 shows a large increase in the electron transfer resistance, R_{et}, when a bare Au electrode was covered with an antigen–thiol monolayer, curve b, whereas a relatively small R_{et} increase was observed upon further binding of the antibodies to the antigen sensing monolayer, curve c, [65]. It has been shown that a mixed self-assembled monolayer composed of the antigen-functionalized thiols and nonderivatized thiols provides better binding of the antibody molecules because the antigen units are diluted and less screened by the neighbor units. Also in the mixed monolayer configuration, the impedance measurements do not suffer from the penetration of the redox probe through pinholes between the antigen–antibody complex because they are blocked by the nonderivatized thiols. Another approach that allows minimizing the redox probe penetration through the pinholes includes the application of bulky redox molecules (e.g., NADH) or even redox en-

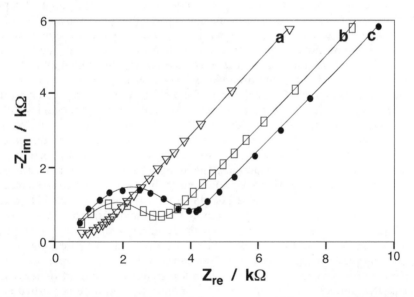

Fig. 3. Faradaic impedance spectra recorded upon sensing of foot-and-mouth disease antibodies at a Au electrode functionalized with a self-assembled monolayer of the thiolated polypeptide (3) corresponding to the foot-and-mouth disease antigen. a) Bare Au electrode, b) (3) monolayer-functionalized Au electrode prior to the reaction with the antibody, c) (3) monolayer-functionalized Au electrode after the reaction with the antibody. The data were recorded in the presence of $[Fe(CN)_6]^{3-/4-}$, 2 mM. (Reproduced from Ref. [65] with permission)

zymes [69]. For example, glucose oxidase (GOx) with tethered ferrocene units was used in the presence of glucose as a redox probe to sense electrode surface blocking upon the antigen–antibody complex formation. Glucose oxidase tethered with ferrocene electron relays was electrically contacted at the electrode surface functionalized with small dinitrophenol (DNP)-antigen molecules, thus providing a bioelectrocatalytic current for glucose oxidation. However, when large DNP-antibody molecules were associated with the electrode surface, the electrical contact between the GOx biocatalyst and the electrode was effectively blocked because the bulky biocatalyst cannot penetrate through small-size pinholes in the antigen–antibody complex monolayer. This allows the Faradaic impedance spectra changes to be especially well observed [75, 76].

Redox polymers (e.g., polypyrrole) were used as sublayers for the immobilization of immunosensing units [77–80]. When the antigen–antibody binding proceeds atop of the redox polymer, the highly charged protein molecules appear at the interface. This results in exchange of the counter ions between the redox polymer support and the protein layer that yields a significant change of the electrochemical properties of the polymer. Faradaic impedance spectroscopy was used to follow this change, thus allowing the impedimetric detection of the antigen–antibody complex formation. The mechanism of the process is, however, not well understood and the quantitative interpretation of the results is difficult.

Impedimetric immunosensors very often suffer from nonspecific adsorption of biomolecules (e.g., proteins) accompanying the analyte [74]. Special attempts were made to minimize the nonspecific adsorption phenomenon and its influence on the impedance measurements. These included special chemical treatment of the immunosensing electrodes, as well as development of special electrochemical cells and instruments. For example, bovine serum albumin (BSA) was coimmobilized in the immunosensing layer in order to decrease nonspecific binding of proteins [65]. Dual electrochemical flow cells were introduced to compare the impedance spectra in the two detection channels: with and without the analyte sample [63, 81]. Experimental [81] and theoretical [63] details on the use of differential impedance spectroscopy for immunosensing were discussed.

Another point that has been addressed is the often-observed poor reversibility of the immunological reactions. The binding between the antigen and the antibody is usually very strong. By applying acidic media or a high concentration of salts or urea, the proteins associated with the antigen–antibody complex are unfolded, leading to separation of antigen and antibody. These severe conditions affect the protein that acts as the receptor component of the sensor and the reactivation of the sensing interface after separation of the antigen–antibody complex is often impossible. Even in the case where the receptor component is not a protein and has high chemical stability (e.g., dinitrophenyl antigen, DNP-antigen), the structure of the monolayer could be perturbed and the thiol anchor groups could be dissociated from the surface upon the conditions required for affinity complex dissociation. Photoisomerizable antigen molecules were suggested as a means to tailor reversible and reusable immunosensor electrodes [75, 76, 82, 83] (Scheme 7A). By this approach the antigen in photoisomer state A enables sensing of the antibody by the electronic transduction of the formation of the antigen-antibody complex at the transducer sur-

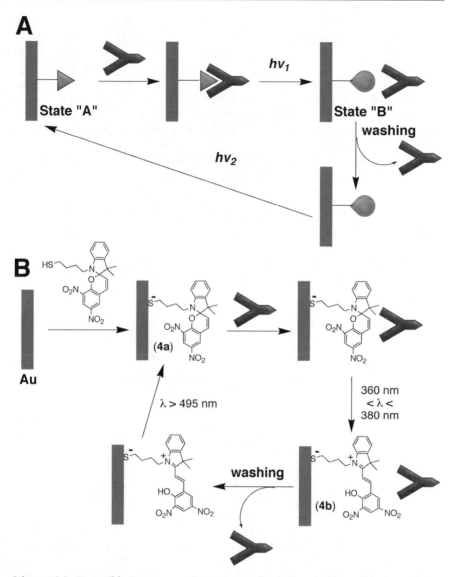

Scheme 7A,B. Reversible immunosensing using a photoisomerizable antigen-monolayer-functionalized electrode. **A** General scheme showing cyclic immunosensing upon an antibody binding-photoisomerization-washing off process. **B** Assembly of the photoisomerizable dinitrospiropyran monolayer onto a Au electrode and its interaction with the DNP-antibody for reversible sensing of the DNP-Ab

face. Upon completion of the sensing process, the active sensing layer is photoisomerized to state B. The latter configuration lacks antigen features for antibody binding and allows one to wash off the associated antibody. By further photoisomerization of the state B to the original photoisomer state A, the antigen layer is regenerated for a secondary sensing cycle. According to this concept a reversible immunosensor electrode for dinitrophenyl antibody, DNP-Ab, was assembled (Scheme 7B). A dinitrospiropyran (**4a**) photoisomerizable monolayer assembled onto a Au electrode acted as the active antigen interface for sensing DNP-Ab. Photoisomerization of the monolayer to the protonated dinitromerocyanine (**4b**) state generated an interface lacking antigen properties for the DNP-Ab. This allowed washing off of the DNP-Ab, and by a subsequent photoisomerization of the monolayer to the dinitrospiropyran state, the active antigen interface was recycled.

Figure 4 shows the Faradaic impedance spectra of the reversible immunosensing system in the different states measured in the presence of a negative-

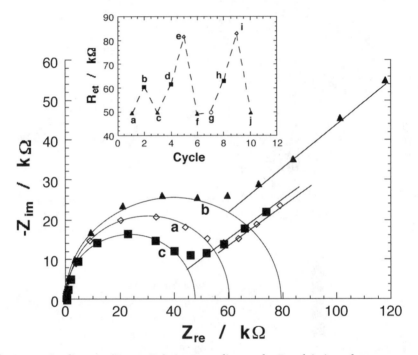

Fig. 4. Nyquist diagram (Z_{im} vs Z_{re}) corresponding to the Faradaic impedance measurements at the photoisomerizable **4a/4b** monolayer-modified Au electrode: a) **4a** monolayer state; b) affinity complex of DNP-Ab with the **4a** monolayer; c) **4b** monolayer state after washing off the DNP-Ab. Inset: Cyclic change of the electron transfer resistance, R_{et}, upon photoisomerization of the monolayer between **4a/4b** states, and association or dissociation of DNP-Ab to and from the monolayer interface: a) and c) monolayer in the **4b** state; b) and d) monolayer in the **4a** state; e) **4a** monolayer state upon addition of the DNP-Ab; f) photoisomerization to **4b** monolayer state and washing off the DNP-Ab; g) **4b** monolayer upon addition of DNP-Ab; h) washing off the DNP-Ab followed by photoisomerization of the monolayer to the **4a** state; i) addition of DNP-Ab to the **4a** monolayer; j) photoisomerization of the monolayer to the **4b** state and washing off DNP-Ab. Impedance measurements were performed in the frequency range from 10 mHz to 20 kHz applying a constant bias potential, 0.6 V; amplitude of alternating voltage, 10 mV, using $[Fe(CN)_6]^{3-/4-}$, 1×10^{-2} M, as the redox probe in 0.01 M phosphate buffer, pH 7.0. (Reproduced from Ref. [75] with permission)

ly charged redox probe, $[Fe(CN)_6]^{3-/4-}$ [75]. The immunosensing was started by following the impedance features of the neutral dinitrospiropyran antigen (4a) monolayer-functionalized Au electrode, curve a. Association of the DNP-Ab with the antigen monolayer resulted in an increase of the electron transfer resistance, R_{et}, curve b, originating from the blocking of the electrode surface with the bulky DNP-Ab molecules. Photoisomerization of the dinitrospiropyran monolayer to the protonated dinitromerocyanine state (4b) and further washing off of the DNP-Ab yields a positively charged monolayer that shows a smaller electron transfer resistance, curve c. It should be noted that the R_{et} value observed for the positively charged dinitromerocyanine state is even smaller than the original R_{et} of the neutral dinitrospiropyran monolayer (4a). This is consistent with the fact that the negatively charged redox probe is electrostatically attracted by the positively charged state of the monolayer, resulting in the lower electron transfer resistance for this state. The next step in the process includes the photoisomerization of the dinitromerocyanine (4b) monolayer to the dinitrospiropyran monolayer (4a), curve a, which is ready again to react with DNP-Ab. Thus, the reversible changes of the electron transfer resistance, R_{et}, were derived from the real part, Z_{re}, of the Faradaic impedance spectra (Fig. 4, inset). These changes reflect the association and dissociation of the antibody to and from the antigen monolayer, respectively, as well as the formation of different charges at the functionalized interface.

One of the major problems of impedimetric immunosensors is the relatively small change of the impedance spectra upon the direct binding of antibodies to the antigen-functionalized electrodes. The generated signals (i.e., the differences in the electron transfer resistance, ΔR_{et}, and in the double-charged layer capacitance, ΔC_{dl}) are especially insignificant when the antibody concentration is low and the surface coverage of the antigen–antibody complex is far from saturation. This major problem was addressed in recent studies directed to the development of amplified impedimetric immunosensors. In general, this approach includes the application of enzyme labels bound to the components of the immunosensing system that perform biocatalytic reactions after the antigen–antibody recognition events in order to amplify the generated signal. It is similar to ELISA immunosensing, but the process is performed at an electrode surface and the results are sensed by Faradaic impedance spectroscopy. In order to increase the interfacial changes at the sensing interfaces and to amplify the resulting impedimetric signal, enzyme labels that precipitate an insoluble product on the sensing interface were employed [44, 84–87].

This concept was exemplified with the sensing of DNP-Ab at the electrode surface functionalized with a DNP-antigen self-assembled monolayer [44] (Scheme 8). A DNP-antigen (5) monolayer was assembled onto the Au electrode surface. Challenging of the functionalized electrode with the complementary analyte antibody, DNP-Ab, results in the formation of the antigen–antibody complex on the transducer. The extent of complex formation is controlled by the concentration of the antibody in the analyzed sample and the time of interaction between the antigen electrode and the antibody solution. Treatment of the resulting electrode with a probe solution consisting of the anti-DNP-antibody-horseradish peroxidase conjugate (anti-DNP-Ab-HRP) yields the biocat-

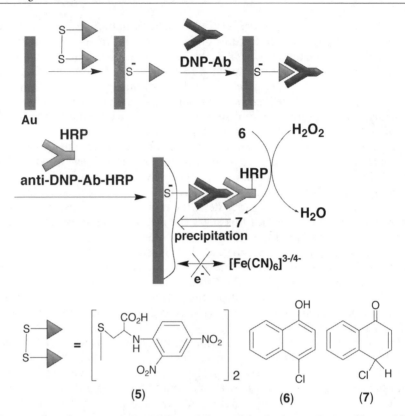

Scheme 8. Impedimetric immunosensing of the DNP-antibody on a DNP self-assembled monolayer-functionalized Au electrode amplified by precipitation of an insoluble product biocatalyzed by anti-DNP-antibody-HRP conjugate bound to the primary sensing interface

alytically labeled antibody/anti-antibody complex on the surface. HRP biocatalyzes the precipitation of an insoluble product (7) upon oxidation of 4-chloro-1-naphthol (6), and thus the formation of the precipitate (7) was used to probe the formation of the tricomponent antigen/antibody/anti-antibody-HRP assembly on the electrode. Note that precipitation of 7 occurs only if the analyte antibody, DNP-Ab, binds to the support and subsequently associates the anti-DNP-Ab-HRP probe. The amount of precipitate formed on the transducer is controlled by the concentration of the parent analyte antibody on the surface and the time interval used for the biocatalyzed production of the insoluble film. That is, the amount of precipitate associated with the electrode surface relates to the concentration of the analyte antibody in the sample. Furthermore, the generation of the insoluble film by the biocatalyzed process represents an amplification route for the formation of low coverage antigen–antibody complexes at the electrode surface.

Substitution of the electrolyte solution with the protein molecules at the electrode surface and further increase of the thickness and density of the organic layer upon the biocatalytic precipitation process results in a significant change of the double-charged layer capacitance, C_{dl}, which is derived from the imaginary part, Z_{im}, of the impedance spectra. The decrease of the effective dielectric constant of the medium at the interface yields the respective decrease of C_{dl}. Figure 5A shows the capacitance at the functionalized electrode upon buildup of the antigen monolayer on the electrode, association of the DNP-Ab and the corresponding anti-DNP-Ab-HRP conjugate to the sensing surface, and upon the HRP-biocatalyzed precipitation of the insoluble product 7 on the electrode. The additional suppression of the interfacial capacitance upon the formation of the precipitate is evident. Nonetheless, the changes in the real part of the impedance are expected to be substantially higher.

The stepwise formation of the antibody and anti-antibody complexes on the electrode is anticipated to yield a hydrophobic layer on the electrode surface that perturbs the interfacial electron transfer processes between the electrode and a redox probe solubilized in the electrolyte solution. The barrier for interfacial electron transfer should be amplified upon precipitation of the insulating product on the electrode. This results in a significant change of the electron transfer resistance values, R_{et}, derived from the impedance spectra. Figure 5B shows the Faradaic impedance spectra of the DNP-antigen-functionalized electrode upon interaction with different concentrations of the analyte DNP-Ab, and amplification of the analysis of the DNP-Ab by means of the anti-DNP-Ab-HRP conjugate and the biocatalyzed precipitation of the insoluble product 7 for 12 min. As the concentration of the DNP-Ab is elevated, the electron transfer resistance, R_{et}, increases as a result of the formation of the insoluble film. Figure 5B, inset, shows the derived calibration plots that correspond to the R_{et} at the electrode before, curve a, and after amplification with the secondary anti-DNP-Ab-HRP conjugate and formation of 7, curve b. A ca. 3-fold amplification of the immunosensing is evident upon comparison of the calibration plots with and without the formation of the precipitate (curves b and a, respectively). The electron transfer resistance increases as the concentration of DNP-Ab is elevated, and then levels off to a constant value at higher concentrations of DNP-Ab. These results are consistent with the fact that the DNP-antigen monolayer is saturated with the DNP-Ab at higher concentrations of the analyte antibody. This results in the saturation of the electrode surface with the biocatalytic conjugate, and the formation of a saturation amount of the precipitant on the electrode.

The extent of the signal amplification is controlled by the time interval for the precipitation of the insoluble product. Figure 6 shows the effect of the time interval used for the precipitation of the insulating layer on the interfacial electron transfer resistances. In this experiment, the DNP-antigen monolayer electrode was treated with a constant concentration of the analyte DNP-Ab, corresponding to 2 ng mL^{-1}, and the biocatalyzed precipitation of 7 onto the electrode was allowed to proceed for different time intervals. The semicircle diameters of the respective Faradaic impedance spectra increase as the biocatalyzed precipitation of 7 occurs for longer time intervals. For example, the electron transfer resistances at the electrode surface after 12 and 180 min of biocatalytic precipitation correspond to R_{et}=2.6 and 12.9 kΩ, respectively. That is, as the biocatalyzed precipitation of 7 yields thicker insulating layers on the electrode, the interfacial

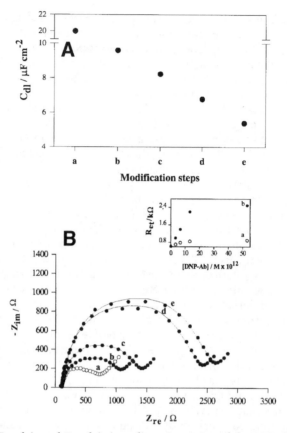

Fig. 5A,B. Non-Faradaic and Faradaic impedimetric sensing of the DNP-Ab amplified by the precipitation of the insoluble product. A Double-layer capacitance, C_{dl}, at the electrode upon the stepwise assembly of: a) the bare Au electrode; b) the DNP-antigen monolayer electrode; c) the DNP-antigen/DNP-Ab layered assembly; d) the DNP-antigen/DNP-Ab/anti-DNP-Ab-HRP layered assembly; e) after the biocatalyzed precipitation of the insoluble product 7 on the electrode. B Nyquist diagrams (Z_{im} vs Z_{re}) for the Faradaic impedance measurements corresponding to the sensing of variable concentrations of the DNP-Ab amplified by biocatalyzed precipitation of the insoluble product 7. The DNP-antigen monolayer electrodes were treated with different concentrations of the DNP-Ab for 5 min. The resulting electrodes were then treated with the anti-DNP-Ab-HRP conjugate, 10 µg mL^{-1}, for 5 min, and the electrodes were then allowed to stimulate the biocatalyzed precipitation of 7; [H$_2$O$_2$]=30 mM, [6]=1 mM. Impedance spectra corresponding to the DNP-antigen-functionalized electrode (a) and to the primary interaction of the DNP-antigen monolayer with the DNP-Ab at concentrations of: b) 0.5 ng mL^{-1}; c) 1 ng mL^{-1}; d) 2 ng mL^{-1}; e) 8 ng mL^{-1}. Inset: Calibration plots corresponding to the electron transfer resistances at the DNP-antigen monolayer-functionalized electrodes upon the analysis of different concentrations of the DNP-Ab: a) R_{et} observed by the direct interaction of the DNP-Ab with the sensing interface; b) R_{et} observed by the amplification of the initial binding of the DNP-Ab with the biocatalyzed precipitation of 7. Impedance measurements were performed in the frequency range from 10 mHz to 20 kHz applying a constant bias potential, 0.17 V; amplitude of alternating voltage, 10 mV, using [Fe(CN)$_6$]$^{3-/4-}$, 1×10^{-2} M, as the redox probe in 0.01 M phosphate buffer, pH 7.0. (Reproduced from Ref. [44] with permission)

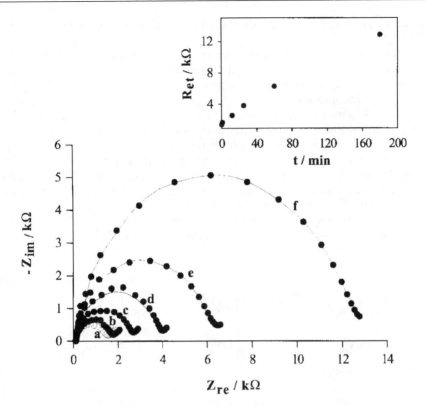

Fig. 6. Nyquist diagrams (Z_{im} vs Z_{re}) corresponding to the Faradaic impedance spectra observed upon the time-dependent precipitation of the insoluble product 7. The DNP-antigen monolayer electrode was treated with the DNP-Ab, 2 ng mL^{-1}, for 5 min, and then interacted with the anti-DNP-Ab-HRP conjugate, 10 µg mL^{-1}, for 5 min: a) prior to the precipitation, b) 1 min of precipitation, c) 12 min of precipitation, d) 40 min of precipitation, e) 60 min of precipitation, f) 180 min of precipitation. The experimental conditions were similar to those in Fig. 5. (Reproduced from Ref. [44] with permission)

electron transfer resistances are enhanced. Figure 6, inset, shows the plot of the R_{et} as precipitation proceeds with time. It is evident that the electron transfer resistance tends to reach a saturation value. This saturation value of R_{et} is reached when the insoluble film insulates the biocatalyst from the substrate that blocks further precipitation of 7.

A similar approach was used for the impedimetric detection of whole cells of *E. coli* bacteria [88]. The primary affinity binding of the *E. coli* cells to the electrode surface functionalized with an *E. coli* antibody was detected by Faradaic impedance measurements. The recognition process was further amplified by the coupling of the *E. coli* antibody-alkaline phosphatase conjugate to the surface, which stimulated the precipitation of an insoluble material onto the electrode surface. This impedimetric immunosensor allowed analysis of the *E. coli* cells with a detection limit of 6×10^3 cells mL^{-1}. A linear response in the electron transfer resistance for the analysis of the *E. coli* cells was found within the range of 6×10^4 to 6×10^7 cells mL^{-1}.

Alternatively, the impedimetric immunosensing could be amplified by building an extended network of noncatalytic but very bulky protein molecules after the primary recognition event. For example, an electrode functionalized with anti-human IgG antibodies (anti-hIgG) was reacted with the analyte hIgG antigens [89] (Scheme 9). After the primary recognition event, the sensing interface was further reacted with biotinylated anti-hIgG and then several steps of the reactions with avidin and a biotinylated protein were performed, resulting in the formation of an extended multilayer of proteins on the electrode surface. The multilayer density relates to the coverage of the sensing surface with the primary analyte antigen, but the formation of the thicker isolating layer results in the amplification of the impedimetric signal (Fig. 7).

An interesting approach for the amplified impedimetric analysis of antigen–antibody interactions has involved the insulation of the electrode with an ex-

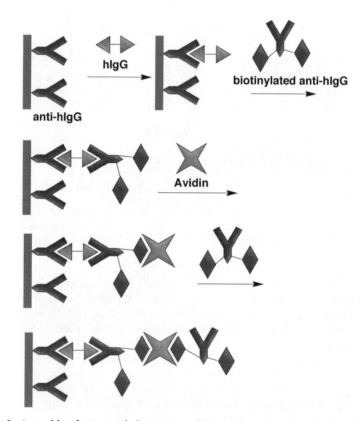

Scheme 9. Assembly of an extended protein multilayer system for impedimetric immunosensing. The primary binding of the analyte hIgG antigens to the sensing interface functionalized with anti-hIgG antibodies is amplified by the secondary stepwise binding of the biotinylated anti-hIgG and avidin

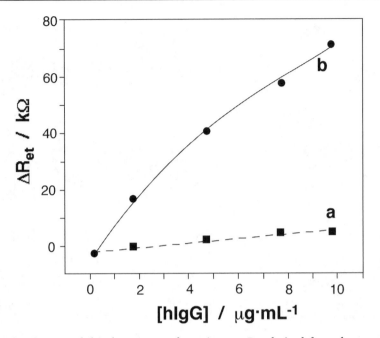

Fig. 7. The changes of the electron transfer resistance, R_{et}, derived from the respective Faradaic impedance spectra upon the sensing of hIgG at the anti-hIgG antibody-functionalized electrode: a) without amplification; b) after amplification by the stepwise binding of the biotinylated anti-hIgG and avidin. (Reproduced from Ref. [89] with permission)

tended array of liposomes [90] or functionalized liposomes that include a biocatalyst which precipitates an insoluble product on the transducer [86]. In the latter system, dual amplification is achieved by the primary insulation of the electrode interface with the liposomes, and the secondary hydrophobic blocking of the electrode surface with enzyme-generated insoluble product. This method was applied for the impedimetric analysis of the cholera toxin [86] (Scheme 10). A Au electrode was primarily modified with protein G that allowed binding of the anti-cholera-toxin antibody, which represents the sensing component for the analysis of cholera toxin. After the sensing event (binding of the cholera toxin to the anti-cholera-toxin antibody), the electrode surface was reacted with liposomes derivatized with ganglioside G_{M1} and HRP. The ganglioside G_{M1} units bind to the cholera toxin and result in the association of the liposomes to the sensing interface. The insulation of the electrode surface by the liposomes perturbs the interfacial impedance features of the electrode and provides the primary amplification path. The HRP-mediated oxidation of 4-chloronaphthol (6) to the insoluble product (7) provides the secondary amplification route, since the precipitation alters the interfacial electron transfer resistance and capacitance of the electrode. Figure 8 shows the amplified impedimetric analysis of cholera toxin. It is evident that elevation of the cholera toxin concentration results in an increase of the electron transfer resistance in the respective Farada-

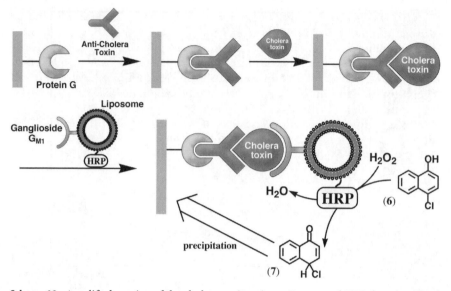

Scheme 10. Amplified sensing of the cholera toxin using a G_{M1}-tagged HRP-functionalized liposome and the biocatalyzed precipitation of the insoluble product

ic impedance spectra. The dual amplification process allows the impedimetric analysis of cholera toxin with a low detection limit equal to ca. 1×10^{-13} M (Fig. 8, inset).

Another amplification route includes enzyme-catalyzed polymer dissolution on the sensing interface [91, 92]. This process results in a decrease of the electron transfer resistance at the sensing interface that is opposite to the formation of the precipitant, which yields a higher electron transfer resistance. According to this method, the immunosensing system was organized on top of an organic polymer membrane. After the primary binding of an analyte antibody to the antigen-functionalized surface, the secondary antibody labeled with a hydrolytic enzyme was attached to the surface, and the hydrolytic breakdown of the polymer support was biocatalyzed. Figure 9 shows an electron micrograph of the polymer layer with pores generated upon the biocatalytic reaction. These pores provide penetration of the redox probe through the membrane that decreases the impedance value. The extent of this impedance decrease relates to the amount of the bound secondary antibody–hydrolytic enzyme conjugate that is proportional to the primary antigen–analyte antibody complex loading on the sensing surface.

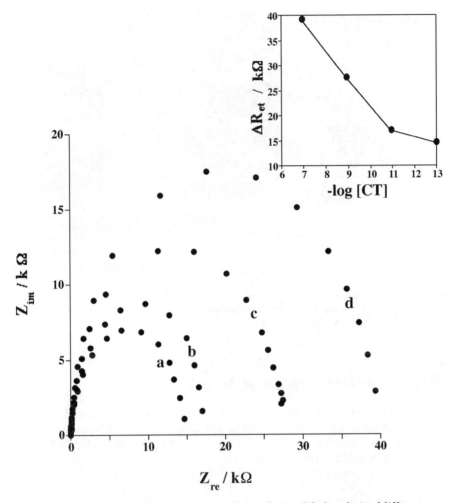

Fig. 8. Faradaic impedance spectra corresponding to the amplified analysis of different concentrations of the cholera toxin (CT) by the biocatalyzed precipitation of 7: a) 1×10^{-13} M of CT, b) 1×10^{-11} M of CT, c) 1×10^{-9} M of CT, and d) 1×10^{-13} M of CT. Inset: Calibration curve corresponding to the changes in the electron transfer resistance at the electrode upon the amplified sensing of different concentrations of CT

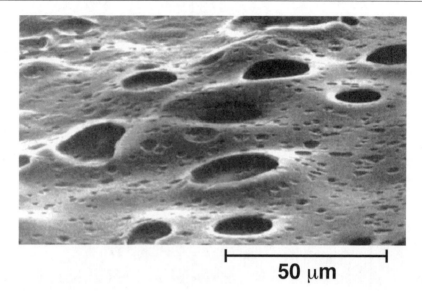

50 μm

Fig. 9. Electron micrograph of a polymer (Eudragit S 100)-coated electrode after partial breakdown due to the enzymatic reaction coupled to the antigen–antibody binding on the sensing interface. (Reproduced from Ref. [91] with permission)

4
DNA Sensors Based on Impedance Spectroscopy

The development of DNA sensor devices [93–95] attracts substantial research efforts directed to gene analysis, detection of genetic disorders, tissue matching, and forensic applications [96–99]. The electronic transduction of the formation of nucleic acid/DNA complexes on electronic transducers such as electrodes or semiconductors could provide quantitative information on the DNA analyte in the sample [94]. The design of bioelectronic DNA sensing devices requires the development of analytical routes that address two basic goals:

(i) The detection of DNA should reveal high sensitivity and eliminate the need for preamplification of the DNA content in the sample by the polymerase chain reaction (PCR). For this purpose, chemical and physical amplification means need to be included in the analytical protocols.

(ii) The detection paths should reveal high specificity and selectivity for the analysis of mutants in the sample. Ideally, analysis schemes for the detection of single-base mismatches are required.

The development of biochemical DNA sensors requires the immobilization of nucleic acids, acting as the sensing interface, on the electronic transducers and the electronic transduction of the hybridization occurring with the analyte DNA on the transducer. As the nucleic acids and the resulting nucleic acid/ DNA complexes are oligoanionic polymers, their immobilization on surfaces generates a negatively charged interface that repels negatively charged redox labels, and particularly multicharged negative ions such as $[Fe(CN)_6]^{3-/4-}$. The repulsion of the redox label is anticipated to inhibit the interfacial electron transfer and thus leads to enhanced electron transfer resistance, R_{et}. Thus, Faradaic impedance spectroscopy seems to be an effective means to follow the adsorption and assembly of nucleic acids on conductive surfaces, and to characterize DNA hybridization processes on the transducers. Indeed, the adsorption of nucleotides [100–102] and oligonucleotides [103–107] on different electrode surfaces was studied by Faradaic impedance spectroscopy. For example, Fig. 10 shows the gradual increase of the interfacial electron transfer resistance, R_{et}, at a Au electrode upon the time-dependent assembly of the thiolated nucleic acid, 5′-TCTATCCTACGCT-$(CH_2)_6$-SH-3′, using $[Fe(CN)_6]^{3-/4-}$ as the redox label for probing the generation of the thiolated nucleic acid monolayer [108]. The increase in the R_{et} value upon buildup of the nucleic acid monolayer was attributed to the enhanced repulsion of the $[Fe(CN)_6]^{3-/4-}$ by the negatively charged interface. Using chronocoulometric and quartz-crystal microbalance measure-

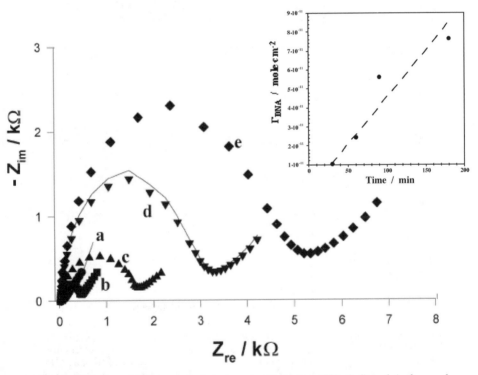

Fig. 10. Faradaic impedance spectra of: a) a bare Au electrode and b), c), d), and e) after modification of the Au electrode with thiolated oligonucleotide, 5′-TCTATCCTACGCT-$(CH_2)_6$-SH-3′, 5 µm, for 30, 60, 90, and 180 min, respectively. Impedance measurements were performed in the frequency range from 10 mHz to 20 kHz applying a constant bias potential, 0.17 V; amplitude of alternating voltage, 10 mV, using $[Fe(CN)_6]^{3-/4-}$, 1×10^{-2} M, as the redox probe in 0.01 M phosphate buffer, pH 7.0. Inset: Time-dependent increase of the DNA coverage on the Au electrode upon self-assembling of the thiolated DNA derived from the complementary microgravimetric measurements

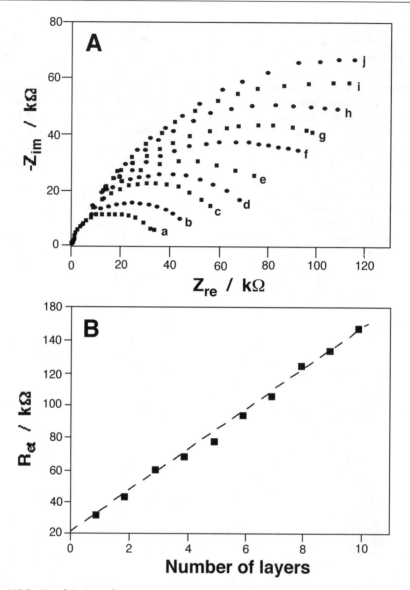

Fig. 11A,B. Faradaic impedance spectroscopy of the layered polyelectrolyte/DNA assemblies. A Nyquist diagrams (Z_{im} vs Z_{re}) of the Faradaic impedance spectra corresponding to layer-by-layer deposition of poly(dimethylammonium chloride) and calf thymus DNA: a)–j) correspond to one to ten layers of the polymer/DNA (DNA layer was always the outer layer). **B** The electron transfer resistances, R_{et}, derived from the Faradaic impedance spectra of the polymer/DNA layered systems. (Reproduced from Ref. [110] with the permission)

ments, the time-dependent surface coverage of the analyzed nucleic acid could be assayed and the interfacial electron transfer resistance could be correlated with the coverage of the nucleic acid. For example, at a surface coverage of ca. 8×10^{-11} mole cm^{-2} the interfacial electron transfer resistance corresponded to R_{et}=5.3 kΩ. Figure 10, inset, shows the gradual increase of the DNA surface coverage upon self-assembling of the thiolated DNA on the Au electrode surface.

The modification of surfaces with different DNA composites such as DNA-carbon nanotubes [109] or oligocationic polymer–DNA layer-by-layer deposited films [110, 111] were studied by Faradaic impedance spectroscopy. For example, the interfacial electron transfer resistance of layer-by-layer deposited films consisting of the cationic polymer, poly(dimethyldiallylammonum chloride), and DNA was found to be controlled by the number of layers associated with the electrode. Figure 11A shows the impedance spectra of deposited films that include increasing numbers of the polymer–DNA bilayers (the DNA is always the outer interface). Figure 11B shows that the interfacial electron transfer resistance ($[Fe(CN)_6]^{3-/4-}$ as redox label) increases with the number of layers. It was also demonstrated that the association of small organic molecules (e.g., drugs) with DNA adsorbed on electrodes could affect the impedance spectrum of the interface, and thus impedance spectroscopy was applied to characterize the binding of these molecules to DNA [112–115].

The dependence of the interfacial electron transfer resistance and capacitance features of DNA-modified electrodes on the amount of the oligonucleic acids at the interface allows the use of Faradaic impedance spectroscopy as a versatile tool for the amplified electronic detection of DNA [90, 116–119]. The analysis of single-base mismatches in DNA [108, 120] and the characterization of biocatalyzed transformations on DNA (or RNA) associated with electrodes (e.g., replication, scission, ligation) were accomplished by Faradaic impedance spectroscopy [121, 122].

4.1
Impedimetric Sensing of DNA Hybridization

The analysis of a target DNA is based on its hybridization with a complementary nucleic acid on the electronic transducer [123]. The fact that the analyte DNA is an oligonucleonic polyelectrolyte suggests that formation of the nucleic acid/DNA complex on an electrode surface would generate a negatively charged interface that electrostatically repels an anionic redox probe such as $[Fe(CN)_6]^{3-/4-}$. The electrostatic repulsion of the redox probe from the electrode introduces a barrier for electron transfer at the electrode support, thus turning impedance spectroscopy into an ideal method to probe DNA hybridization at electrodes [116, 124]. That is, in the presence of an anionic redox probe, DNA hybridization with a nucleic acid-functionalized electrode will be accompanied by an increase in the interfacial electron transfer resistance, R_{et}, that will be reflected in the Z_{re} component of the impedance spectrum.

A model system that involves the amplified impedimetric detection of the mutant characteristic to the Tay-Sachs (TS) genetic disorder was reported [116]. The nucleic acid primer (**8**) was assembled onto a Au electrode surface, using a five-base thiophosphate-T tag in the oligonucleotide as an anchoring group (Scheme 11A). The 13-mer oligonucleotide (8) included a 12-base sequence that is complementary to a part of the TS mutant (9). The resulting 8-functionalized electrode was interacted with the TS mutant. Figure 12 shows the impedance fea-

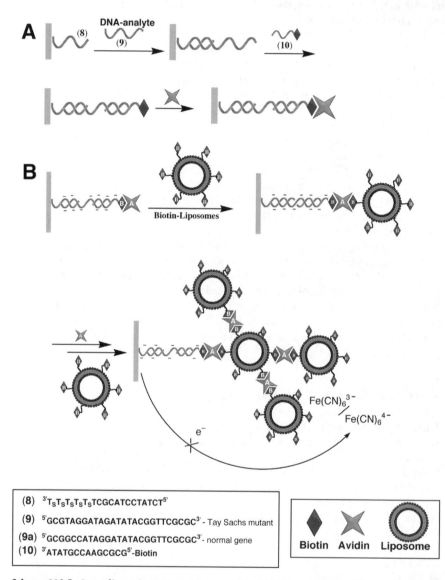

(8)	$^{3'}T_sT_sT_sT_sT_s$TCGCATCCTATCT$^{5'}$
(9)	$^{5'}$GCGTAGGATAGATATACGGTTCGCGC$^{3'}$ - Tay Sachs mutant
(9a)	$^{5'}$GCGGCCATAGGATATACGGTTCGCGC$^{3'}$- normal gene
(10)	$^{3'}$ATATGCCAAGCGCG$^{5'}$-Biotin

Biotin   Avidin   Liposome

Scheme 11A,B. Impedimetric DNA sensing amplified by the assembly of secondary super-structures. **A** Primary DNA hybridization amplified by the secondary binding of a bioti-nylated oligonucleotide and avidin. **B** Primary DNA hybridization amplified by secondary binding of a biotinylated oligonucleotide followed by the formation of a multilayer dendritic superstructure composed of avidin units and biotinylated liposomes

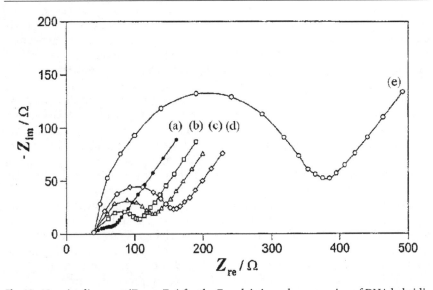

Fig. 12. Nyquist diagrams (Z_{im} vs Z_{re}) for the Faradaic impedance sensing of DNA hybridization amplified by secondary hybridization with a biotinylated oligonucleotide followed by avidin binding: a) the bare Au electrode; b) the DNA probe (**8**)-functionalized Au electrode; c) the (**8**)-functionalized Au electrode after 30 min of incubation with the DNA analyte (**9**), 30 ng mL^{-1}; d) the ds-DNA complex (**8/9**) after 30 min of incubation with the biotinylated oligonucleotide (**10**), 40 ng mL^{-1}; e) after 10 min of incubation with avidin, 5 ng mL^{-1}. The impedance spectra were recorded within the frequency range 0.1 Hz–50 kHz at the bias potential, 0.17 V; amplitude of alternating voltage, 5 mV, using [Fe(CN)$_6$]$^{3-/4-}$, 1×10^{-2} M, as the redox probe in 0.01 M phosphate buffer, pH 7.0. (Reproduced from Ref. [116] with permission)

tures presented as Nyquist plots of the bare Au electrode (curve a), the **8**-functionalized electrode (curve b), and the functionalized electrode after treatment with the complementary mutant (**9**) (curve c), in the presence of [Fe(CN)$_6$]$^{3-/4-}$ as the redox probe. Significant differences in the impedance spectra were observed upon the stepwise formation of the double-stranded (ds) oligonucleotide (**8/9**) complex. The respective electron transfer resistances, R_{et}, were derived from the semicircle diameters of the impedance spectra: R_{et} for the bare Au electrode was ca. 60 Ω and it increased to ca. 80 and 105 Ω upon functionalization of the surface with **8** and further formation of the ds-oligonucleotide complex with **9**, respectively. This is consistent with the fact that the negatively charged interface, formed upon the assembly of **8** on the electrode, repels the negatively charged redox probe. This introduces a barrier for interfacial electron transfer and results in the increased electron transfer resistance. Formation of the ds-oligonucleotide complex (**8/9**) enhances the repulsion of the redox probe and results in a further increase in the electron transfer resistance at the electrode.

In order to confirm and amplify the formation of the ds-oligonucleotide complex, the sensing interface was interacted with the biotinylated oligonucleotide (**10**) (Scheme 11A). The latter oligonucleotide was complementary to the residual sequence of the analyte DNA that was left nonhybridized in the previous modification step. The hybridization process between the 8/9 complex and 10 results in further increase of the electron transfer resistance (R_{et}=134 Ω) in the impedance spectrum (Fig. 12, curve d), that confirms the formation of 8/9 complex in the previous step and amplifies the signal. As the oligonucleotide 10 is labeled by biotin, the subsequent reaction of the oligonucleotide 8/9/10 superstructure with avidin is anticipated to further inhibit electron transfer due to the hydrophobic insulation of the sensing interface. Figure 12, curve e, shows the impedance spectra after the reaction of the sensing interface with avidin, demonstrating a substantial increase in the diameter of the semicircle domain, which translates to R_{et}=340 Ω. The last amplification step allowed the analysis of the oligonucleotide 9 with a detection limit of 3.5×10^{-12} mole mL^{-1}. A control experiment demonstrated that the oligonucleotide 9a, which includes the normal gene sequence, does not bind to the 8-functionalized interface and, therefore, does not result in an impedance spectrum change. Thus, the impedimetric method allowed discrimination between the mutant 9 and normal 9a gene sequence and provided the amplified analysis of 9 with a low detection limit.

Additional steps were introduced in the sensing process in order to reach higher amplification of the impedimetric signal. After the sensing process outlined in Scheme 11A was finished, the interface was further stepwise reacted with biotinylated liposomes and avidin yielding a dendritic multilayer structure on the sensing surface [108] (Scheme 11B). The electrode surface was significantly isolated by the liposomes, thus the interfacial electron transfer process was further perturbed and the R_{et} value was further increased. This allowed additional amplification of the sensing signal for the primary DNA hybridization process. The liposomes also decreased the nonspecific adsorption of foreign proteins on the sensing interface. An even higher extent of the impedimetric signal amplification was reached when a biocatalytic process was used as amplification means. For this purpose, after the sensing interface was reacted with a biotinylated oligonucleotide according to the reaction sequence outlined in Scheme 11A, it was further reacted with avidin–HRP conjugates [118] (Scheme 12A) or liposomes functionalized with avidin and HRP units [90] (Scheme 12B). The HRP biocatalytic moieties were used to catalyze formation of the insoluble product (7) that precipitated on the sensing surface and resulted in a growing barrier for the interfacial electron transfer process. Thus, the sensing signal amplification was also controlled by the time interval of the precipitation process.

A further route for the amplified detection of DNA by the biocatalyzed precipitation of an insoluble product on the electrode has involved the intercalation of doxorubicin (11) into a ds-DNA assembly at a Au electrode surface [119] (Scheme 13). Doxorubicin intercalates only into double-stranded DNA, and thus its association with the electrode indicates the hybridization event between the sensing interface and the analyte DNA. The electrochemical reduction of the quinonoid intercalator stimulates electrocatalyzed O_2 reduction resulting in the generation of H_2O_2, which participates in the formation of the insoluble precipitant 7 in the presence of HRP and 6 in the solution. The impedimetric detection of the DNA hybridization event was amplified by the insulation of the electrode as a result of the precipitation process.

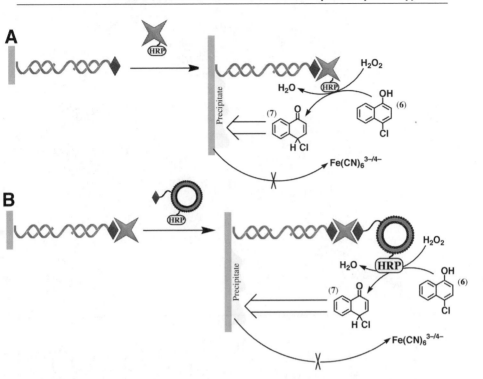

Scheme 12A,B. Impedimetric DNA sensing amplified by biocatalytic reactions that follow the primary sensing events. **A** Primary DNA hybridization amplified by the avidin-HRP-biocatalyzed precipitation of the insoluble product on the electrode. **B** Primary DNA hybridization amplified by the biocatalyzed precipitation of the insoluble product by an avidin/biotinylated liposome-HRP superstructure

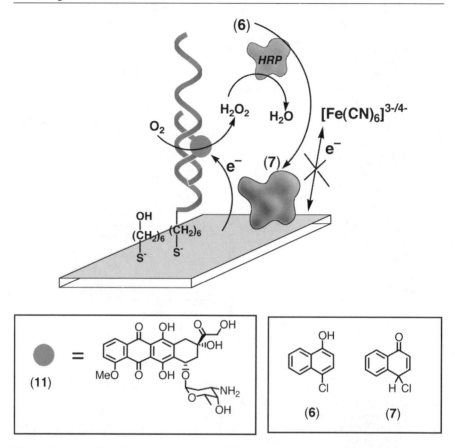

Scheme 13. Impedimetric DNA sensing amplified by precipitation of the insoluble product using electrochemical generation of H_2O_2 catalyzed by the redox active intercalator

4.2
Impedimetric Sensing of Single-Base Mismatches in DNA Sequences

The specific detection of DNA mutants is a basic challenge in DNA analysis. Partial selectivity in DNA analysis was achieved by the application of a sensing interface that is capable of generating a single double-helix turn with the analyte DNA. The introduction of four-to-six mismatches between the sensing interface and the analyte DNA is sufficient to prevent hybridization and discriminate the mutants from fully complementary DNA [116]. A different approach to tailor specificity in DNA analysis involves the use of the different melting temperature of double-stranded DNA that includes variable numbers of base mismatches. As the number of base mismatches increases, the melting temperature of the double-stranded DNA is lowered. This property has been used for the specific electrochemical analysis of DNA by controlling the hybridization temperature between the sensing nucleic acid and the analyzed DNA [125].

A method for the amplified detection of single-base mismatches in DNA using Faradaic impedance spectroscopy as transduction means was developed [120]. The method is schematically presented in Scheme 14, where the detection of a 41-base mutant oligonucleotide (12) that includes a single mutation, the substitution of a guanine for an adenine in the normal sequence (12a), is illustrated. The thiolated oligonucleotide (13), which is complementary to the mutant or the normal gene 12 or 12a up to the point of mutation, is assembled on a Au electrode and used as the active sensing interface. Interaction of the sensing interface with the mutant (12) or the oligonucleotide representing the normal sequence (12a) generates the respective double-stranded assembly on the electrode surface. The resulting interface is then reacted with the biotinylated base complementary to the mutation site in the mutant (e.g., biotinylated cytosine triphosphate, b-dCTP) in the presence of DNA polymerase I (Klenow fragment). In the presence of the double-stranded assembly that includes the mutant, the surface coupling of b-dCTP to the probe oligonucleotide proceeds. The resulting assembly is then interacted with an enzyme-linked avidin conjugate that catalyzes the precipitation of an insoluble product on the electrode surface. Specifically, avidin–alkaline phosphatase conjugate (avidin-AlkPh) was used, and its association with the sensing interface catalyzed the oxidative hydrolysis of 5-

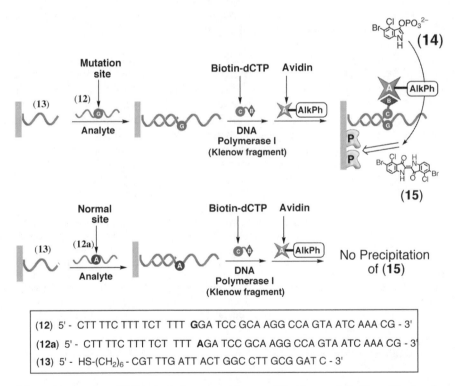

Scheme 14. Sensing of DNA single-base mutation amplified by precipitation of the insoluble product

bromo-4-chloro-3-indoyl phosphate (14) to an insoluble indigo derivative (15). Precipitation of 15 occurs only if the single-base mutant hybridizes to the probe linked to the sensing surface. The biocatalyzed precipitation of 15 provides a means to amplify the detection process, and the extent of precipitate formed on the electrode surface is controlled by the amount of DNA mutant associated with the sensing interface and the time interval employed for the biocatalyzed precipitation of 15. The insoluble product formed on the electrode alters the capacitance and electron transfer resistance at the conductive support.

Figure 13A, curve a, shows the Faradaic impedance spectrum of the sensing interface modified with the probe oligonucleotide (13) using $[Fe(CN)_6]^{3-/4-}$ as redox label. Curve b shows the Faradaic impedance spectrum of the sensing interface after interaction with the mutant (12), 3×10^{-9} mole mL^{-1}, and curve c shows the spectrum after the sensing interface treatment with b-dCTP and polymerase. Curve d shows the spectrum of the resulting surface after treatment with avidin-AlkPh conjugate, and curves e and f show the spectra after the biocatalyzed precipitation of the insoluble product (15) was allowed to proceed for 10 and 40 min, respectively. The electron transfer resistance increases from 1.6 to ca. 4.1 kΩ upon the formation of the double-stranded assembly between the probe (13) and the mutant (12) oligonucleotides. This is consistent with the electrostatic repulsion of the redox label from the electrode surface upon formation of a double-stranded assembly that introduces a barrier for interfacial electron transfer. Although the treatment of the electrode with the polymerase and b-dCTP does not affect the interfacial electron transfer resistance, the association of the hydrophobic avidin-AlkPh conjugate introduces a barrier for the electron transfer, R_{et}=6.2 kΩ. Biocatalyzed precipitation of 15 onto the electrode insulates the sensing surface and increases the electron transfer resistance to ca. 8.4 and 16 kΩ after 10 and 40 min of precipitation, respectively. Thus, the primary sensing event was amplified by the binding of avidin-AlkPh conjugate and even further amplified by the biocatalyzed precipitation process.

The Faradaic impedance spectra corresponding to similar experiments performed with the oligonucleotide representing the normal sequence (12a) are

▶

Fig. 13A–C. Impedimetric sensing of DNA single-base mutation amplified by the precipitation of an insoluble product on the electrode. A Faradaic impedance spectra corresponding to: a) the (13)-modified electrode; b) the (13)-modified electrode after interaction with the mutant DNA (12), 3×10^{-9} mole mL^{-1}; c) the ds-(13)/(12)-functionalized electrode after interaction with the biotinylated dCTP in the presence of Klenow fragment, 20 U mL^{-1} in Tris buffer, pH=7.8; d) after the interaction of the biotin-labeled ds-DNA assembly with avidin–alkaline phosphatase conjugate, 100 nmole mL^{-1} in 0.1 M phosphate buffer solution for 15 min; e) after the precipitation of 15 biocatalyzed by alkaline phosphatase for 10 min, [14]=2×10^{-1} M f) after the precipitation of 15 biocatalyzed by alkaline phosphatase for 40 min. B Faradaic impedance spectra corresponding to: a) the (13)-functionalized electrode; b) after the interaction of the (13)-functionalized electrode with the normal gene DNA (12a), 3×10^{-9} mole mL^{-1}; c)–f) repetition of the steps outlined in A. C Calibration curve corresponding to the changes in the electron transfer resistances of the electrode as a result of the precipitation of 15 upon the detection of different concentrations of 12. (Reproduced from Ref. [120] with permission)

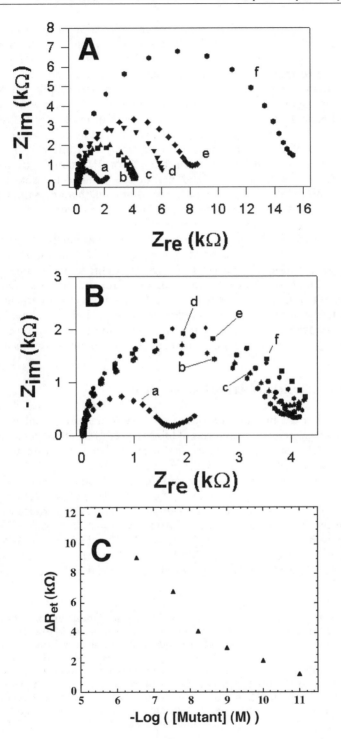

shown in Fig. 13B. It is evident that after the formation of the double-stranded assembly, R_{et}=3.9 kΩ, no further increase in the electron transfer resistance is observed upon treatment of the surface with avidin-AlkPh conjugate and an attempt to precipitate the insoluble product (15). Thus, the successful analysis of 12 is attributed to the specific polymerase-mediated coupling of b-dCTP to the mutant assembly, resulting in further binding of the avidin-AlkPh conjugate and the biocatalyzed precipitation of 15. A calibration plot for the analysis of the single-base mismatched DNA mutant was derived from the Faradaic impedance spectra obtained for different concentrations of the analyte (Fig. 13C). The method was successfully applied to the analysis of DNA samples from blood for the Tay-Sachs genetic disorder. The method has enabled the detection of the mutants with the discrimination of homocygotic and heterocygotic carriers of the genetic disorder with no need for pre-PCR amplification.

Using a related approach, single-base mismatches in DNA were identified by applying negatively charged biotinylated liposomes as labels and Faradaic impedance spectroscopy as transduction means [108]. The method involves coupling of the biotin-labeled base complementary to the mutation site as outlined in Scheme 14. The subsequent association of avidin and the biotin-labeled negatively charged liposomes yields an interface that blocks the interfacial electron transfer to the negatively charged redox probe $[Fe(CN)_6]^{3-/4-}$. The inhibition of the electron transfer process resulted in a high interfacial electron transfer resistance, R_{et}, which was monitored by Faradaic impedance spectroscopy.

4.3
Impedimetric Sensing of DNA and RNA Replication

The control of the interfacial properties of electrodes functionalized with nucleic acids has enabled the analysis of biocatalyzed transformations on surface-confined nucleic acids by means of Faradaic impedance spectroscopy [122]. Enzyme-induced ligation, replication, and specific scission of DNA have been examined on electrode surfaces (Scheme 15). The thiolated 18-mer oligonucleotide primer (16) was assembled on a Au electrode and ligated to oligonucleotide (17) in the presence of ligase. The interfacial electron transfer resistance of the 16-modified electrode (Fig. 14, curve a) is R_{et}=0.44 kΩ and it is increased upon ligation with 17 to R_{et}=1.33 kΩ (curve b). This is consistent with the fact that negative charge was accumulated on the electrode upon ligation and the fact that $[Fe(CN)_6]^{3-/4-}$ was used as a redox probe. Hybridization of oligonucle-

▶

Fig. 14. Faradaic impedance spectra corresponding to the biocatalytic transformations of nucleic acid-functionalized electrode: a) 16-functionalized electrode; b) after ligation of 17, 3×10^{-5} M, with the 16-functionalized electrode in the presence of ligase, 20 units, 37 °C, 30 min; c) after hybridization of the resulting assembly with 18, 2.5×10^{-5} M, 2 h; d) after replication of the double-stranded assembly in the presence of dNTP, 1×10^{-3} M, and polymerase, 3 units, 37 °C, 30 min; e) after scission of the resulting assembly with endonuclease, CfoI, 10 units, 37 °C, 1 h. The impedance spectra were recorded within the frequency range 0.1 Hz–50 kHz at the bias potential, 0.17 V; amplitude of alternating voltage, 5 mV, using $[Fe(CN)_6]^{3-/4-}$, 1×10^{-2} M, as the redox probe in 0.01 M phosphate buffer, pH 7.0

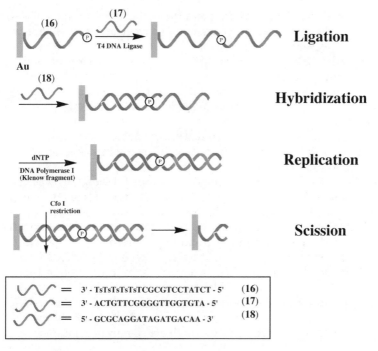

Scheme 15. Biocatalyzed ligation, replication, and scission of single- and double-stranded DNA on electronic transducer (Ts=thymine thiophosphate)

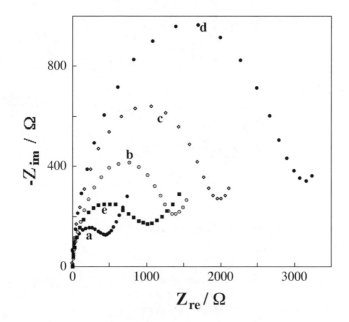

otide (**18**) with the ligated product and the replication of the double-stranded assembly further increased the interfacial electron transfer resistance to $R_{et}=1.9$ and 3.1 kΩ, respectively (curves c and d), due to the accumulation of negative charges on the surface in the respective two processes. Treatment of the surface with the endonuclease Cfo I that is specific to the double-stranded sequence 5'GCG/C3' resulted in the decrease of the interfacial electron transfer resistance to $R_{et}=0.9$ kΩ (Fig. 14, curve e) due to the removal of nucleic acid residues upon hydrolytic scission.

An ultrasensitive detection of viral DNA was achieved with an amplified impedimetric sensing technique that does not need the PCR amplification process prior to analysis. The method for analysis of the target viral DNA that involves the surface replication and concomitant labeling of the analyzed DNA is depicted in Scheme 16 [121]. The primer thiolated oligonucleotide (**19**), complementary to a segment of the target M13ϕ mp8 DNA, was assembled on a Au electrode through a thiol functional group. The sensing interface was then treated with the analyte DNA of M13ϕ mp8 (+)strand, and the resulting complex on the electrode surface was treated with a mixture of bases: dATP, dGTP, dTTP, dCTP, and biotinylated-dCTP (ratio 1:1:1:$^2/_3$:$^1/_3$, nucleotide concentration of 1 mM) in the presence of DNA polymerase I, Klenow fragment (20 U mL^{-1}). Polymerization and the formation of a double-stranded assembly with the target DNA provides the first amplification step in the analysis of the viral DNA. Furthermore, the biotin tags into the replicated double-stranded assembly provide docking sites for

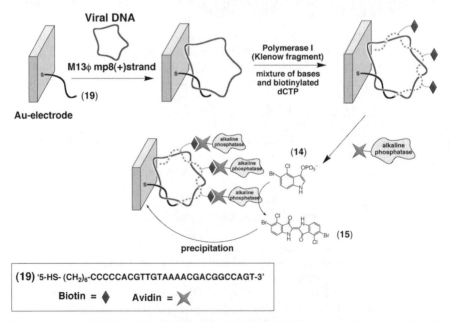

Scheme 16. Amplified impedimetric transduction of viral DNA by the polymerase-induced replication of the analyte and the biocatalyzed precipitation of the insoluble product on the electrode (broken line shows replicated nucleic acid)

the binding of the avidin-AlkPh conjugate. The associated enzyme biocatalyzed the precipitation of the insoluble material (15) on the electrode surface, thus providing the second amplification step for the analysis of the target DNA. The process was monitored by Faradaic impedance spectroscopy.

Figure 15, curve a, shows the impedance spectrum of the Au electrode functionalized with the primer oligonucleotide (19). The electron transfer resistance, R_{et}, increased upon binding of the viral DNA from 3 to ca. 20 kΩ (curve b). This is consistent with the fact that binding of the high molecular weight DNA electrostatically repels the negatively charged redox label, $[Fe(CN)_6]^{3-/4-}$.

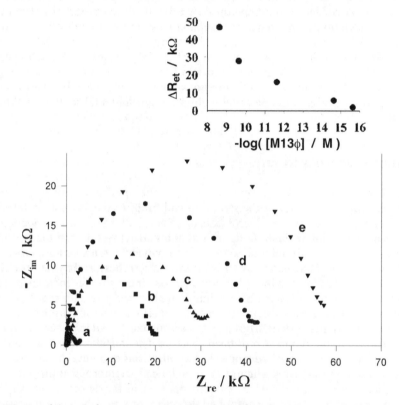

Fig. 15. Faradaic impedance spectra (Nyquist plots) of: a) the DNA probe (16)-functionalized electrode; b) after hybridization with the viral DNA, M13φ mp8 (+)strand, 2.3×10^{-9} M; c) after the polymerase-induced replication and formation of the double-stranded assembly for 45 min; d) after the binding of the avidin–alkaline phosphatase conjugate to the surface; e) after the biocatalyzed precipitation of the insoluble product (15) for 20 min in the presence of 14, 2×10^{-3} M. Inset: The changes in the electron transfer resistance, ΔR_{et}, upon the sensing of different concentrations of M13φ by the amplified biocatalyzed precipitation of 15 onto the electrode. The ΔR_{et} is the difference in R_{et} at the electrode after the precipitation of 15 for 20 min and R_{et} at the 16-functionalized electrode. The impedance spectra were recorded within the frequency range 0.1 Hz–50 kHz at the bias potential, 0.17 V; amplitude of alternating voltage, 10 mV, using $[Fe(CN)_6]^{3-/4-}$, 1×10^{-2} M, as the redox probe in 0.01 M phosphate buffer, pH 7.0. (Reproduced from Ref. [121] with permission)

The DNA polymerase-mediated synthesis of the complementary strand further increased the electron transfer resistance to R_{et}=33 kΩ, due to the accumulation of the negative charges on the surface (curve c). Note that the polymerization does not double the R_{et} value, consistent with the partial replication of the complementary strand on the target DNA molecule. The binding of avidin-AlkPh conjugate (curve d), and the subsequent biocatalyzed precipitation of 15 onto the electrode surface, resulted in an insulating layer that introduced a high barrier for the interfacial electron transfer, R_{et}=58 kΩ (curve e).

A calibration plot was derived from the Faradaic impedance measurements with different concentrations of the viral DNA (Fig. 15, inset). The amplification process involved in the impedimetric analysis of the target viral DNA allowed its detection with a sensitivity limit corresponding to 2×10^{-16} M. A similar approach was used for the amplified sensing of the 11,161-base RNA of vesicular stomatitis virus (VSV) using reverse transcriptase as the replication biocatalyst [121]. The RNA detection limit was found to be 1×10^{-17} M (ca. 60 copies in 10 µL). These impedimetric methods for DNA and RNA analyses would find broad applicability in clinical diagnostics, environmental control, and food analysis.

5
Conclusions and Perspectives

This review has addressed recent advances in the applications of impedance spectroscopy to characterize the structure and functional operation of various kinds of immunosensors and DNA sensors. Specifically, the use of impedance spectroscopy to follow the buildup of biological structures on conductive supports, to observe the affinity interactions of biomolecules (such as antigen–antibody binding, oligonucleotide hybridization on electrodes), and to monitor biocatalytic reactions (e.g., DNA replication, ligation, or enzymatic reactions resulting in the formation of insoluble products) were discussed. Different kinds of impedance spectroscopy (including non-Faradaic capacitance measurements, Faradaic impedance spectroscopy in the presence of an external redox probe) were utilized in the different biosensor systems. The theoretical background of the different methods and their practical applications in analytical procedures were outlined in the paper. The use of impedance spectroscopy as an electronic readout signal of novel biosensor protocols, such as the detection of a single-base mutation in DNA or the amplified detection of viral DNA via replication, were demonstrated. Furthermore, impedance spectroscopy offers a very sensitive method to follow time-dependent protein binding to electrodes, e.g., the reversible association/dissociation of antigen–antibody complexes on electrodes. The combination of impedance spectroscopy with other physical methods (such as surface plasmon resonance, ellipsometry, quartz-crystal microbalance, FTIR spectroscopy, photoelectrochemistry) is especially productive to characterize the interfacial properties of sensing and bioelectronic devices.

Acknowledgements: The support of our research activities in the area of bioelectronics by the Israel Ministry of Science and the German-Israeli Program (DIP) is acknowledged.

References

1. Willner I, Willner B, Katz E (2002) Rev Mol Biotechnol 82:325
2. Willner I, Katz E (2000) Angew Chem Int Ed 39:1180
3. Willner I, Katz E, Willner B (2000) Layered functionalized electrodes for electrochemical biosensor applications. In: Yang VC, Ngo TT (eds) Biosensors and their applications. Kluwer, New York, p 47
4. Heller A (1990) Acc Chem Res 23:128
5. Göpel W, Heiduschka P (1995) Biosens Bioelectron 10:853
6. Zhong C-J, Porter MD (1995) Anal Chem 67:709A
7. Bourdillon C, Demaille C, Gueris J, Moiroux J, Savéant J-M (1993) J Am Chem Soc 115:12264
8. Bourdillon C, Demaille C, Moiroux J, Savéant J-M (1994) J Am Chem Soc 116:10328
9. Bourdillon C, Demaille C, Moiroux J, Savéant J-M (1996) Acc Chem Res 29:529
10. Rogers KR (2000) Mol Biotechnol 14:109
11. Yang M, McGovern ME, Thompson M (1997) Anal Chim Acta 346:259
12. Willner I, Katz E, Willner B (2002) Amplified and specific electronic transduction of DNA sensing processes in monolayer and thin-film assemblies. In: Brajter-Toth A, Chambers JQ (eds) Electroanalytical methods of biological materials. Marcel Dekker, New York, p 43
13. Campbell CN, Gal D, Cristler N, Banditrat C, Heller A (2002) Anal Chem 74:158
14. Bardea A, Katz E, Bückmann AF, Willner I (1997) J Am Chem Soc 119:9114
15. Anzai J, Takeshita H, Kobayashi Y, Osa T, Hoshi T (1998) Anal Chem 70:811
16. Willner I, Rubin S, Cohen Y (1993) J Am Chem Soc 115:4937
17. Sethi RS (1994) Biosens Bioelectron 9:243
18. Bartlett PN, Tebbutt P, Whitaker RC (1991) Prog React Kinet 16:55
19. Ianniello RM, Lindsay TJ, Yacynych AM (1982) Anal Chem 54:1980
20. Ghindilis AL, Kurochkin IN (1994) Biosens Bioelectron 9:353
21. Lion-Dagan M, Ben-Dov I, Willner I (1997) Colloids Surf B 8:251
22. Ben-Dov I, Willner I, Zisman E (1997) Anal Chem 69:3506
23. Bardea A, Dagan A, Ben-Dov I, Amit B, Willner I (1998) Chem Commun 839
24. O'Sullivan CK, Vaughan R, Guilbault GG (1999) Anal Lett 32:2353
25. Kazanskaya N, Kukhtin A, Manenkova M, Reshetilov AN, Yarysheva L, Arzhakova O, Volynskii A, Bakeyev N (1996) Biosens Bioelectron 11:253
26. Zayats M, Kharitonov AB, Katz E, Bückmann AF, Willner I (2000) Biosens Bioelectron 15:671
27. Pardo-Yissar V, Katz E, Wasserman J, Willner I (2003) J Am Chem Soc 125:622
28. Amador SM, Pachence JM, Fischetti R, McCauley JP Jr, Smith AB, Blasic JK (1993) Langmuir 9:812
29. Hobara D, Niki K, Zhou C, Chumanov G, Cotton TM (1994) Colloids Surf A 93:241
30. Liedberg B, Nylander C, Lundström I (1995) Biosens Bioelectron 10:1
31. Jordan CE, Corn RM (1997) Anal Chem 69:1449
32. Raitman OA, Katz E, Bückmann AF, Willner I (2002) J Am Chem Soc 124:6487
33. Raitman OA, Patolsky F, Katz E, Willner I (2002) Chem Commun 1936
34. Kharitonov AB, Alfonta L, Katz E, Willner I (2000) J Electroanal Chem 487:133
35. Athey D, Ball M, McNeil CJ, Armstrong RD (1995) Electroanalysis 7:270
36. Nahir TM, Bowden EF (1996) J Electroanal Chem 410:9
37. Bresler HS, Lenkevich MJ, Murdock JF, Jr, Newman AL, Robbin RO (1992) Application of capacitive affinity biosensors; HIV antibody and glucose detection. In: Biosensor design and application. ACS Symp Ser 511, p 89
38. Bard AJ, Faulkner LR (1980) Electrochemical methods: fundamentals and applications. Wiley, New York
39. Stoynov ZB, Grafov BM, Savova-Stoynova BS, Elkin VV (1991) Electrochemical impedance. Nauka, Moscow

40. Randles JEB (1947) Discuss Faraday Soc 1:11
41. Ershler BV (1947) Discuss Faraday Soc 1:269
42. Champagne GY, Belanger D, Fortier G (1989) Bioelectrochem Bioenerg 22:159
43. Savitri D, Mitra CK (1999) Bioelectrochem Bioenerg 48:163
44. Bardea A, Katz E, Willner I (2000) Electroanalysis 12:1097
45. Janata J (2002) Crit Rev Anal Chem 32:109
46. Pravda M, Kreuzer MP, Guilbault GG (2002) Anal Lett 35:1
47. Sapsford KE, Charles PT, Patterson CH, Ligler FS (2002) Anal Chem 74:1061
48. Luppa PB, Sokoll LJ, Chan DW (2001) Clin Chim Acta 314:1
49. Mullett WM, Lai EPC, Yeung JM (2000) Methods 22:77
50. Fare TL, Cabelli MD, Dallas SM, Herzog DP (1998) Biosens Bioelectron 13:459
51. Berggren C, Bjarnanson B, Johansson G (2001) Electroanalysis 13:173
52. Laureyn W, Nelis D, Van Gerwen P, Baert K, Hermans L, Magnée R, Pireaux J-J, Maes G (2000) Sens Actuators B 68:360
53. Van Gerwen P, Laureyn W, Laureys W, Huyberechts G, De Beeck MO, Baert K, Suls J, Sansen W, Jacobs P, Hermans L, Mertens R (1998) Sens Actuators B 49:73
54. Valdes JJ, Wall JG, James J, Chambers P, Eldefrawi ME (1988) Johns Hopkins APL Tech Digest 9:4
55. Eldefrawi ME, Sherby SM, Andreou AG, Mansour NA, Annau Z, Blum NA, Valdes JJ (1988) Anal Lett 21:1665
56. Pak SC, Penrose W, Hesketh PJ (2001) Biosens Bioelectron 16:371
57. Taylor RF, Marenchic IG, Spencer RH (1991) Anal Chim Acta 249:67
58. Bresler HS, Lenkevich MJ, Murdock JF, Newman AL, Robin RO (1992) ACS Symp Ser 511:89
59. Sergeyeva TA, Lavrik NV, Piletsky SA, Rachkov AE, El'skaya AV (1996) Sens Actuators B 34:283
60. Sergeyeva TA, Lavrik NV, Rachkov AE, Kazantseva ZI, El'skaya AV (1998) Biosens Bioelectron 13:359
61. DeSilva MS, Zhang Y, Hesketh PJ, Maclay GJ, Gendel SM, Stetter JR (1995) Biosens Bioelectron 10:675
62. Hardeman S, Nelson T, Beirne D, DeSilva M, Hesketh PJ, Maclay GJ, Gendel SM (1995) Sens Actuators B 24–25:98
63. Sadik OA, Xu H, Gheorghiu E, Andreescu D, Balut C, Gheorghiu M, Bratu D (2002) Anal Chem 74:3142
64. Finklea HO (1996) Electrochemistry of organized monolayers of thiols and related molecules on electrodes. In: Bard AJ, Rubinstein I (eds) Electroanalytical chemistry, vol 19. Marcel Dekker, New York, p 109
65. Knichel M, Heiduschka P, Beck W, Jung G, Göpel W (1995) Sens Actuators B 28:85
66. Rickert J, Göpel W, Beck W, Jung G, Heiduschka P (1996) Biosens Bioelectron 11:757
67. Taira H, Nakano K, Maeda M, Takagi M (1993) Anal Sci 9:199
68. Ameur S, Martelet C, Jaffrezic-Renault N, Chovelon JM (2000) Appl Biochem Biotechnol 89:161
69. Blonder R, Katz E, Cohen Y, Itzhak N, Riklin A, Willner I (1996) Anal Chem 68:3151
70. Mirsky VM, Riepl M, Wolfbeis OS (1997) Biosens Bioelectron 12:977
71. Ouerghi O, Touhami A, Jaffrezic-Renault N, Martelet C, Ben Ouada H, Cosnier S (2002) Bioelectrochemistry 56:131
72. Feng C-D, Ming Y-D, Hesketh PJ, Gendel SM, Stetter JR (1996) Sens Actuators B 35:431
73. Ameur S, Martelet C, Jaffrezic-Renault N, Chovelon JM, Plossu C, Babier D (1997) Proc Electrochem Soc 19:1019
74. Jie M, Ming CY, Jing D, Cheng LS, Huaina L, Jun F, Xiang CY (1999) Electrochem Commun 1:425
75. Patolsky F, Filanovsky B, Katz E, Willner I (1998) J Phys Chem B 102:10359
76. Willner I, Willner B (1999) Biotechnol Prog 15:991

77. Sargent A, Loi T, Gal S, Sadik OA (1999) J Electroanal Chem 470:144
78. Lillie G, Payne P, Vadgama P (2001) Sens Actuators B 78:249
79. Farace G, Lillie G, Hianik T, Payne P, Vadgama P (2002) Bioelectrochemistry 55:1
80. Ouerghi O, Senillou A, Jaffrezic N, Martelet C, Ben Ouada H, Cosnier S (2001) J Electroanal Chem 501:62
81. Maupas H, Soldatkin AP, Martelet C, Jaffrezic N, Mandrand B (1997) J Electroanal Chem 421:165
82. Willner I, Blonder R, Dagan A (1994) J Am Chem Soc 116:9365
83. Blonder R, Levi S, Tao G, Ben-Dov I, Willner I (1997) J Am Chem Soc 119:10467
84. Alfonta L, Bardea A, Khersonsky O, Katz E, Willner I (2001) Biosens Bioelectron 16:675
85. Katz E, Alfonta L, Willner I (2001) Sens Actuators B 76:134
86. Alfonta L, Willner I, Throckmorton DJ, Singh AK (2001) Anal Chem 73:5287
87. Yoon HC, Yang H, Kim YT (2002) Analyst 127:1082
88. Ruan CM, Yang LJ, Li YB (2002) Anal Chem 74:4814
89. Pei R, Cheng Z, Wang E, Yang X (2001) Biosens Bioelectron 16:355
90. Alfonta L, Singh AK, Willner I (2001) Anal Chem 73:91
91. McNeil CJ, Athey D, Ball M, Ho WO, Krause S, Armstrong RD, Wright JD, Rawson K (1995) Anal Chem 67:3928
92. Ho WO, Krause S, McNeil CJ, Pritchard JA, Armstrong RD, Athey D, Rawson K (1999) Anal Chem 71:1940
93. Wang J (2000) Nucleic Acid Res 28:3011
94. Wang J, Palecek E, Nielsen PE, Rivas G, Cai X, Shiraishi H, Dontha N, Luo D, Farias PAM (1996) J Am Chem Soc 118:7667
95. Homs WCI (2002) Anal Lett 35:1875
96. Mikkelsen SR (1996) Electroanalysis 8:15–19
97. Wilson EK (1998) Chem Eng News 76:47
98. Millan KM, Mikkelsen SR (1993) Anal Chem 65:2317
99. Yang MS, McGovern ME, Thompson M (1997) Anal Chim Acta 364:259
100. Oliveira-Brett AM, Brett CMA, Silva LA (2002) Bioelectrochemistry 56:33
101. Oliveira-Brett AM, Silva LA, Brett CMA (2002) Langmuir 18:2326
102. Lust E, Jänes A, Lust K (1998) J Electroanal Chem 449:153
103. Brett CMA, Brett AM, Serrano SHP (1999) Electrochim Acta 44:4233
104. Zhao Y-D, Pang D-W, Hu S, Wang Z-L, Cheng J-K, Qi Y-P, Dai H-P, Mao B-W, Tian Z-Q, Luo J, Lin Z-H (1999) Anal Chim Acta 388:93
105. Saoudi B, Despas C, Chehimi MM, Jammul N, Delamar M, Bessière J, Walcarius A (2000) Sens Actuators B 62:35
106. Lisdat F, Ge B, Krause B, Ehrlich A, Bienert H, Scheller FW (2001) Electroanalysis 13:1225
107. Strasák L, Dvoak J, Haso S, Vetterl V (2002) Bioelectrochemistry 56:37
108. Patolsky F, Lichtenstein A, Willner I (2001) J Am Chem Soc 123:5194
109. Wang G, Xu J-J, Chen H-Y (2002) Electrochem Commun 4:506
110. Luo L, Liu J, Wang Z, Yang X, Dong S, Wang E (2001) Biophys Chem 94:11
111. Pei R, Cui X, Yang X, Wang E (2001) Biomacromolecules 2:463
112. Hason S, Dvorák J, Jelen F, Vetterl V (2002) Crit Rev Anal Chem 32:167
113. Yan F, Sadik OA (2001) Anal Chem 73:5272
114. Yan F, Sadik OA (2001) J Am Chem Soc 123:11335
115. Haso S, Dvoak J, Jelen F, Vetterl V (2002) Talanta 56:905
116. Bardea A, Patolsky F, Dagan A, Willner I (1999) Chem Commun 21
117. Patolsky F, Lichtenstein A, Willner I (2000) Angew Chem Int Ed 39:940
118. Patolsky F, Katz E, Bardea A, Willner I (1999) Langmuir 15:3703
119. Patolsky F, Katz E, Willner I (2002) Angew Chem Int Ed 41:3398
120. Patolsky F, Lichtenstein A, Willner I (2001) Nat Biotechnol 19:253
121. Patolsky F, Lichtenstein A, Kotler M, Willner I (2001) Angew Chem Int Ed 40:2261

122. Alfonta L, Willner I (2001) Chem Commun 1492
123. Gooding JJ (2002) Electroanalysis 14:1149
124. Vagin MY, Karyakin AA, Hianik T (2002) Bioelectrochemistry 56:91
125. Caruana DJ, Heller A (1999) J Am Chem Soc 121:769

Chapter 5

"Voltohmmetry" – a New Transducer Principle for Electrochemical Sensors

Michael J. Schöning

Abstract. The lateral resistance of a thin metal film (<30 nm) depends on the presence or absence of other species at its surface. This phenomenon is investigated in terms of analytical applications for the determination of heavy metals in aqueous solutions. Therefore, microstructured polycrystalline noble metal electrodes have been fabricated by means of thin-film techniques; their resistance–potential behaviour has been recorded during the electrochemical deposition and dissolution of various heavy metals such as Cd, Pb, Ni, Tl and Zn. During the deposition of the heavy metals, an increase of the electrode's lateral resistance has been observed, whereas the subsequent dissolution of the deposited heavy metals decreases the resistance to its initial value. The magnitude of the resulting resistance change linearly depends on the heavy metal content of the solution (from ppb (μg/l) to ppm (mg/l)). The resistance measurements show a specific selectivity in the electrode potentials of thermodynamic equilibrium between the deposited (i.e. reduced) and the dissolved (i.e. oxidised) form of the species. Corresponding to voltammetric methods, this new principle is called *voltohmmetry*.

Keywords: Voltohmmetry, Thin-film electrode, Heavy metal determination, Silicon technology

1
Introduction

Nowadays, electrochemical methods for the determination of the heavy metal content in aqueous solutions are well established within the armoury of analytical techniques [1, 2]. Such analytical techniques at the interface "solid/liquid" are mainly based on interfacial properties and processes and make use of the measurement of current, charge, potential or capacity. As one example, voltammetric techniques offer reliable and cost-effective possibilities for the trace analysis of heavy metal ions in aqueous solutions [3]. When, in addition, electrochemical detection principles are combined with batch preparation processes, like thin-film, semiconductor and micromachining technologies, then powerful miniaturised analytical detection systems can be developed [4]. In such µTAS (micro total analysis systems) [5], chemical sensors are utilised as detection unit, and include several advantages such as well-defined and reduced size as well as small sample volume. Furthermore, the integration of sensors to sensor arrays simplifies a simultaneous multi-analyte determination. Nonetheless, voltammetric techniques can suffer from several drawbacks, like disturbances by capacitive currents and the lack of selectivity between adsorbing and non-adsorbing reactants. From this background, alternative detection principles are always desirable with regard to practical sensor applications (e.g. environmental monitoring, medical analysis, food control) for the purpose of enhanced selectivity and improved quality assurance.

On the other hand, it has been known from the literature for more than 100 (!) years that the lateral d.c. resistance of a thin metal film depends on its surface properties [6]. The adsorption of ions and molecules can lead to an increase of the lateral d.c. resistance, which has been discussed in different theoretical and experimental works [7–16]. In surface physics, this phenomenon has been investigated intensively with respect to the adsorption of gases at the metal/vacuum interface and the deposition of dissolved ions at the metal/liquid interface. The theory of surface resistance changes reaches from semi-classical models up to quantum-mechanical considerations of state density functions. All these models allow this phenomenon to be described. For example, different groups investigated the adsorption of gases at metal surfaces under vacuum conditions in order to validate the theoretical consumptions. The most significant and accepted model, however, is based on the adsorption-induced increase of the d.c. resistance due to diffuse scattering of conduction electrons of a thin metallic film (d<30 nm) by adatoms and -molecules on its surface. The Fuchs–Sondheimer theory distinguishes the surface resistance of a single crystalline, perfectly smooth metal film exposed to the vacuum into two regions: a "regular" film resistance in the case that no foreign particles are present, and an "increased" resistance if particles with suitable properties are present at the metal surface, and can act as centres for diffuse scattering of the metal's film conduction electrons.

Therefore, the measurement of the surface resistance or surface resistance change should be a suitable parameter for analytical methods at the metal (electrode)/liquid (analyte) interface if one can realise appropriate electrode structures and experimental conditions. The resistance change of thin epitaxial metal films by means of adsorption and dissolution of cations and anions from the electrolyte solution has been reported in electrochemical experiments [17]. A practical analytical application for such chemical sensors, however, is mainly imaginable in basic research studies, because the preparation of single crystalline, thin metal films is not always compatible with conventional batch processes. An efficient mass production of the required electrode structures on a chip basis would be necessary for a widespread use of this new electrochemical principle in analytical laboratories.

The progress of modern silicon planar technology now allows the well-controlled preparation of very thin, well-defined polycrystalline metal layer structures with high reproducibility. Based on such polycrystalline working electrodes, recently the application of surface resistance measurements for the investigation of the potential-controlled deposition and dissolution of various heavy metals at platinum and gold electrodes has been proposed [18–26]. To employ this sensing principle for trace metal analysis in aqueous solutions, the thin polycrystalline metal films are used as working electrode in a "three-electrode" setup. For measurement of the resistance change during deposition and dissolution, a constant current is forced through this working electrode. An additional analytical selectivity can be achieved by controlling the potential of the working electrode. When scanning the electrode potential, the deposition of heavy metal species causes an increase, whereas the subsequent dissolution leads to a decrease of the d.c. resistance to the initial value. Because of the registration of resistance–voltage (potential) curves, this measurement principle is called "voltohmmetry" in analogy to voltammetry. Certainly, the sensor can be classified as an electrochemical one and it can be utilised both as an alternative and/or a supplement to traditional voltammetric techniques.

This article describes the theory, experimental setup, principles and requirements of voltohmmetry with the purpose of demonstrating its capabilities for electroanalytical applications. Characteristics of specifically designed thin-film working electrodes with thicknesses of <30 nm, integrated in a three-electrode chip design, are presented. Although the cited theoretical descriptions mainly imply the use of single crystalline, atomically smooth metal films, silicon-based polycrystalline metal films can also be applied as chemical sensing layers with almost the same resistance behaviour due to adsorption and deposition of foreign atoms and molecules. Different measurement and evaluation procedures for qualitative and quantitative analysis as well as the characteristics of this sensor type towards trace concentrations of lead, nickel, thallium, zinc, cadmium and Cd-EDTA complexes are discussed.

2
Theory and Experiment

2.1
Theory of Surface Resistance Change

The resistance of metallic layers will be influenced by both the properties of the bulk of the solid state and its surface. In both cases, the reason for that is the scattering of conduction electrons at traps of the lattice. When considering very thin films, the contribution of the surface (i.e. surface resistance) can be considerable. Therefore, in order to describe these scattering processes of conduction electrons in solids, different classical and semi-classical models can be applied, even if the phenomenon is a quantum-mechanical one. Different theories and models are discussed in the literature that deal with this phenomenon: the Drude model of electrical resistance, the transport of conduction electrons in crystalline solids, the Fuchs–Sondheimer theory of surface resistance (including its extended version) and the microscopic theory of surface resistance [27–46]. The most significant and accepted model is the Fuchs–Sondheimer theory, which will be explained in the following.

An indication that the surface properties of a solid influence the electrical resistance was found when experiments demonstrated the dependence of a thin metal film from its specific resistivity. This effect can be qualitatively explained by the scattering model shown schematically in Fig. 1. It is based on the adsorp-

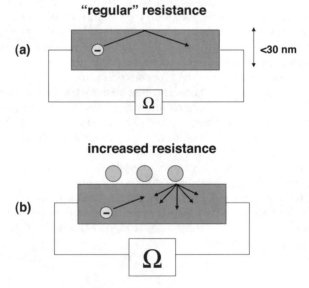

Fig. 1a,b. Sensing principle of voltohmmetry: specular reflection of conduction electrons at smooth metal surface (**a**); diffuse scattering of conduction electrons at metal surface due to adatoms (**b**)

tion-induced increase of the d.c. resistance due to diffuse scattering of conduction electrons of a thin metallic film (d<30 nm) by adatoms and -molecules on its surface. At a perfectly smooth surface of the metal film, the conduction electrons running through it will be specularly reflected at its surface. In this case, the electron momentum parallel to the plane of the metal film is conserved and consequently, the thin-film surface does not contribute to the lateral film resistance (Fig. 1a). If, however, there are any foreign atoms or ions adsorbed (deposited) on the film surface, the conduction electrons are diffusely scattered due to additionally shielded Coulomb potentials, which have been formed inside the metallic film. Now, the electron momentum parallel to the plane of the thin film is changed, resulting in an additional contribution and yielding an increase in the lateral d.c. film resistance (Fig. 1b). If, as a prerequisite, the thickness of the metal thin film is in the order of the mean free pathway of the conduction electrons (typically some 10 nm) this contribution is significant compared to the overall film resistance and thus, it can serve as a measuring signal. For low adsorbate coverages, up to 10% of a monolayer, the resistance increase is proportional to the coverage.

The qualitative description of the behaviour of the conduction electrons by this simple model mirrors the experimental results with good conformity. For an exact theoretical description, however, one has to solve Boltzmann's transport equation with the boundary values of coplanar surfaces of the metallic film, a parabolic conduction level of quasi-free electrons and a steric isotropic electron scattering at the film surface. These assumptions allow one to draw a relationship between the specific resistance of a thin metallic film and the bulk material. A deeper interpretation takes into account the polycrystalline structure of the metallic thin film together with the lateral and vertical extension of the crystallites. Due to the quantum-mechanical nature of the scattering process of electrons, an original, exact description of the surface resistance is only possible with the help of quantum-mechanical models (see work of Ishida et al. [35–37]). Self-consistent calculations of the density function assume the influence of surface roughness, adsorbates and crystallographic surface orientation on the surface scattering of the electrons and thus, finally, on the specific resistance of thin metallic films. Nonetheless, when considering the spatial isotropic and diffuse scattering of conduction electrons at adsorbates, this theoretical, quantum-mechanical-based model agrees with the results demonstrated by Fuchs and Sondheimer.

2.2
Fabrication of Thin-Film Electrodes

The major prerequisite for significant contributions of the surface resistance to the measuring signal is the reproducible fabrication of very smooth metal layers of a few nanometers thickness. Therefore, the sensor chip, including the working electrode as the most critical part of the experimental setup, has been prepared by means of silicon planar technology. The actual working electrode material exposed to the solution consists of a polycrystalline Au layer. In contrast to previous studies at single crystal silver electrodes [17] and polycrystalline Ag and Pt electrodes [22], Au was the most stable and favoured material found for the deposition of the thin-film electrodes.

As basic raw material, highly ohmic and oxidised Si wafers with a (100)-surface orientation and a resistivity of 2–100 kΩ cm have been used. Figure 2 depicts a scheme of the single fabrication steps: after depositing a photoresist by spin-coating, the patterning of the metal structures is achieved by photolitho-

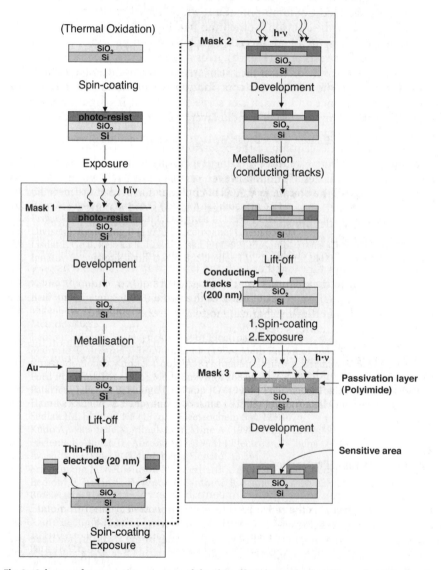

Fig. 2. Scheme of preparation process of the thin-film electrodes by silicon planar technology (photolithographic patterning and lift-off process)

graphic patterning (mask 1, first box), subsequent metallisation by electron-beam evaporation and a lift-off process. For stabilising the 15-nm-thick Au layer, a 5-nm-thick Cr layer served as adhesion promoter between the Si substrate and the Au layer. With this first mask step, the area of the thin-film working electrode is defined. The second mask step, including the same process sequence, leads to the areas of the conducting tracks that are thickened up to 200 nm (mask 2, second box). Finally, the whole wafer with the exception of the thin-film electrodes and the bond pads is covered with a 1-μm-thick electrically insulating polyimide layer. This step is performed by mask step 3 (third box). The sensor chips are cut into chip sizes of 15×15 mm^2. Each sensor chip carries up to eight pairs of Au thin-film electrodes, having dimensions from 10×20 μm^2 up to 400×500 μm^2. Each sensor chip is glued on a printed circuit board (PCB) and electrically connected by means of ultrasonic wire bonding.

2.3
Instrumentation and Chemicals

Figure 3 shows the experimental setup of the voltohmmetry and the cross-sectional scheme of the thin-film working electrode. Actually, each Au working electrode consists of two identical parts (see Fig. 3a), which are connected in series by a metallic pad (conducting track). The pad also serves as the connection of the electrode to the external potentiostat. Such an arrangement ensures that interfering electrical signals for the resistance measurement, like faradaic currents arising from the electrode during the deposition and dissolution of species in the test sample, are eliminated. In the chosen "two-finger" arrangement of the working electrode, electrochemical currents do not influence the d.c. resistance measurements: the potential drop caused by the current flowing in one half of the electrode (e.g. upper part) is compensated by a potential drop of the same magnitude but with opposite sign that is generated by the current flowing through the other half (e.g. lower part) of the thin-film electrode. On the other end of the experimental setup, a constant current I_R of 0.5 mA is applied between both parts of the working electrode and the d.c. resistance is measured by recording the voltage V.

In the cross-sectional view in Fig. 3b, the sensor is operated in a three-electrode configuration. A pair of the Au working electrodes is connected on one side to the common outlet of a potentiostat, and the other end of the working electrodes is connected to a constant-current source. A voltmeter is utilised to measure the d.c. resistance of the thin-film electrodes. A Pt rod electrode serves as counter electrode; the Ag/AgCl reference electrode is filled with saturated KNO$_3$ electrolyte and employed as potential reference (except for the current source and the voltmeter, the setup exactly corresponds to a conventional voltammetric experiment). For the measurements, the sensor chip is mounted in a batch cell of PTFE and O-ring sealed. Thus, only the thin-film electrode has contact with the test sample.

Figure 4a shows a photograph of a single sensor chip on a PCB. The cut-out magnification of the sensor chip (Fig. 4a, bottom) clearly depicts the area of the two-finger arrangement of the Au thin-film electrode. In Fig. 4b, the microscop-

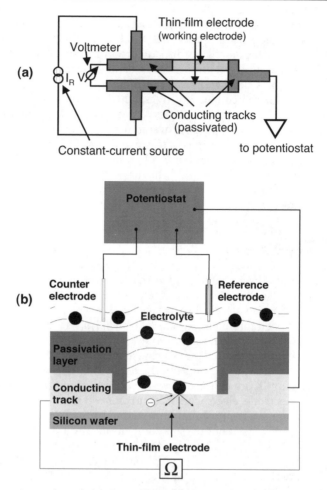

Fig. 3a,b. Experimental setup for d.c. resistance measurements at the two-finger thin-film (working) electrode (**a**), and cross-sectional scheme of the thin metal film electrode (**b**)

ic pictures of freshly prepared Au thin-film electrodes are presented. The thin-film electrodes of 5/15 nm Cr/Au appear homogeneous (left picture), whereas the enclosing passivation layer of 1 μm polyimide looks slightly wavelike. Each electrode is built up of two anti-parallel connected, identical thin-film layers (right picture).

2.4
Measurement Procedure

The deposition and dissolution of analyte species is realised for the voltohm-metric experiment under potential control as for voltammetry. According to voltammetry, the measurement of the resistance of such thin-film electrodes

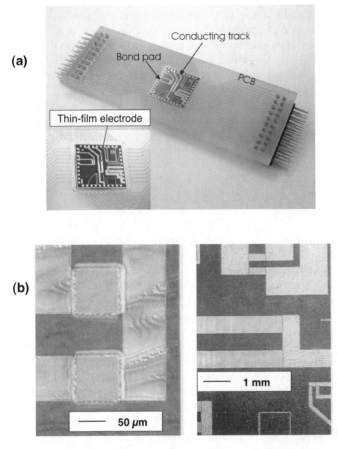

Fig. 4a,b. Photograph of a fabricated sensor chip, glued and bonded on a printed circuit board (a), and microscopic pictures of freshly prepared thin-film electrodes (b); the electrodes appear metallic grey and homogeneous, surrounded by an organic, slightly wave-like passivation layer

can be specified by the waveform of the excitation signal such as cyclic voltohmmetry, linear-sweep voltohmmetry or square-wave voltohmmetry. In the case of a separate preceding accumulation period, the resulting method can even be utilised as stripping voltohmmetry.

In this work, two kinds of measurement procedures will be discussed. In the so-called cyclic experiment (*cyclic voltohmmetry*), the electrode potential was periodically varied with a constant sweep rate from an anodic potential (E_{max}) towards a cathodic potential (E_{min}). The experiment was started and stopped at E_{max}, at which the studied analyte species was not reduced and deposited on the electrode surface. Then, within the cathodic scan, the reactive species was reduced and deposited onto the thin-film electrode. In the subsequent anodic scan, the deposited species was then oxidised and dissolved again from the surface of the thin-film electrode. In the so-called stripping experiment (*stripping voltohmmetry*), the electrode potential was switched from an anodic value to a cathodic one, at which the electroactive species was deposited onto the electrode. The cathodic potential was kept for a certain deposition time t_i. During this time period, a diffusion-controlled deposition of the species from solution takes place. After t_i, the potential was swept at a constant rate towards anodic values until it reached a requested final potential. During the anodic sweep, the deposited species was oxidised, and consequently dissolved. For both measurement procedures, the lateral potential drop over the thin-film electrode was recorded as a function of the electrode potential at a constant adjusted d.c. current.

3
Characterisation and Applications of Voltohmmetric Sensors

For the development and application of *voltohmmetry* as an interfacial method at thin-film electrodes in solution one has to consider several fundamental prerequisites:
- The contribution of the surface resistivity has to be substantial in relation to the bulk resistivity of the working electrode; this has been realised by the preparation of the Au layers with a thickness in the mean free pathway of the conduction electrons in the metal.
- The surface roughness of the thin-film electrodes has to be approximately small and the relevant interfacial characteristics of the Au layer have to be homogeneous enough to allow the observation of identical interfacial processes, like adsorption or deposition of species from the test sample; this has been checked by AFM (atomic force microscopy) measurements to determine the mean roughness and the density of the deposited thin films.
- Interfering electrical signals, like faradaic currents from redox potentials at the electrode during deposition and dissolution of species in the analyte, have to be eliminated; this is done by utilising the two-finger arrangement of the working electrode.

3.1
Physical Characterisation of Thin-Film Electrodes

During the fabrication of the Au thin-film electrode, the thickness was carefully controlled and optimised in the range between 15 and 20 nm for the layer system Cr/Au. Because of the necessity of a very smooth and flat surface of the working electrode, the surface roughness was checked by AFM and optical reflectance measurements. Figure 5a gives a comparison of both methods: in the diagram the nominal values of the thin-film thickness are compared with the respective measured film thickness. The obtained linear fit validated the good agreement between both methods with a maximum variation of less than 10% from the mean value with a slope factor of 1.

Figure 5b shows an AFM top view picture of the surface of a 5/15 nm Cr/Au thin-film electrode. In the figure, the dark black areas correspond to a rough-

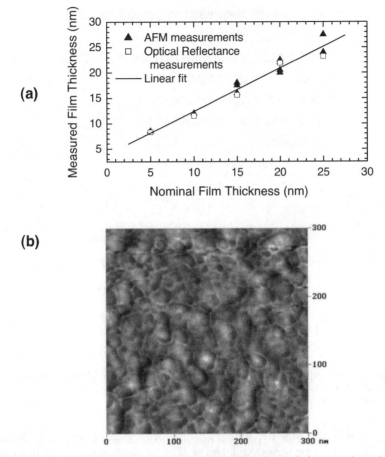

Fig. 5a,b. Proof of thickness of freshly prepared thin-film electrode by AFM (atomic force microscopy) and optical reflection measurement (**a**); morphological characterisation of thin-film electrode by AFM (**b**)

ness of 0 nm, whereas the bright areas define a maximum roughness of 2 nm. For the investigation, a surface area of 300×300 nm^2 was scanned with a resolution of 256×256 pixels. As can be seen, the maximum roughness does not exceed 2 nm in height.

The resistance values of eight freshly prepared working electrodes of 5/15 nm Cr/Au with an electrode size of $2 \times 75 \times 75$ μm^2 are presented in Fig. 6a. For the exact determination of the surface resistance values, the thin-film electrodes are not in contact with the analyte solution. The maximum resistance value amounts 35.1 Ω, the minimum resistance 25.92 Ω; the mean value (see dotted line in diagram) is 29.52 Ω with a standard deviation of 2.74 Ω, which is marked as an error bar for each working electrode. The resulting standard deviation of about 10% perfectly fits the variations of the film thickness (roughness), obtained by AFM and optical reflectance measurements in Fig. 5.

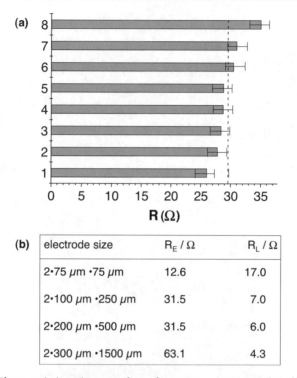

Fig. 6a,b. Characteristic resistance values of N=8 Cr/Au (5/15 nm) thin-film electrodes with Cr/Au (5/195 nm) conducting tracks (**a**), and mean values of thin-film electrodes (R_E) and conducting tracks (R_L) depending on their electrode size (**b**)

Figure 6b displays the dependency of the mean values of the thin-film electrodes (R_E) and their conducting tracks (R_L) on the respective electrode size. When considering the measurements of N=14 sensor chips, the specific resistivity for the thin-film electrodes can be calculated as ϱ_E=12.6 µΩ cm±0.8 µΩ cm, and for the conducting tracks as ϱ_L=2.0 µΩ cm±0.4 µΩ cm [22]. Since the conducting tracks were prepared in two steps (first mask: thin-film electrode of 15 nm; second mask: conducting tracks of 180 nm in addition), the specific resistivity of the complete Au layer of about 200 nm has a mean value of 1.1 µΩ cm, which is in good agreement with values discussed in the literature [47].

3.2
Analytical Approach

3.2.1
Resistance/Voltage Curve and Evaluation Procedure

A typical measurement curve for cyclic voltohmmetry in a 2 ppm-containing Pb solution is shown in Fig. 7 (upper diagram). During the sweep of the Au thin-film working electrode from 0.5 to –0.9 V with a sweep rate of 50 mV/s, the re-

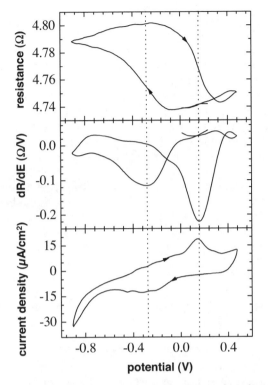

Fig. 7. Comparison of cyclic voltohmmetry (upper diagram) with cyclic voltammetry (lower diagram) of a Au thin-film electrode in 2 ppm lead-containing solution; the middle diagram belongs to the first derivative of the cyclic voltohmmetry

sistance changed significantly. In the beginning of the cycle, i.e. starting from anodic to cathodic potentials, the surface resistance slightly decreases with lowering of the potential until about –0.1 V. This behaviour is similar to that observed in pure electrolytes and is due to the dependence of the potential on the electrical double layer as well as its influence on the surface conductivity of the metal. For potential values <–0.1 V, the surface resistance increases, which is a result of the deposition of Pb atoms on the surface of the thin-film electrode: in this potential range, the adatoms on the electrode surface are acting as scattering centres for conduction electrons in the thin metal film.

During the backwards potential scan, the electrochemical deposition of Pb is continued up to –0.2 V. Later, the reoxidation, i.e. the dissolution of Pb, takes place and is accompanied by a decreasing resistance. At the end of the experiment, the resistance value at 0.4 V is almost identical with the initial starting value. This behaviour not only demonstrates the reversibility of the deposition (reduction) and dissolution (oxidation) of Pb, but also the reversibility of the interface structure of the metallic thin film/solution. Therefore, the registration of the surface resistivity change is an appropriate method for the determination of species not only in the gas phase or at single crystalline electrodes in solution, but also for polycrystalline thin-film materials, prepared by silicon-compatible batch processing and immersed in electrolyte solutions.

A convenient way of data evaluation for cyclic voltohmmetry consists in the transformation of the resistance/voltage curve into its first derivative (Fig. 7, middle). The resulting peak-shaped signals allow a more precise determination of the values of maximum resistance change in the cathodic and anodic cycle, respectively (the exact evaluation procedure will be discussed in Fig. 8).

In addition, Fig. 7 allows a direct comparison of cyclic voltohmmetry (upper diagram) and cyclic voltammetry (lower diagram) in terms of the Pb/Pb^{2+} process at the thin-film electrode. As for voltohmmetry, in voltammetry the reductive deposition of Pb also leads to a signal below –0.1 V. There is a good correlation to the point of inflection around –0.3 V. However, voltammetry only yielded weakly pronounced signals. Nonetheless, a similar signal behaviour, with a well-shaped cyclic voltammogram at 0.2 V and the following resistance decrease were caused by the anodic dissolution of Pb within the reverse potential scan.

When discussing voltohmmetry and voltammetry, the main advantage of voltohmmetry is due to the fact that no electron-transfer reaction between the Au electrode and the Pb species is necessary. The resulting sensor signal solely originates from the effect of foreign surface particles (e.g. Pb) on the conduction electrons of the Au thin-film electrode itself. Therefore, voltohmmetry is not restricted to determining analyte species that undergo heterogeneous faradaic reactions. In contrast, an interaction between the adsorbed analyte and the surface of the thin-film electrode is already sufficient.

Figure 8 exemplifies a cyclic voltohmmogram for Tl determination at the Au thin-film electrode. In the left diagram, significant resistance changes have been observed in the presence of 600 ppb Tl in comparison to the background curve (no Tl, dashed line), measured in the pure electrolyte solution. During the experiment, the electrode potential was swept from –0.1 to –0.9 V and backwards. The sweep rate was 10 mV/s, adjusting a constant current of 0.5 mA. In the case

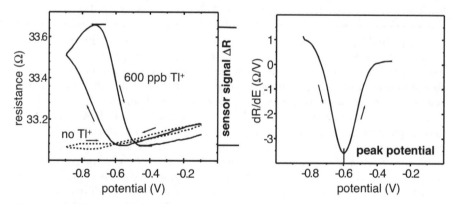

Fig. 8. Cyclic voltohmmetry in a 600 ppb-containing Tl solution in comparison to a Tl-free solution: evaluation of the sensor signal ΔR by maximum resistance change (left) and peak potential by its first derivative (right)

of the Tl-containing solution, the strong increase of the film resistance in the cathodic sweep is caused by the deposition of Tl adatoms from the electrolyte onto the surface of the Au working electrode. Then, during the anodic scan, the resistance decreases to its initial value when the deposited Tl atoms are oxidised and dissolved in the electrolyte.

For the quantitative evaluation of the sensor signal, two parameters of the surface resistance measurements have been examined: the amplitude of the resistance change (Fig. 8, left) and the peak potential of the respective species to be detected (Fig. 8, right). The amplitude of the resistance decrease during the anodic scan caused by the dissolution of the deposited species (here, Tl) is taken as the sensor signal ΔR. The inflection point in the anodic cycle that corresponds to the peak potential is determined by plotting the first derivative dR/dE of the thin-film resistance versus the electrode potential.

For the purpose of quantitative analysis, the concentration-dependent parameter is given by the relative change between the surface resistance of an analyte-free solution and the maximum ΔR in presence of deposited analyte species. The experiments did show that usually stripping voltohmmetry (similar to voltammetry) yields better-defined signals together with a lower detection limit than cyclic voltohmmetry. Therefore, cyclic voltohmmetry is not mainly envisaged for the determination of very low concentrations, but is more favourable in achieving a general survey on the surface resistance–potential behaviour of analytes at the thin-film electrode/electrolyte interface.

3.2.2
Sensitivity and Selectivity

As can be seen from Fig. 9, depending on the applied scan rate, a linear relationship was found between ΔR and the Cd concentration from 4 to 1,000 ppb. Thus, the principal feasibility of a quantitative analysis of heavy metal species in aqueous solutions by using voltohmmetry has been proven. The measurements were performed between –0.6 and –0.1 V with a d.c. current of 0.5 mA. All concentration measurements were carried out with sweep rates of 100 mV/s (circles), 50 mV/s (triangles) and 10 mV/s (squares). The three calibration curves are a result of the different scan rates. By variation of this experimental parameter one can tune the sensitivity of voltohmmetry: according to the concentration range of interest, the residence time of the Au working electrode within the potential window of analyte deposition is defined, and thus also the number of adatoms at the electrode surface.

For Cd concentrations from 10 to 200 ppb a linear relationship was achieved for a sweep rate of 10 mV/s with a sensitivity of 34 mΩ/ppb. By using this scan rate, however, for concentrations higher than 200 ppb Cd, the resistance changes deviate from the linear fit (dotted line). This can be explained by an increase of the specularity parameter for large amounts of reduced Cd atoms, i.e. high surface coverages. To circumvent this drawback for higher Cd concentrations, a further potential sweep rate can be applied, which leads to a smaller amount of deposited Cd species and thus, to a lower surface coverage. For scan rates of 50 and 100 mV/s, linear sensitivities for Cd of 9 and 7 mΩ/ppb, respectively, exist

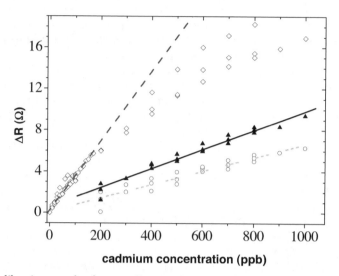

Fig. 9. Calibration graphs from cyclic experiment in the concentration range from 4 to 1,000 ppb Cd for different scan rates (circles: 100 mV/s; triangles: 50 mV/s; squares: 10 mV/s)

in the concentration range from 200 to 1,000 ppb. On the other hand, for such high scan rates, no measurable resistance changes were observed for Cd concentrations below 200 ppb.

For cyclic experiments with a particularly small scan rate of 1 mV/s and a 100-fold amplification, even Cd concentrations of some parts per billion yield a measurable signal that is proportional to the Cd content with a sensitivity of 11 mΩ/ppb down to 4 ppb. This experiment demonstrates that slower scan rates allow the determination of smaller analyte concentrations by cyclic voltohmmetry. Moreover, the detected small concentrations of Cd (but also for Pb in other experiments) become relevant for concentration values that are required by the World Health Organization for maximum residues in drinking water (Cd: 5 ppb; Pb: 50 ppb). By using cyclic voltohmmetry, the detection limits of the heavy metal species can be summarised as 4 ppb for Cd, 20 ppb for Pb and 200 ppb for Tl.

To utilise the presented method of voltohmmetry for the determination of different heavy metals in solution, their corresponding potential ranges for deposition and dissolution at the thin-film electrode have to be studied. In this way, the selectivity behaviour can be proven. First, by performing cyclic experiments, the suitable potential ranges were estimated for Cd, Pb, Ni, Tl, Zn and Cd-EDTA complexes. By the following stripping experiments, the corresponding peak potentials of the respective species were exactly calculated. As a result, it was found that due to the different peak potentials in the anodic scan of the cyclic experiment (first derivative of dR/dE), a distinction between various heavy metals and even heavy metal complexes is possible. The peak potentials for the investigated species are listed in Fig. 10a.

(a)

species	peak potential
Ni	-252±30 mV
Cd	-352±22 mV
Pb	-367±26 mV
Cd-EDTA	-452±16 mV
Tl	-566±12 mV
Zn	-868±19 mV

(b)

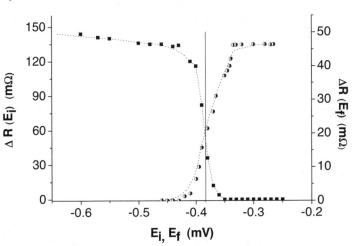

Fig. 10a,b. Determination of the peak potentials for different species, evaluated from the dR/dE plot by means of cyclic voltohmmetry (**a**), and stripping experiment for the determination of the deposition (reduction: squares) and dissolution (oxidation: circles) potential in a Cd-containing solution (**b**)

For the stripping experiment, the resistance changes can be described by a sigmoidal curve fit, which corresponds to a Boltzmann distribution. Figure 10b shows the stripping experiment for the determination of the deposition potential (squares) as well as the dissolution potential (circles) in a 200 ppb Cd-containing electrolyte solution. The stripping sweep was 5 mV/s, the accumulation time 120 s and the adjusted d.c. current 0.5 mA. The Boltzmann distribution describes the probability of heavy metal atoms occupying the reduced and oxidised states that correspond to the dissolved and deposited heavy metal species at a given potential. The measured values can be fitted to two sigmoid functions of the form:

$$\Delta R = \frac{\Delta R_{max}}{1 + e^{\left(E_i - E_0\right)/\delta}} \tag{1}$$

$$\Delta R = \frac{\Delta R_{max}}{1 + e^{\left(E_0 - E_f\right)/\delta}} \tag{2}$$

In the equations, ΔR_{max} represents the maximum resistance change, E_i is the deposition (accumulation) potential, E_f is the final value in the stripping experiment, δ is the fitting parameter that describes the half width of the Boltzmann distribution and E_0 is the equilibrium potential. At E_0, the ratio of the concentrations of reduced and oxidised Cd atoms near the surface of the thin-film electrode is unity.

From Fig. 10b it can be seen that a maximum signal is obtained for $E_i = -380$ mV (straight line), because here, all Cd ions arriving at the surface of the thin-film electrode are deposited. This corresponds to the fact that for trace metal analysis it is advantageous to deposit a maximum amount of heavy metal species at given measurement conditions. On the other hand, it is not necessary to chose a more negative deposition potential, since this will not increase the measured signal. As a negative effect, it may even further initiate the evolution of hydrogen, which can damage the thin-film electrode. For the subsequent dissolution of the deposited heavy metal species it is favourable to select a more anodic potential than the peak potential that is listed in Fig. 10a. Now, it is ensured that all deposited species are dissolved again and a maximum sensor signal ΔR can be achieved.

3.2.3
Reproducibility

One of the most important parameters to realise a new electrochemical analytical method by means of voltohmmetry is defined by the reliability, i.e. the reproducibility of the thin-film sensor device. Especially from the analogue – voltammetry – it is well known that the varying surface properties of the electrode material can have a distinct influence on the general sensor behaviour. In order to check the reproducibility of the surface resistivity technique, both single measurements of the same thin-film electrode (Fig. 11a) and a series of measurements (e.g. calibration plots, see Fig. 11b) have been performed.

Figure 11a shows a typical bare plot for the change of resistivity ΔR for N=11 subsequent measurements in a Cd test sample containing 200 ppb Cd salt. In this cyclic experiment, the potential was swept from –0.6 to –0.2 V with a scan rate of 5 mV/s, an adjusted current of 0.5 mA and an amplification of 100. The mean value of resistance change amounts to 70±5 mΩ and is marked by the dotted line in the diagram. The resulting standard deviation is less than 10% of the evaluated mean value, indicating a good reproducibility of the fabricated Au thin-film electrodes. The slight rise in the absolute value of resistance change with increasing number of measurements can be explained by a starting, slow degradation of the only 5/15-nm-thick Cr/Au working electrode that is accompanied by an increase of the surface roughness.

Even more interesting are the repeated measurements of different sensor chips or measurements made at different times. As one example, Fig. 11b depicts a comparison of calibration plots of two sensors after several weeks of operation

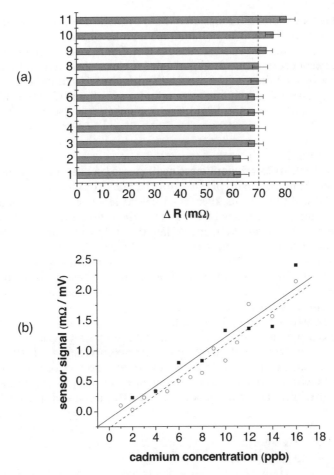

Fig. 11a,b. Change of resistivity ΔR for N=11 subsequent measurements in 200 ppb-containing Cd solution (**a**), and comparison of calibration plots of two sensors with varying cadmium concentration (**b**)

with varying Cd concentration from 0 to 16 ppb (for all experimental details, see [22]). The two calibration plots possess an excellent conformity with a slope of 0.13 ± 0.01 mΩ/mV; only a parallel shift of 0.1 mΩ/mV exists. The somewhat slower sensitivity in contrast to freshly prepared sensors is also probably due to the degradation behaviour of the thin-film sensors with operation time.

3.2.4
Heavy Metal Determination in Complex Analytes

To utilise the developed thin-film sensor for real applications such as the determination of Cd in lake water samples, one has to consider that the composition and the purity of the investigated sample may completely differ from laboratory test samples. In particular, matrix effects (e.g. organic complexes of heavy metals) can have a strong influence on the expected sensor signal.

Figure 12 presents the calibration plot of a "real" lake water sample when adding trace concentrations of Cd, evaluated by means of cyclic voltohmmetry. Before successive adding of Cd from 4 to 10 ppb, the test sample was characterised by an additional, highly sensitive reference method: ICP-MS (inductively coupled plasma mass spectrometry). The original sample had a composition of 0.12 ppb Cd, 0.21 ppb Pb, 0.26 ppb Tl, 4.5 ppb Ni and 42.4 ppb Zn.

During the cyclic experiment from −0.6 to −0.2 V, the scan rate was 1 mV/s, the current 0.5 mA and the amplification 100. The calibration plot shows a linear slope of 0.54 ± 0.05 mΩ/ppb. For detecting higher Cd concentrations, the analyte was acidified from originally pH 7.67 to pH 3. As a consequence, the peak potential was also shifted from -498 ± 20 to -398 ± 5 mV (for experimental details, see [22]). In this case, the absolute change of the sensor signal in milliohms is adequate for values obtained in "pure" laboratory test solutions. Thus, the experiments demonstrated the principal feasibility to determine heavy metal species also in real samples. However, further detailed investigations are necessary to study the influence of other disturbing anions, cations and also complexes, which may lead to the shift of the peak potential and change of the slope of the calibration curve of the sensor.

Since the inflection point, obtained from the anodic cycle, represents a characteristic peak potential for each heavy metal species, in a further experiment the simultaneous determination of more than one species in an analyte was investigated. As an example, Fig. 13 shows the simultaneous determination of Tl and Pb (each with a concentration of 300 ppb) by means of stripping voltohmmetry. The background solution was 10^{-3} M HNO$_3$, the sweep rate 10 mV/s,

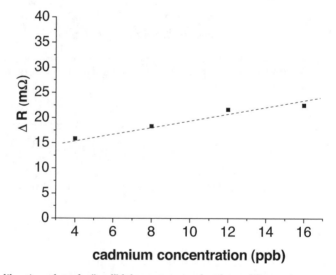

Fig. 12. Calibration plot of a "real" lake water sample when adding ppb concentrations of Cd by means of cyclic voltohmmetry

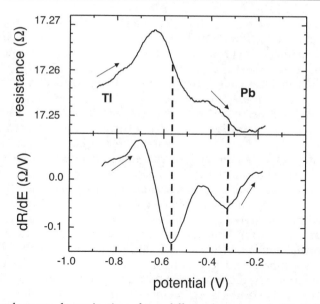

Fig. 13. Simultaneous determination of two different heavy metals (Tl, Pb) by means of stripping voltohmmetry: resistance/voltage curve (top) and its first derivative (bottom)

the adjusted d.c. current 0.5 mA, the amplification 100 and the deposition time 120 s.

In the upper curve, there are two regions where the resistance reaches a maximum, followed by a strong resistance decrease. In the first derivative of the resistance/voltage curve (Fig. 13, bottom), the peak potentials of –566 mV for Tl and –352 mV for Cd exactly correspond to the values obtained in Fig. 10. For Tl, the resistance change is about 20 mΩ, whereas for Pb it is about 7 mΩ. The measurement showed that the resistance changes can be clearly distinguished from each other by means of their characteristic peak potentials. In analogy to anodic stripping voltammetry, a few different metal ions can be analysed (see also Fig. 10a). As a prerequisite, an appropriate selection of the accumulation potential is given, if the stripping potentials in voltohmmetry differ by at least 100 mV.

4
Conclusion and Perspectives

The application of polycrystalline gold thin-film electrodes to trace metal analysis in aqueous solutions has been demonstrated by means of the principle of voltohmmetry. The surface resistance (change) turned out to present a highly sensitive parameter for the quantification of immobilised species, like heavy metals (Cd, Tl, Ni, Zn and Pb). As shown here, the sensors can be easily prepared with high accuracy in mass production processes by means of silicon planar technology. Simple polycrystalline Au electrodes with a layer thickness of about 20 nm serve as the sensor layer. The morphology and topography of the deposited noble metal films of Au, Pt and Ag show a high reproducibility (layer thickness: 20 nm±10%) of the polycrystalline materials (crystallite size: 15–20 nm); the specific resistance of the thin-film electrodes is about 12 µΩ cm. Although the measured resistance changes are relatively small, they can be determined with a very high precision in the order of several milliohms, which implies low detection limits down to a few parts per billion. The best sensor performance has been obtained with Au electrodes, whereas Pt as sensitive layer only allows cycling in a small potential region and the Ag layer is not stable in permanent contact with the analyte. Consequently, the surface resistance is a much more sensitive parameter in comparison to capacity measurements, because a very small electrode coverage is sufficient for a detectable voltohmmetric signal. First experiments even pointed out that Cd-EDTA complexes can be distinguished from hydrated cadmium ions.

The utilised measuring setup, i.e. the three-electrode arrangement, matches a conventional voltammetric analyser, thus voltohmmetry gives an alternative as well as a supplement to traditional electroanalytical methods. Moreover, as for voltammetry, the voltohmmetry can be carried out as cyclic (cyclic voltohmmetry) or as stripping (stripping voltohmmetry) experiment. In spite of the fact that this method cannot compete with anodic stripping voltammetry at the mercury electrode for trace analysis, voltohmmetry offers great potential for the quantification of species at solid metal/solution interfaces. This could be a chance for legislative action to remove mercury electrodes from analytical laboratories.

A significant improvement of the analytical quality parameters of voltohmmetry, like detection limit, precision, selectivity and response time, can be expected from a systematic know-how transfer between voltammetry and voltohmmetry. In this direction, voltohmmetry is envisaged to be an alternative method in electrochemistry for different fields of practical analytical purposes such as waste or drinking water analysis, speciation analysis, medical analysis and food control. This seems to be especially promising for non-electroactive species (adsorbates) that can be deposited or dissolved under defined potential control at the polycrystalline thin-film electrode. In contrast to voltammetry, here no electron transfer or catalytic reactions are required.

Our current investigations concern the suitability and modification of the thin-film microelectrodes [48], the performance of the sensor in an analyte which contains more than one species as well as in natural waste and drinking water, and its corrosion stability [49]. Extensive studies on the range of chemical species which can be qualitatively and quantitatively detected by this technique are intended. Moreover, the feasibility of realising a miniaturised sensor array combined with microsystem technology including micropumps, -valves, -channels and an integrated counter and reference electrode opens the possibility to build up a "lab-on-chip" sensor to broaden the spectrum for electroanalytical purposes. In order to further improve the selectivity as well as the simultaneous determination of compounds, the developed voltohmmetric sensor is planned to be coupled as a detector to a separation system, like chromatography or electrophoresis [50].

Acknowledgements: The author gratefully acknowledges H. Emons and O. Glück for valuable discussion. The author would like to thank the Ministerium für Wissenschaft und Forschung des Landes Nordrhein-Westfalen for financial support.

References

1. Kellner R, Otto M, Mermet JM, Widmer M (1998) Analytical Chemistry. Wiley-VCH, Weinheim
2. Strobel HA, Heinemann WR (1989) Chemical instrumentation: a systematic approach. Wiley, New York
3. Henze G (1994) Analytical voltammetry and polarography; Ullman's encyclopaedia of industrial chemistry. VCH, Weinheim
4. Suzuki H (2000) Electroanal 12:703
5. Manz A, Graber N, Widmer H (1990) Sens Actuators B 1:244
6. Thompson JJ (1901) Proc Camb Phil Soc 11:120
7. Fuchs K (1938) Proc Camb Phil Soc 34:100
8. Sondheimer EH (1952) Adv Phys 1:1
9. Lucas MPS (1965) J Appl Phys 36:1632
10. Wißmann P (1975) The electrical resistivity of pure and gas-covered metal films, vol 77, Springer Tracts in Modern Physics. Springer, Berlin Heidelberg New York
11. Schumacher D (1992) Surface scattering experiments with conduction electrons, vol 128, Springer Tracts in Modern Physics. Springer, Berlin Heidelberg New York
12. Ishida H (1995) Phys Rev B 52:10819
13. Romeo FM, Tucceri RI, Posadas D (1988) Surf Sci 203:186
14. Anderson WJ, Hansen WN (1973) J Electroanal Chem 43:329
15. Rath DL, Hansen WN (1984) Surf Sci 136:195
16. Winkes H (1996) PhD thesis, Heinrich-Heine-Universität Düsseldorf
17. Hanewinkel C, Winkes H, Schumacher D, Otto A (1997) Electrochim Acta 42:3345
18. Glück O, Schöning MJ, Lüth H, Emons H, Hanewinkel C, Otto A (1997) In: Proceedings of the 11th European conference on solid-state transducers, Warsaw, 21–24 Sept 1997
19. Glück O, Schöning MJ, Lüth H, Emons H, Otto A (1999) Adv Sci Technol 4:79
20. Glück O, Schöning MJ, Lüth H, Otto A, Emons H (1999) Electrochim Acta 44:3761
21. Schöning MJ, Glück O, Kordos P, Lüth H, Emons H (1999) SPIE Reprint 3857:135
22. Glück O (1999) Untersuchungen von Widerstandsänderungen polykristalliner Dünnschichtelektroden bei Schwermetallabscheidungen. Berichte des Forschungszentrums Jülich 3713, Jülich, Germany
23. Schöning MJ, Hüllenkremer B, Glück O, Lüth H, Emons H (2001) Sens Actuators B 76:275
24. Emons H, Hüllenkremer B, Schöning MJ (2001) Fresenius J Anal Chem 369:42
25. Emons E, Glück O, Hüllenkremer B, Schöning MJ (2001) Electroanal 13:677
26. Emons H, Glück O, Schöning MJ (2001) Proc Electrochem Soc, vol 2001-18. Electrochem Soc, San Francisco
27. Ibach H, Lüth H (1990) Festkörperphysik. Springer, Berlin Heidelberg New York
28. Dayal D, Finzel HU, Wißmann P (1987) Thin metal films and gas chemisorption, vol 32. Elsevier, Amsterdam
29. Juretschke HJ (1965) J Appl Phys 37:435
30. Mayadas AF, Shatzkes M (1970) Phys Rev B 1:1382
31. Tellier CR, Tosser AJ (1982) Size effects in thin films, vol 2. Elsevier, Amsterdam
32. Liebsch A (1997) Electronic excitations at metal surfaces. Plenum, New York
33. Löwdin PO (1999) Density function theory, vol 33. Academic, San Diego
34. Derouane EG, Lucas AA (1976) Electronic structure and reactivity of metal surfaces, vol 16. Plenum, New York

35. Ishida H (1996) Phys Rev B 54:10905
36. Ishida H (1998) Phys Rev B 57:4140
37. Ishida H (1994) Phys Rev B 49:14610
38. Persson BNJ (1991) Phys Rev B 44:3277
39. Persson BNJ (1993) J Chem Phys 98:1659
40. Lessie D (1979) Phys Rev B 20:2491
41. Lessie D, Crosson ER (1989) J Appl Phys 59:504
42. Krim J, Watts ET, Digel J (1990) J Vac Sci Technol A 8:3417
43. Dayal D, Wißmann P (1989) Vakuum-Tech 38:121
44. Körwer D, Schumacher D, Otto A (1991) Ber Bunsenges Phys Chem 95:1484
45. Alefeld G, Völkl J (1978) Hydrogen in metals I, vol 28. Springer, Berlin Heidelberg New York
46. Wedler G (1987) Thin metal films and gas chemisorption, vol 32. Elsevier, Amsterdam
47. Pietsch EHE (1954) Gmelins Handbuch der analytischen Chemie, vol 62. Verlag Chemie, Weinheim
48. Buß G, Schöning MJ, Lüth H, Schultze JW (1999) Electrochim Acta 44:3899
49. Schmitt G, Schultze JW, Faßbender F, Buß G, Lüth H, Schöning MJ (1999) Electrochim Acta 44:3865
50. Kissinger PT, Heinemann WR (1996) Laboratory techniques in electroanalytical chemistry, 2nd edn. Marcel Dekker, New York

Electrochemical Sensors for the Detection of Superoxide and Nitric Oxide – Two Biologically Important Radicals

F. Lisdat

Abstract. Electrodes are favourable sensing elements for the detection of the short-lived species nitric oxide and superoxide in the medical field. They can provide access to a time- and spatially resolved analysis allowing a deeper insight into the physiological role and interaction of both radicals. Superoxide sensing uses mainly electrodes modified with cytochrome c or superoxide dismutase as recognition elements. Operation is based on direct protein electrochemistry or on the detection of superoxide decomposition products. For NO quantification mainly electrodes modified with gas-permeable membranes have been developed. Transition metal complexes have been applied for the enhancement of NO electrocatalysis lowering the applied electrode potential. More recently haem proteins have also been combined with electrodes. Based on direct protein-electrode contacts, NO sensing is feasible exploiting specific NO interaction with the haem group.

Keywords. Superoxide, Nitric oxide, Radicals, Sensors, Protein electrodes

1
Introduction

Oxygen is the final electron acceptor in aerobic metabolism. Four-electron uptake results in water, however partial reduction leads to cytotoxic species such as superoxide or peroxide. The enzymatic release of these species is not just a result of imperfection in the evolution of biocatalysis; these species have to perform a role in controlled cell death (apoptosis) as well as in the defence against viral and bacterial attacks. The view of these reactive species as signal molecules is also supported by the recognition of nitric oxide – another related radical – as the endothelium-derived relaxing factor and the discovery of the involvement of this radical in several biochemical key processes such as inhibition of platelet aggregation, neurotransmission and inflammatory processes [1–4]. However, high concentrations of reactive species are toxic for the organism, thus a comprehensive antioxidative system counterbalances the generation processes of reactive species. It comprises enzymatic and non-enzymatic compounds as well as signal pathways to trigger gene and protein expression. Thus, the presence of reactive species is an ambivalent situation and the regulation of their in vivo concentration is a sophisticated process.

Superoxide was in the focus of biomedical research as long ago as the late 1960s. It has seen a renaissance in the 1990s since it can be considered as a counterpartner of NO, the biological role of which was discovered at that time. Superoxide reacts with NO under diffusion control and thus peroxynitrite is liberated. Since this species is even more reactive than both radicals the mechanisms discussed for oxidative damage by the radicals has had to be re-evaluated. One example is the oxidative conversion of the Fe-S enzyme aconitase into the iron regulatory protein (IRP1), which triggers the expression of ferritin – an iron storage protein – and the transferrin receptor [3, 5, 6]. Further objectives are DNA and protein damage [7–9] and operation of the soxRS regulon [10]. Ischaemia/reperfusion injuries where superoxide is liberated by NADPH oxidase, mitochondrially or by the xanthine dehydrogenase/xanthine oxidase pathway, have been intensively investigated [11–17]. The concentration rises rapidly and overwhelms the antioxidative defence within the timescale of minutes. Further concerns are brain injury [18, 19], apoptosis [20, 21], carcinogenesis [22] and superoxide involvement in ageing [9, 23, 24]. The interplay of superoxide and NO for such conditions has captured more and more attention [4, 17, 25–30], which has been supported not least by the finding that NO synthase may also act as a source of superoxide [31–33].

Along with the biomedical research, reliable techniques for radical detection have been developed. In the scope of this chapter are methods based on a sensor approach which allow the spatially resolved, on-line detection of radical concentrations. To start with the specific properties of these short-lived species and the demands for their sensor construction are examined; superoxide and NO sensors will be treated separately.

2
Superoxide and Nitric Oxide: Short-Lived Reactive Species

Both superoxide and NO are species with an unpaired electron. Reactivity is based on the tendency to stabilise the electron configuration by electron uptake or release. Therefore, both species appear to be redox amphoteric: one-electron reduction leads to nitrosyl (NO^-) or peroxide and oxidation to nitrosonium (NO^+) or oxygen, respectively. Thus, reaction pathways in physiological media depend strongly on the reaction partners available.

For the superoxide radical, oxidation and reduction potentials are pH-dependent. At neutral pH the values are –0.33 V vs. NHE for the O_2^-/O_2 couple and

+0.89 V vs. NHE for the O_2^-/H_2O_2 couple (pH 7, [34]). Superoxide is therefore a moderate reducing agent. This applies for example to the known reactions with ferricytochrome c or nitroblue tetrazolium. Substrate oxidation occurs with reducing substances, such as ascorbic acid, reduced flavins, free iron(II) etc.

For NO the picture is more diverse since the one-electron reaction is followed by a variety of reaction schemes which may lead to N_2O or NH_2OH, nitrosation of alcohols, thiols and DNA for NO reduction (E=0.25 V NO/NO$^-$) and to NO_2^-, N_2O_3 or ONOO$^-$ for NO oxidation.

For sensor development and calibration two opposite processes in aqueous solution have to be considered for both radicals: generation and decomposition. Superoxide is not only reactive with respect to other substances but it also shows spontaneous dismutation according to a second order kinetics:

$$2\,O_2^- + 2\,H^+ \rightarrow H_2O_2 + O_2 \tag{1}$$

Once liberated, superoxide thus decomposes in a concentration- and pH-dependent manner. The half-life of 1 µM superoxide at pH 6.5 is about 0.26 s. A much higher stability is observed in strong alkaline solutions and even in organic solvents such as DMSO (see also Table 1).

A similar behaviour is observed for NO in the presence of oxygen. Whereas in oxygen-free solution NO is quite stable, it reacts in solution with oxygen in a third order kinetics:

$$4\,NO + O_2 + 2\,H_2O \rightarrow 4\,NO_2^- + 4\,H^+ \tag{2}$$

The rate constant was determined to be k=2.1×10^6 M^{-2}s^{-1} resulting in an apparent rate constant for the NO decomposition of 8.4×10^6 M^{-2}s^{-1} (4×k=k$_{app}$) [35]. With an oxygen concentration of 200 µM the rather pH-independent reaction results in a half-life for 5 µM NO of about 120 s and for 500 nM of 1,200 s. For NO sensor calibration the problem of limited lifetime of the species to be detected is therefore much smaller than with superoxide. This is especially relevant for applications with low oxygen and NO concentrations. Thus, for NO sensor calibration NO stock solutions can be used elegantly. For superoxide and higher NO concentrations (micromolar) however, steady-state situations in which radical decomposition is counterbalanced by an enzymatic synthesis or a radical-liber-

Table 1. pH dependency of the half- life time of the superoxide radical in solution

pH of aqueous solution	Half-life time[a] (s)
6.5	0.26 s
7.0	0.8 s
7.5	2.5 s
8.5	24.5 s

[a] Calculated from dismutation rate constants according to Eq. 6 in [135]

ating compound can be applied. Additionally, measurements determining the rate of radical liberation can be used for calibration.

In Fig. 1 a typical superoxide sensor response to the enzymatic generation of superoxide in solution is shown. For sensor calibration the response rate after the start of the measurement can be used as described by Tammaveski et al. [36]. The steady-state sensor output after reaching a steady-state radical concentration due to a kinetic balance between the enzymatic generation and the spontaneous dismutation of the radical can also be mathematically analysed [37]. Increasing enzyme activity results in increasing steady-state superoxide concentrations in solution. This allows the use of the current output as a calibration term.

For the accuracy of the current–concentration relation with both approaches careful determination of the rate constant of enzymatic radical generation, the stoichiometry of dismutation and control of the pH during the measurement are important. The time period in which a steady-state superoxide concentration can be observed is shortened by increasing enzyme activity and prolonged by increasing substrate concentration (at higher concentrations, however, enzyme inhibition has to be considered). This period is not decisive for sensor calibration but important for in vitro radical interaction studies and also for the analysis of antioxidant efficiency.

In vitro, superoxide is mainly generated by the oxidative conversion of xanthine or hypoxanthine by xanthine oxidase [38]. Alternative approaches are the

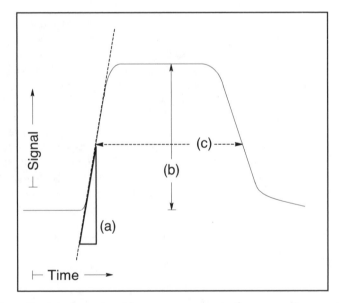

Fig. 1. Schematic plot of a superoxide sensor response to the enzymatic generation of the radical. Concentration is increasing because of the continuous biocatalytic radical generation; increasing spontaneous dismutation results in a steady-state situation and finally the consumption of the substrate leads to a concentration decrease. For sensor calibration, generation rate (a) or the steady-state concentration (b) can be used. The time period (c) is important when interaction of the superoxide radical with other species is studied

use of diamine oxidase [39], fumarate reductase [40] or NAD(P)H:flavin oxidoreductase [41]. Besides these enzymatic sources there are some reports using modified electrodes for the one-electron reduction of oxygen to form superoxide [42–44]. Another approach was introduced by additions of a stock solution of the potassium salt of the radical in DMSO into the aqueous solution [37]. A peak-shaped behaviour was found at the sensor electrode resulting from the spontaneous dismutation of the radical. However, mathematical modelling is not trivial and conditions such as pH and stirring during calibration as well as the quality of the stock solution have to be controlled.

NO can be synthesised by NO synthase [45, 46], generated chemically (nitrite+acid) or used as a gas to prepare a stock solution. Alternatively a NO donor is used (diazeniumdiolates, S-, C- and N-nitroso compounds, organic nitrates, oxatriazoles etc.). Since the liberation profile of NO donors may vary by orders of magnitude, different concentration–time profiles can be provided [35].

From the high reactivity of the oxygen-derived radicals several demands for sensor-based detection can be derived:

(i)   the detection has to be fast in order to follow actual concentration changes, in particular the interplay with other reactive species;

(ii) rather low concentrations have to be analysed since the range under physiological conditions covers the nano- and micromolar concentration level; and

(iii) the sensor configuration should be stable under repeated radical bursts in order to be reusable several times. Other demands are miniaturisation of the sensor element for spatially resolved detection and high selectivity of the sensor signal for the reactive species under investigation.

3
Sensors for Superoxide

Superoxide is the starting point of a chain of reactive oxygen species. Along with the elucidation of its production rates and time profiles the importance of antioxidative therapies has increased [47, 48]. However, the discussion about the efficiency is quite controversial. This seems to be mainly connected with the lifetime and membrane permeability of the antioxidant and in some cases the concentration and matrix dependence of their action. For example, it was found that the well-known antioxidant α-tocopherol can even show pro-oxidant properties and may induce tumour formation under special conditions [49]. Another finding in this direction is the change of the superoxide scavenging ability of ascorbic acid if incorporated into different hydrophobic matrices [50].

The complexity of the mechanisms of radical generation, the overwhelming of the natural defence system and the interaction with external antioxidants formed the background for the development of direct and on-line detection methods with the help of sensors. Depending on the measuring principle two main groups can be distinguished. The first group uses an enzymatic conversion of superoxide and focuses on the detection of reaction products. The second group exploits the direct electrochemical conversion of a protein at a modi-

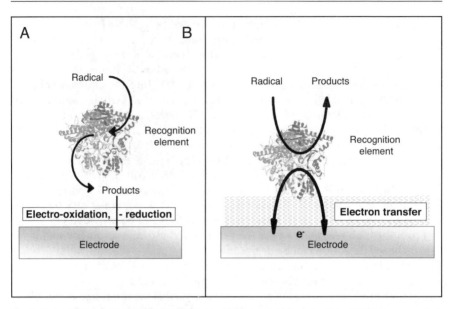

Fig. 2. Principles of electrochemical sensors for superoxide detection. Within approach A reaction products (O_2, H_2O_2) of an enzymatic conversion are detected. Approach B uses an electron transfer chain from the radical via a redox protein to the electrode

fied electrode and detects the changes of electron transfer in the presence of the radical interacting with the protein (see Fig. 2).

3.1
Sensors Based on Enzymatic Product Detection

For the first approach, superoxide dismutase (SOD) is used exclusively as a highly specific enzyme for superoxide radicals. It generates oxygen and hydrogen peroxide during the conversion. The latter can be detected at the electrode and thus, the H_2O_2 oxidation current is taken as a measure of the superoxide level in solution. SOD immobilisation was done by simply embedding SOD in gelatine and sandwiching it between two membranes which were spanned over a Pt electrode [51]. More elegantly SOD can be immobilised in a polypyrrole layer which was prepared by electropolymerisation [52–54]. This also allows the preparation of thin microelectrodes useful for stereotactic measurements.

The detection of H_2O_2 as a product of superoxide conversion, however, causes problems with H_2O_2 present in the test medium and other interferents which can be detected at about +0.5 to +0.7 V vs. Ag/AgCl [52, 53]. This can elegantly be solved by the introduction of a radical permeable membrane separating the SOD-catalysed conversion from the test medium [51]. Problems arise, however, in miniaturised sensor fabrication and response rate. Alternatively a simultaneous H_2O_2 and O_2^- measurement and differential analysis can be used [54, 55].

Another approach combines the SOD-catalysed dismutation with an amperometric oxygen electrode [56]. Although it is not a real superoxide sensor since the electrode reacts also to external oxygen variations, it is valuable for the characterisation of antioxidants.

3.2
Sensors Based on Direct Protein-Electrode Contacts

Here the sensor concept is based on an efficient electrochemical communication of the redox centre of a suitable redox protein with an electrode. Bare metal electrodes are seldom suited to a stable direct protein electrochemistry and thus promoter molecules are used to modify the electrode surface in a "protein-friendly" way. In addition, these promoter layers can offer the possibility of covalent fixing of the protein in order to avoid protein loss by desorption. Ideally they may also form a kind of barrier against potentially interfering substances and thus enhance selectivity. An efficient electron transfer through the thin promoter layer will allow the detection of redox reactions of the protein with superoxide radicals and subsequently the sensor-based analysis of the radical (see Fig. 2B).

For sensors of this group communication between the protein and the electrode is decisive. Here the extensive research on direct protein electrochemistry has resulted in one of the first practical applications with the focus on cytochrome c. Cytochrome c is a small redox protein showing rather fast reduction by superoxide ($k=6\times10^5$ $M^{-1}s^{-1}$ [57]). This is also the basis for the standard photometric test to prove radical generation. For the electrochemical detection an efficient re-oxidation of the reduced protein results in a current which can be correlated with the superoxide concentration in solution. Since electron transfer from cytochrome c to the electrode is influenced by the thickness of the promoter layer on the gold electrode, the early developments used a short-chain thiol for cytochrome c immobilisation [36, 58–60]. With these electrodes several measurements at cell cultures could be successfully shown [61]. Additionally to the promoter thickness, protein orientation influences electron transfer to the electrode. Layers from mixed short-chain thiols have been shown to improve the electrochemical communication between the immobilised protein and the electrode surface [62]. This was also used for sensitive superoxide analysis [63].

However, quasi-reversible redox transformation was also found with longer thiols, which proved to be valuable for interferent rejection since they can form rather dense layers on gold [64]. With these electrodes in vivo measurements of the reperfusion syndrome were feasible [65]. Recent developments have shown that by means of a mixed long-chain thiol layer the advantages of interference rejection and fast electron transfer can be combined [37]. A mixed promoter layer of hydroxy- and carboxy-terminated thiols results in a significantly enhanced electron transfer (35 s^{-1} in the covalently fixed state compared to 4 s^{-1} for only carboxy-terminated thiol layers) and still effectively rejects interfering substances from the electrode surface. The mixed thiol approach proved to be valuable not only for an optimum orientation of the protein with respect to efficient electron transfer but also for an enhanced reaction rate with superoxide radicals compared to a promoter layer using only carboxy-groups (C11-hydroxy-, carboxy-terminated: $k_{mixed}=3.1\times10^4$ $M^{-1}s^{-1}$ compared to C11-carboxy-terminated: $k=1.4\times10^4$ $M^{-1}s^{-1}$). Furthermore, the amount of electroactive protein can be drastically increased after the covalent linkage of cytochrome c to the promoter layer. With these electrodes the dependence of radical production during reperfusion on conditions of ischaemia was shown in an animal model [66].

More recently direct protein–electrode contacts have also been established with superoxide dismutase which was immobilised on modified gold (Cu/Zn-SOD [67], Cu/ZnSOD, FeSOD [68]). Once more self-assembled monolayers (SAMs) have been applied to create surface properties at metal electrodes, which are suited for protein immobilisation and effective electron transfer to the electrode. Mercaptopropionic acid and cysteine have been used for these electrode modifications. The reaction rate constant of superoxide with SOD is much higher compared to cytochrome c and the interaction is more specific. This can facilitate a more sensitive detection of superoxide in the nanomolar concentration range.

The direct protein electrochemistry approach uses surface reactions for radical detection – usually based on monolayers of the promoter molecule as well as the protein. Thus, the measurement is not only sensitive but also fast (response time in the range of seconds), allowing the on-line detection of radical concentration profiles. In addition, this approach is also suited for the preparation of small sensing areas of different shape. However, it has to be noted that miniaturisation is limited by the immobilisation density of the biomolecule.

Alternatively to the direct protein oxidation, a mediator can be used in combination with a carbon electrode; however, rather high potentials have been reported [69]. The redox behaviour of superoxide can also be exploited directly at bare or modified electrodes (mercury, carbon etc.) to generate and detect this oxygen radical. Because of interferences these investigations are of minor importance for superoxide quantification in physiological media but may have a potential for the evaluation of antioxidative substances [70–72].

3.3
Sensors for Antioxidants

Sensitive and stable sensors for superoxide offer the possibility to characterise antioxidants if they are combined with a stable radical source. Antioxidants are substances which can effectively remove reactive oxygen from solution by catalytic conversion (e.g. SOD) or chemical reaction (e.g. ascorbic acid). The evaluation of the antioxidative potential of individual substances or mixtures has to be seen in the light of antioxidative therapies in the medical area, and the use of such substances in the food and cosmetics industries for product persistence, health and beauty effects (e.g. anti-ageing).

Cytochrome c electrodes based on SAMs have been used to detect SOD as well as non-enzymatic radical scavengers [73, 74]. The measuring principle is based on the analysis of steady-state superoxide concentrations in the presence and absence of potentially antioxidative substances as shown in Fig. 3. An IC50 value, which represents the antioxidant concentration that can diminish the radical concentration to 50%, can serve as a measure for the efficiency of the antioxidant action. More recently it has been shown that the antioxidant analysis

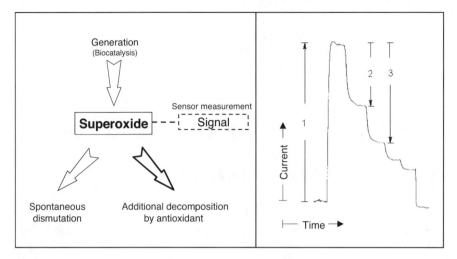

Fig. 3. Scheme of antioxidant detection using a superoxide sensor. The steady-state superoxide concentration resulting from a steady state between enzymatic generation and spontaneous dismutation is shifted because of an additional decomposition process by the scavenging reaction with the added antioxidant. The relative signal decrease detected at the sensor electrode can be evaluated for the quantification of the antioxidant efficiency

can be extended to mixtures of aqueous buffer and organic solvents, thus broadening the applicability of the measuring system [50].

The combination of an oxygen electrode with the SOD-catalysed dismutation and a gas-permeable membrane separating the test medium from the sensor compartment can also be applied to organic solvents [56]. When KO_2 is used as superoxide source, only the spontaneous and SOD-catalysed dismutation (and thus oxygen production) and their change after addition of radical scavengers have to be evaluated.

4
Nitric Oxide Sensors

For NO detection a variety of sensor systems have been developed. The amount of literature during the last decade is considerable and several comparative studies have been published [75–78]. Here the focus is on the different principles of sensor operation. Most sensors are electrochemical, using the redox transformation of NO at metal or carbon electrodes [79–83] or even semiconductor electrodes [84]. Because of the redox-amphoteric behaviour of NO it can be reduced or oxidised at the electrode surface. To ensure selectivity against other constituents of the solution, the peculiarities of a small gas molecule can be exploited and the electrode compartment can be separated from the test media by a gas-permeable membrane (see also Fig. 4). For NO reduction, however, there is still an interference by oxygen which can also permeate the membrane. Thus, most sensor developments have used the oxidation of NO. In the classical approach the sensor reflects the construction of a Clark electrode [79]. More re-

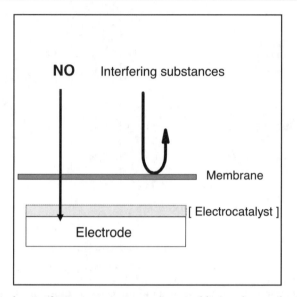

Fig. 4. General scheme of NO sensors using gas-permeable membranes for the separation of the detection compartment from the test solution. Additionally the electrochemical detection of the NO molecule can be enhanced by means of an electrocatalyst – mainly transition metal complexes applied for the surface modification of electrodes

liable for sensor miniaturisation are systems with an integrated membrane on top of the electrode surface. Membranes which are often prepared from a casting solution are cellulose acetate [75, 85], nitrocellulose [86], Nafion [85, 87–98], polymer combinations [75, 85, 87, 92, 99, 100] or Teflon-type membranes [101]. Electropolymerisation, in contrast, provides an in situ sensor preparation method. Examples are non-conducting polymers from diaminobenzene (DAB) [75, 92], dihydroxybenzophenone (DHB) [75], resorcinol [92] and phenylenediamine [87, 92]. But even conducting polymers such as polypyrrole have been used for sensor construction [93]. Electropolymerised films can be prepared reproducibly and rather thinly, ensuring a rapid response of the sensor. However, quite often they have to be combined with Nafion to avoid interference from ascorbic acid and other anionic interferents. A rather high selectivity has been reported with electropolymerised eugenol [102]. Another approach has used ascorbate oxidase within an additional interference layer [99].

Concomitant with membrane development, attempts have been made to reduce the high potential for NO oxidation (+0.9 V at Pt electrodes). Since the pioneering work of Malinski et al. [94, 103] several nickel complexes have been used for the electrocatalysis [91], including salens [104] and phthalocyanines [88, 95], but mainly porphyrins [87, 89, 96, 99, 102, 105] – not least because of their preparation potential by electropolymerisation. The basic idea exploits the fact that NO can coordinate several transition metal complexes. Thus, it is activated for a redox transformation and catalytic effects for both oxidation and reduction have been reported (copper complexes [106], iron centres [95, 97, 107–

110], manganese porphyrins [111], chromium complexes [98, 112] and ruthenium chelates [113]). Furthermore, palladium and iridium oxide were used for electrocatalysis of NO [90].

The electrodes using transition metal complexes and thin membranes are well suited for sensor miniaturisation. Sensors with diameters in the micrometre scale can be used for spatially resolved NO detection. In addition, the thin-layer design provides a high response rate (up to milliseconds), thus providing access to time-resolved observation of releasing processes [114, 115]. Sensitivity can be provided in the nanomolar range with lower detection limits of, e.g., 85 nM [102], 35 nM [87] or even 0.5 nM [99]. Sensitivity is not only influenced by the kind of electrode modification but can be improved by differential normal-pulse amperometry [76, 90] or interdigitated electrode arrays [116].

Alternatively to metal complexes attempts have been made to use haem proteins as recognition elements for the electrochemical detection of NO. The detection is based on the direct protein electrochemistry at the modified electrode. Myoglobin was incorporated in a thin surfactant film on a pyrolytic graphite electrode [117] and investigated with respect to NO reduction by the surface-immobilised protein. The study focused mainly on mechanistic effects with micromolar NO concentrations. Recently, clay-modified glassy carbon electrodes with an efficient protein–electrode contact have also been found to be sensitive to low NO concentrations [118]. Haemoglobin can spontaneously be adsorbed onto glassy carbon electrodes and an electrochemical communication with the haem centre can be established. These electrodes were shown to give a reduction current in the presence of NO [119]. Alternatively the protein can be incorporated into a DNA [120] or surfactant [121, 122] film on carbon electrodes. However, sensitivity was reported in the micromolar concentration range. More recently an electrode modification strategy using cysteamine-modified gold nanoparticles and haemoglobin resulted in a much more sensitive detection of nanomolar NO concentrations [123].

Compared to the haem proteins myoglobin and haemoglobin, the interaction of cytochrome c with NO is not very pronounced. Denaturation has been reported to be a precondition for sensor construction [124, 125]. Cytochrome c′ (prime) as a high-spin and pentacoordinated haem protein, in contrast, was found to be stably immobilised on SAM-modified gold electrodes and to show a pronounced reaction with NO. Interaction of the protein with mercaptosuccinic acid-modified gold was found to be dominated by non-electrostatic interactions and resulted in dissociation of the protein dimer. The reaction of the immobilised cytochrome c′ with NO produced a reduction current which is suited for amperometric detection in the nanomolar concentration range [126, 127]. In general, however, protein-based sensors still suffer mainly from lack of stability and interference problems in comparison to the use of NO oxidation at membrane-modified electrodes.

Detection of radicals by optical sensors, which shall be mentioned here only briefly, uses either metalloproteins for specific interaction with NO or fluorescent dyes and chemiluminescence reactions. Naturally the classical haem proteins such as cytochrome c, haemoglobin and myoglobin were used in combination with photometric detection [128, 129]. An in vivo receptor of NO, the enzyme guanylate cyclase, can also be used. The fluorescent dye-coupled enzyme provides access to measurements in the micromolar concentration range [130]. Another example with a slightly lower sensitivity is the use of cytochrome c' which shows a rather high specificity for NO [131, 132]. Alternatively to the interaction of NO with biomolecules, the adsorption to a fluorescent dye-modified gold surface can be monitored [133]. However, the detection limit is still high in the micromolar range. The same applies to the use of the chemiluminescence reaction of NO with H_2O_2 and luminol, which can be followed at 425 nm [134].

5
Conclusion

Electrochemical transduction is feasible for superoxide as well as NO sensing. Electrodes are applied to measurements of physiological relevance, exploiting their high sensitivity and response rate. While for superoxide analysis biological recognition elements such as cytochrome c and SOD are combined with the electrode, for NO detection mainly combinations of transition metal complexes and membranes on electrodes have been used. Alternatively haem protein electrodes, which make use of the selective interaction of the messenger molecule with the biomolecule, have been developed. However, their applicability is still limited because of stability and interference problems.

From the sensor construction point of view, the combination of membrane systems with electrodes is being replaced more and more by surface modifications allowing the in situ preparation of the sensing layer on top of the electrode. Electropolymerisation, self-assembling and nanoparticles are the most straightforward direct techniques. They are also valuable for miniaturised sensor fabrication, which is necessary for spatially resolved radical analysis in vivo.

Acknowledgements: The author cordially wants to thank Professor Frieder W. Scheller for helpful discussions, Frau Edith Micheel for editorial assistance and Cheuk Hung Kwan for graphical support. Financial support by the "Fonds der Chemischen Industrie" (VCI, Germany) and the Ministry of Research, Science and Culture of Brandenburg, Germany, is acknowledged (BLV project 24#2598-04/327).

References

1. Torreilles J (2001) Front Biosci 6:D1161
2. Ramamurthi A, Robson SC, Lewis RS (2001) Thromb Res 102:331
3. West AR, Galloway MP, Grace AA (2002) Synapse 44:227
4. Schwentker A, Vodovotz Y, Weller R, Billiar TR (2002) Nitric Oxide–Biol Chem 7:1
5. Bouton C, Hirling H, Drapier J-C (1997) J Biol Chem 272:19969
6. Castro L, Rodriguez M, Radi R (1994) J Biol Chem 269:29409
7. Grisham MB, Jourd'Heuil D, Wink DA (2000) Aliment Pharm Therap 14:3 Suppl
8. Cadenas E, Davies KJA (2000) Free Radic Biol Med 29:222
9. Linton S, Davies MJ, Dean RT (2001) Exp Gerontol 36:1503
10. Pomposiello PJ, Demple B (2001) Trends Biotechnol 19:109
11. Meneshian A, Bulkley GB (2002) Microcirculation 9:161

12. Fabian RH, Kent TA (1999) Free Radic Biol Med 26:355
13. Jassem W, Fuggle SV, Rela M, Koo DDH, Heaton ND (2002) Transplantation 73:493
14. Ushiroda S, Maruyama Y, Nakano M (1997) Jap Heart J 38:91
15. Granger DN, Stokes KY, Shigematsu T, Cerwinka WH, Tailor A Krieglstein CF (2001) Acta Physiol Scand 173:83
16. Chung HY, Baek BS, Song SH, Kim MS, Huh JI, Shim KH, Kim KW, Lee KH (1997) Age 20:127
17. Kobayashi T, Seguchi H (2001) Acta Histochem Cytochem 34:85
18. Shohami E, BeitYannai E, Horowitz M, Kohen R (1997) J Cereb Blood Flow Metab 17:1007
19. Chan PH (1996) Stroke 27:1124
20. Warner HR (1999) Mech Cell Death–Annals New York Acad Sci 887:1
21. Wolin MS, Gupte SA, Oeckler RA (2002) J Vasc Res 39:191
22. Brown NS, Bicknell R (2001) Breast Cancer Res 3:323
23. Casteilla L, Rigoulet M, Penicaud L (2001) IUBMB Life 52:181
24. Camougrand N, Rigoulet M (2001) Resp Physiol 128:393
25. Kumura E, Yoshimine T, Iwatsuki K, Yamanaka K, Tanaka S, Hayakawa T, Shiga T, Kosaka H (1996) Am J Physiol Cell Physiol 270:C748
26. Thomson L, Trujillo M, Telleri R, Radi R (1995) Arch Biochem Biophys 319:491
27. Beinert H, Kiley PJ (1999) Curr Opin Chem Biol 3:152
28. Tschudi MR, Mesaros S, Luscher TF, Malinski T (1996) Hypertension 27:32
29. van der Vliet A, 'tHoen PA, Wong PS, Bast A, Cross CE (1998) J Biol Chem 273:30255
30. Maruyama W, Kato Y, Yamamoto T, Oh-hashi K, Hashizume Y, Naoi M (2001) J Am Aging Assoc 24:11
31. Mayer B, Hemmens B (1997) Trends Biochem Sci 22:477
32. Vasquez-Vivar J, Martasek P, Hogg N, Karoui H, Masters BS, Pritchard KA Jr, Kalyanaraman B (1999) Meth Enzymol 301:169
33. Vasquez-Vivar J, Joseph J, Karoui H, Zhang H, Miller J, Martasek P (2000) Analusis 28:487
34. Sawyer DT, Nanni EJ Jr, Roberts JL Jr (1982) In: Biological redox components. American Chemical Society, Washington, p 585
35. Feelisch M, Stamler JS (eds) (1996) Methods in nitric oxide research. Wiley, New York
36. Tammeveski K, Tenno TT, Mashirin AA, Hillhouse EW, Manning P, McNeil CJ (1998) Free Radic Biol Med 25:973
37. Ge B, Lisdat F (2002) Anal Chim Acta 454:53
38. Rubbo H, Radi R, Prodanov E (1991) Biochim Biophys Acta 1074:386
39. Silva IJ, Azevedo MS, Manso CF (1996) Free Radic Res 24:167
40. Imlay JA (1995) J Biol Chem 270:19767
41. Gaudu P, Touati D, Niviere V, Fontecave M (1994) J Biol Chem 269:8182
42. Divisek J, Kastening B (1975) J Electroanal Chem 65:603
43. Ohki N, Uesugi S, Murayama K, Matsumoto F, Koura N, Okajima T, Ohsaka T (2002) Electrochemistry 70:23
44. Matsumoto F, Tokuda K, Ohsaka T (1996) Electroanal 8:648
45. Wie G, Dawson VL, Zweier JL (1999) BBA–Mol Basis Dis 1455:23
46. Xia Y, Zweier JL (1997) Proc Natl Acad Sci USA 94:12705
47. Gilgun-Sherki Y, Rosenbaum Z, Melamed E, Offen D (2002) Pharmacol Rev 54:271
48. Youdim KA, Spencer JPE, Schroeter H, Rice-Evans C (2002) Biol Chem 383:503
49. Yamashita N, Murata M, Inoue S, Burkitt MJ, Milne L, Kawanishi S (1998) Chem Res Toxicol 11:855
50. Beisenhirtz M, Ge B, Scheller FW, Lisdat F (2003) Electroanal 15(18):1425
51. Song MI, Bier FF, Scheller FW (1995) Bioelectrochem Bioenerg 38:419
52. Mesaros S, Vankova Z, Grunfeld S, Mesarosova A, Malinski T (1998) Anal Chim Acta 358:27

53. Mesaros S, Vankova Z, Mesarosova A, Tomcik P, Grunfeld S (1998) Bioelectrochem Bioenerg 46:33
54. Lvovich V, Scheeline A (1997) Anal Chim Acta 354:315
55. Lvovich V, Scheeline A (1997) Anal Chem 69:454
56. Campanella L, De Luca S, Favero G, Persi L, Tomassetti M (2001) Fresenius J Anal Chem 369:594
57. McCord JM, Crapo JD, Fridovich I (1977) Superoxide dismutase assays: a review of methodology. In: Superoxide and superoxide dismutases. Academic, London
58. McNeil CJ, Smith KA, Bellavite P, Bannister JV (1989) Free Radic Res Commun 7:89
59. Cooper JM, Greenough KR, McNeil CJ (1993) J Electroanal Chem 347:267
60. McNeil CJ, Athey D, Ho WO (1995) Biosens Bioelectron 10:75
61. Manning P, McNeil CJ, Cooper JM, Hillhouse EW (1998) Free Radic Biol Med 24:1304
62. El Kasmi A, Wallace JM, Bowden EF, Binet SM, Linderman RJ (1998) J Am Chem Soc 120:225
63. Gobi KV, Mizutani F (2000) J Electroanal Chem 484:172
64. Lisdat F, Ge B, Ehrentreich-Förster E, Reszka R, Scheller FW (1999) Anal Chem 71:1359
65. Scheller W, Jin W, Ehrentreich-Förster E, Ge B, Lisdat F, Büttemeier R, Wollenberger U, Scheller FW (1999) Electroanal 11:703
66. Büttemeyer R, Philipp AW, Mall JW, Ge B, Scheller WF, Lisdat F (2002) Microsurgery 22:108
67. Tian Y, Mao L, Okajima T, Ohsaka T (2002) Anal Chem 74:2428
68. Ge B, Scheller FW, Lisdat F (2003) Biosens Bioelectron 18:295
69. Campanella L, Favero G, Tomassetti M (1997) Sens Actuators B Chem 44:559
70. Rigo A, Viglino P, Rotilio G (1975) Anal Biochem 68:1
71. Privat C, Trevin S, Bedioui F, Devynck J (1997) J Electroanal Chem 436:261
72. Divisek J, Kastening B (1975) J Electroanal Chem 65:603
73. Lisdat F, Ge B, Reszka R, Kozniewska E (1999) Fresenius J Anal Chem 365:494
74. Ignatov S, Shishniashvili D, Ge B, Scheller FW, Lisdat F (2002) Biosens Bioelectron 17:191
75. Pallini M, Curulli A, Amine A, Palleschi G (1998) Electroanal 10:1010
76. Bedioui F, Trevin S (1998) Biosens Bioelectron 13:227
77. Villeneuve N, Bedioui F, Voituriez K, Avaro S, Vilaine JP (1998) J Pharmacol Toxicol Methods 40:95
78. Bedioui F, Trevin S, Devynck J (1996) Electroanal 8:1085
79. Malinski T, Mesaros S, Tomboulian P (1996) Meth Enzymol 268:58
80. Christodoulou D, Kudo S, Cook JA, Krishna MC, Miles A, Grisham MB, Murugesan R, Ford PC, Wink DA (1996) Meth Enzymol 268:69
81. Pataricza J, Penke B, Balogh GE, Papp JG (1998) J Pharmacol Toxicol Methods 39:91
82. Hara K, Kamata M, Sonoyama N, Sakata T (1998) J Electroanal Chem 451:181
83. Sedlak JM, Blurton KF (1976) J Electrochem Soc 123:1476
84. Kudo A, Watanabe K, Minakata Y, Mine A (1998) Chem Lett 5:391
85. Pariente F, Alonso JL, Abruna HD (1994) J Electroanal Chem 379:191
86. Iwashita E (1998) J Gastroenterol Hepatol 13:391
87. Friedemann MN, Robinson SW, Gerhardt GA (1996) Anal Chem 68:2621
88. Bedioui F, Trevin S, Devynck J, Lantoine F, Brunet A, Devynck M-A (1997) Biosens Bioelectron 12:205
89. Lantoine F, Trevin S, Bedioui F, Devynck J (1995) J Electroanal Chem 392:85
90. Xian YZ, Sun WL, Xue J Luo AM, Jin LT (1999) Anal Chim Acta 381:191
91. Tu HP, Mao LQ, Cao XN, Jin LT (1999) Electroanal 11:70
92. Park JK, Tran PH, Chao JKT, Ghodadra R, Rangarajan R, Thakor NV (1998) Biosens Bioelectron 13:1187
93. Fabre B, Burlet S, Cespuglio R, Bidan G (1997) J Electroanal Chem 426:75
94. Malinski T, Taha Z (1992) Nature 358:676

95. Raveh O, Peleg N, Bettleheim A, Silberman I, Rishpon J (1997) Bioelectrochem Bioenerg 43:19
96. Bedioui F, Trevin S, Devynck J (1994) J Electroanal Chem 377:295
97. Leung E, Cragg PJ, O'Hare D, O'Shea M (1996) Chem Commun 1:23
98. Maskus M, Pariente F, Wu Q, Toffanin A, Shapleigh JP, Abruna HD (1996) Anal Chem 68:3128
99. Mitchell K M, Michaelis EK (1998) Electroanal 10:81
100. Ishida Y, Hashimoto M, Fukushima S, Masumura S, Sasaki T, Nakayama K, Tamura K, Muratami E, Isokawa S, Momose K (1996) J Pharmacol Toxicol Methods 35:19
101. Broderick MP, Taha Z (1995) World Precision Instruments, 4th IBRO world congress of neuroscience, Kyoto, Japan
102. Ciszewski A, Milczarek G (1998) Electroanal 10:791
103. Malinski T, Taha Z, Grunfeld S, Burewicz A, Tomboulian P, Kiechle F (1993) Anal Chim Acta 279:135
104. Mao L, Tian Y, Shi G, Liu H, Jin L (1998) Anal Lett 31:1991
105. Trevin S, Bedioui F, Devynck J (1996) J Electroanal Chem 408:261
106. Tran D, Skelton BW, White AH, Laverman LE, Ford PC (1998) Inorg Chem 37:2505
107. Smith SR, Thorp HH (1998) Inorg Chim Acta 273:316
108. Bonakdar M, Yu J, Mottola HA (1989) Talanta 36:219
109. Zhang J, Lever ABP, Pietro WJ (1994) Inorg Chem 33:1392
110. Hayon J, Ozer D, Rishpon J, Bettelheim A (1994) J Chem Soc Chem Commun 5:619
111. Yu Ch-H, Su YO (1994) J Electroanal Chem 368:323
112. Maskus M, Abruna HD (1996) Langmuir 12:4455
113. Chen Y, Shepherd RE (1997) J Inorg Biochem 68:183
114. Malinski T, Mesaros S, Patton SR, Mesarosova A (1996) Physiol Res 45:279
115. Hyre CE, Unthank JL, Dalsing MC (1998) J Vasc Surg 27:726
116. Toda K, Hashiguchi Sh, Oguni S, Sanemasa I (1997) Anal Sci 13:981
117. Bayachou M, Lin R, Cho W, Farmer PJ (1998) J Am Chem Soc 120:9888
118. Kröning S, Scheller FW, Wollenberger U, Lisdat F (2004) Electroanal 16(4):253
119. Yu AM, Zhang HL, Chen HY (1997) Anal Lett 30:1013
120. Fan CH, Li GX, Zhu JQ, Zhu DX (2000) Anal Chim Acta 423:95
121. Mimica D, Zagal JH, Bedioui F (2001) Electrochem Commun 3:435
122. Fan CH, Pang JT, Shen PP, Li GX, Zhu DX (2002) Anal Sci 18:129
123. Gu HY, Yu AM, Yuan SS, Chen HY (2002) Anal Lett 35:647
124. Haruyama T, Shiino S, Yanagida Y, Kobatake E, Aizawa M (1998) Biosens Bioelectron 13:763
125. Miki K, Ikeda T, Kinoshita H (1994) Electroanal 6:703
126. Lisdat F, Ge B, Stöcklein W, Scheller FW, Meyer T (2000) Electroanal 12:946
127. Ge B, Meyer T, Schöning MJ, Wollenberger U, Lisdat F (2000) Electrochem Commun 2:557
128. Blyth DJ, Aylott JW, Richardson DJ, Russell DA (1995) Analyst 120:2725
129. Lan EH, Dave BC, Fukuto JM, Dunn B, Zink JI, Valentine JS (1999) J Mater Chem 9:45
130. Barker SLR, Zhao Y, Marletta MA, Kopelman R (1999) Anal Chem 71:2071
131. Barker SLR, Clark HA, Swallen SF, Kopelman R, Tsang AW, Swanson JA (1999) Anal Chem 71:1767
132. Barker SLR, Kopelman R, Meyer TE, Cusanovich MA (1998) Anal Chem 70:971
133. Barker SLR, Kopelman R (1998) Anal Chem 70:4902
134. Zhou X, Arnold MA (1996) Anal Chem 68:1748
135. Behar D, Czapski G, Rabani J, Dorfman LM, Schwarz HA (1970) J Phys Chem 74:3209

Part III
Noninvasive Electrical Monitoring of Living Cells

Chapter 7

Living Cells on Chip: Bioanalytical Applications

Martin Brischwein, Helmut Grothe, Angela M. Otto,
Christoph Stepper, Thomas Weyh, Bernhard Wolf

Abstract: Cells are able to respond in an extremely sensitive way to changes in their environment. Electronic microstructures on sensor chips can be applied to analyse such responses by recording properties of cell metabolism and cell morphology. These parameters are linked intimately to the cellular signalling network and can thus be used as sensitive indicators for any perturbation of cellular physiology like toxic insults or receptor activation events. Sensor chips might be able to fill a gap by providing functional and dynamic assays related to cell metabolism and cell morphology. Arraying of sensor chips is a prerequisite in order to make such novel types of cell-based assays work efficiently. It depends on technological efforts in chip packaging and electrical connections, electronic sensor control and specific modes of automated liquid handling. A presentation of solutions developed in the authors' laboratory is given but other approaches in the field of sensor-based cellular analysis will also be included.

Keywords: Whole cell biosensors, Intracellular signalling, Sensor chips, Cell metabolism, Cell morphology

Abbreviations

MOSFET   Metal oxide semiconducting field-effect transistor
ISFET   Ion-sensitive field-effect transistor
IDES   Interdigitated electrode structure
LAPS   Light-activated potentiometric sensor

1
Introduction

Animal and plant cells are dynamic microstructures with functional units located in distinct compartments which are interconnected by a complex signalling network. For sensing specific input signals from the outside, the cell uses specific receptor proteins, transmitting and amplifying the signals either to the nucleus in order to change gene activities or to further functional units in order to elicit an immediate response. Figure 1 is a schematic representation of the complexity of interacting signal transduction and metabolic pathways. In order to better understand the dynamic properties of this network and its input–output relationship, approaches of system analysis have to be employed taking into account the parallel structure of cellular signal computing [2, 3].

A biochemical stimulation of the cellular signal perception apparatus is amplified by intracellular signalling cascades in order to perform an appropriate functional reaction. This response, i.e. the cellular "output signal", is usually accompanied by a change in the rates of energy metabolism, by changes in electrical activity or by cell morphological alterations. If cells are successfully interfaced with adequate microsensors, these changes can be monitored in real time and non-invasively. This interfacing is usually achieved by growing cells directly on the surface of silicon- or glass-based sensor chips with structures of noble metals, i.e. materials which are non-toxic and accepted by most cells as substrate for stable adhesion and growth. The chips may integrate different types of sensor structures with either potentiometric, amperometric or impedimetric principles of function. The setup can be regarded as a cell–transducer hybrid applicable in pharmaceutical drug discovery, individualized clinical testing, or in whole-cell-based biosensors.

The state of the art in functional analysis of living cells is the use of micro test plates in combination with various optical readout systems. Components such as optical microscopes, fibre optics and CCD cameras can be used to detect visible, fluorescent or luminescent signals from cells or tissues. The advent of cellular engineering to include reporter elements such as green fluorescent protein has allowed the detection of specific cellular signalling events. Although the ability to array and to screen cells in high-density formats meets the demands of many users, such optical screening seems less suitable for long-term monitoring of cells and tissues, mostly due to the toxic or phototoxic properties of many dyes. Besides, no truly appropriate dyes are available for time-resolved studies concerning cell adhesion, cell metabolic activity or electrical activity, which is the domain of sensor-based functional monitoring of cells (Fig. 2).

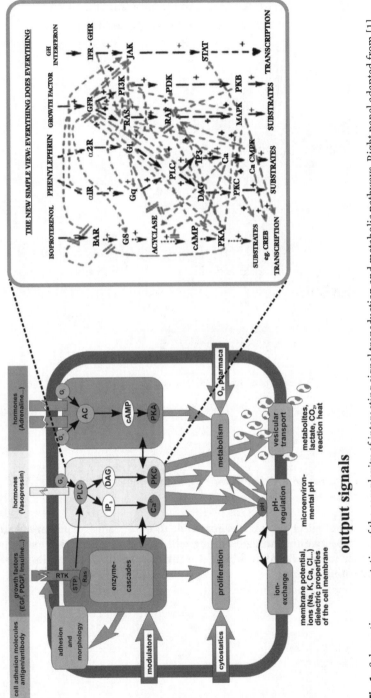

Fig. 1. Schematic representation of the complexity of interacting signal transduction and metabolic pathways. Right panel adapted from [1]

Parameters currently accessible
to microsensor detection

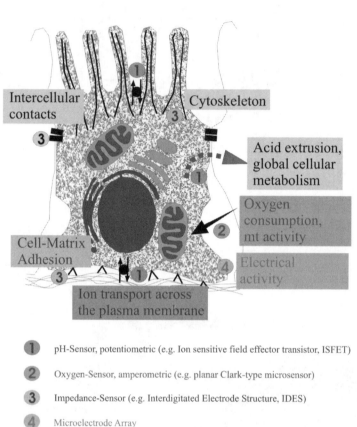

1 pH-Sensor, potentiometric (e.g. Ion sensitive field effector transistor, ISFET)

2 Oxygen-Sensor, amperometric (e.g. planar Clark-type microsensor)

3 Impedance-Sensor (e.g. Interdigitated Electrode Structure, IDES)

4 Microelectrode Array

Fig. 2. Parameters of living cells, which are currently monitored by chip-based sensors

The cell biological parameters amenable to monitoring on planar substrates and the respective technologies, along with some representative examples of their realization, are listed in Table 1 (for further explanations see Sect. 4 of this contribution).

Table 1. Cellular parameters accessible for sensor structures on planar substrates

	Technology	Examples, literature
Extracellular acidification	ISFETs	Baumann WH et al (1999) Microelectronic sensor system for microphysiological application on living cells. Sens Actuators B 55:77–89
	LAPS	(1) McConnell HM et al (1992) The cytosensor microphysiometer: biological applications of silicon technology. Science 257:1906–1913 (2) Metzger R et al (2001) Towards in vitro prediction of an in vivo cytostatic response of human tumour cells with a fast chemosensitivity assay. Toxicology 166:97–108
Oxygen consumption	Amperometric sensors	(1) Amano Y et al (1998) Measuring respiration of cultured cell with oxygen electrode as a metabolic indicator for drug screening. Hum Cell 12:3–10 (2) Brischwein M et al (2002) Microphysiological testing for chemosensitivity of tumour cells with multiparametric microsensor chips. 2nd European medical and biological engineering conference, Vienna, 4–8 December 2002. IFMBE Proceedings, vol I, pp 320–321
	Optical sensors	(1) Timmins MR, Guarino RD Monitoring the oxygen consumption rates of cells in culture. Technical brief, www.bdbiosciences.com (2) Key G, Katerkamp A (2002) On-line estimation of cell viability and cytotoxicity. Screening 1:30–32
Changes in cellular morphology	Electrical impedance sensors	(1) Ehret R et al (1997) Monitoring of cellular behaviour by impedance measurements on interdigitated electrode structures. Biosens Bioelectron 12:29–41 (2) Wegener J et al (2000) Electric cell–substrate impedance sensing (ECIS) as a non-invasive means to monitor the kinetics of cell spreading to artificial surfaces. Exp Cell Res 259:158–166
Electrical activity	Microelectrode arrays	(1) Gross GW et al (1995) The use of neuronal networks on multielectrode arrays as biosensors. Biosens Bioelectron 10:553–567 (2) Egert U et al (1998) A novel organotypic long-term culture of the rat hippocampus on substrate-integrated multielectrode arrays. Brain Res Prot 2:229–242
	FET arrays	(1) Ingebrandt S et al (2001) Cardiomyocyte–transistor hybrids for sensor application. Biosens Bioelectron 16:565–570 (2) www.infineon.com/bioscience

2
Chip Design and Fabrication

In the following, a brief description of the basics of sensor chip design and processing is provided. It is confined to the technology used in the authors' clean-room facilities and treats the construction of ISFETs, planar oxygen sensors and interdigitated electrode structures on silicon and glass substrates.

For the silicon chip a layout was developed to include different types of sensors for multiparametric readout (Fig. 3). The outer chip dimensions are 7.5×7.5 mm^2; the central sensor and cell culture area is 6 mm in diameter. On this area, four ISFETs, two pO$_2$-FETs, an amperometric pO$_2$ sensor, an IDES, a temperature diode and a reference transistor are processed.

Most of the process steps are standard MOS. Due to the particular application, however, special materials and a different order of process steps must be applied. Silicon chip fabrication is started with diffusion of the source and drain areas (n-channel MOSFETs). Afterwards the gate dielectric material is processed by dry oxidation and additional deposition of a thin silicon nitride layer (10 nm SiO$_2$, 20–30 nm Si$_3$N$_4$). This non-standard step is included to obtain better cell adhesion on the nitride layer. Further materials with other physical parameters (dielectric constant, proton binding characteristics, surface charge), such as Al$_2$O$_3$ and TaO$_5$, are currently being tested. After defining the active areas by local oxidation and deposition of a polysilicon protection layer, a titanium layer of a few nanometres thickness has to be deposited to ensure better adhe-

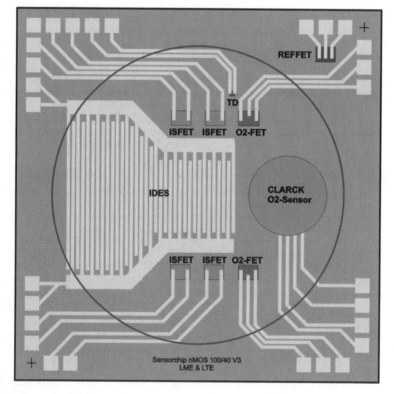

Fig. 3. Silicon chip layout

sion of the following sputtered platinum or palladium (200–300 nm). The metallization layer yields the electrode structures for IDES, pO_2-FET and amperometric pO_2-sensor by a lift-off photoresist step. The use of these metals implements another non-standard step in the fabrication process. They are, however, necessary for catalysing electrochemical reactions and for reasons of biocompatibility. On the ISFETs, two passivation layers (silicon nitride and/or silicon oxide) are deposited and opened on the gate areas. It is important that these passivation layers provide good adhesion and sealing to liquids in order to prevent any leakage current.

The completed silicon chips are mounted and bonded on printed wire boards (24.5×24.5 mm²), fitting into standard PLCC 68 sockets. They are encapsulated

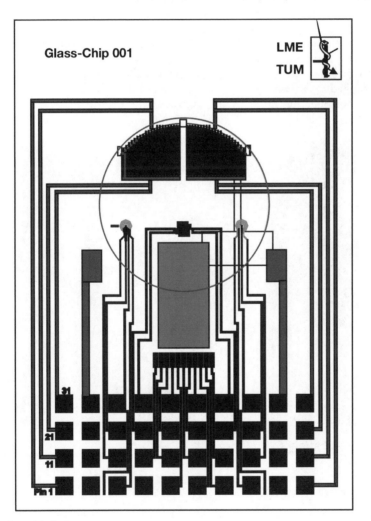

Fig. 4. Glass chip layout

using a mould that is filled with a biocompatible epoxy resin. Prior to cell culture, the packaged chips are disinfected or sterilized either with 70% ethanol or by autoclaving.

The glass chip (Fig. 4) has the outer dimensions 24×33.8 mm², and the sensor and cell culture area is 12 mm in diameter. It includes two amperometric pO_2 sensors and two IDES. Furthermore, a glass–silicon hybrid technology is used to insert a small silicon chip (7.3×3.6 mm²) with four ISFETs into an opening of the glass chip. The metal structures are fabricated using the same platinum lift-off process as described above. Afterwards, a single passivation layer is deposited (Si_3N_4, about 500 nm). For insertion of the silicon chip, an opening is etched into the glass with hydrofluoric acid. A special passivation layer (hard mask) is deposited on the glass prior to etching in order to protect the surface.

For many applications, the use of mathematical methods is necessary in order to obtain optimal electrode geometries. For example, a glass chip was constructed with a novel sensor structure for measuring the concentration of bacteria in an aqueous solution using dielectrophoretic forces. The four sensor electrodes have a sandwich structure. In each layer two similar modelled electrodes are arranged in two insulated layers. The electrodes of one single layer are acting for the collection mode of the sensor. In the measuring mode two of the

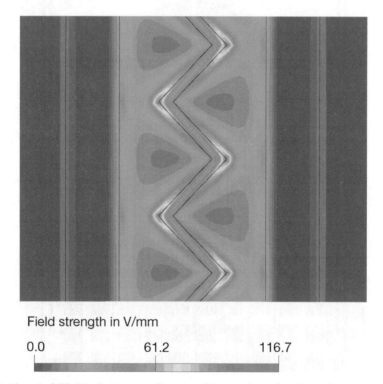

Field strength in V/mm

0.0   61.2   116.7

Fig. 5. Electric field distribution at a distance of 20 μm above the electrode surface. These data were used to optimize the electrode geometry

back-to-back lying electrodes are used to measure the impedance. The collecting property can be achieved by a combination of high electric field strength and a large field gradient. A minimal capacity is necessary for high sensitivity of the sensor. The electrode structure was therefore optimized by numerical field calculations. With this program the dielectric properties of the glass substrate, the electrodes and the aqueous solution were simulated. The result of optimization was an electrode arrangement consisting of one even and two zigzag electrodes (Fig. 5).

3
Cell Culture on Chips and Accessory Devices for Biohybrid Systems

Cells require a defined environment in order to survive and to respond reproducibly to signals from the outside. As a general rule it is desirable to mimic the cell's physiological conditions. The major properties that have to be maintained are (1) physicochemical properties of the cell culture medium, (2) temperature and (3) aseptic conditions.

The physicochemical properties of the cell culture medium include pH, oxygen partial pressure, osmolarity and defined concentrations of nutrients. In the course of cell culture, metabolic waste products and secreted compounds (e.g. proteins) become additional constituents. The composition of the medium is maintained by a liquid handling system connected to the sensor chips. Other important functions of the liquid handling system are the delivery of defined concentrations of bioactive compounds to the cells and, in the case of sensitive measurements of metabolic rates, the adjustment of a small microvolume of culture medium surrounding the cell culture. Only if a sufficiently high ratio of cell number to medium volume can be achieved can those rates can be recorded precisely [4, 5].

In the most convenient setup, the liquid transport in flexible tubing is driven by a pump and controlled by valves (perfusion system) (Fig. 6). The tubing is sterilized prior to connecting the chips using disinfectants such as hypochloric solutions. Engineering advances in microfluidics, also referred to as lab-on-a-chip, have allowed the construction of devices transporting substances through microchannels of a glass or plastic chip by electrokinesis and even the creation of individually addressable array elements (Fig. 7). The technology has evolved from microelectromechanical systems research. In combination with detection systems, the field also became known as micro total analysis systems (μTAS). Cell-based assays in this format are already available [6] and the use of microsensor technology is being worked on [7]. However, there is a limit to the miniaturization of microchannels since beyond a critical size, aggregation of cells and debris may clog up the system.

There may be reasons not to choose a perfusion setting for liquid handling. A problem often encountered after several hours of perfusion is the formation of gas bubbles within the tubing or the microchannels. Although special debubbling devices can improve the situation partially, complete degassing is not a solution with regard to cellular physiology. An alternative approach for long-term monitoring would be to combine sensor chips with existing liquid handling ro-

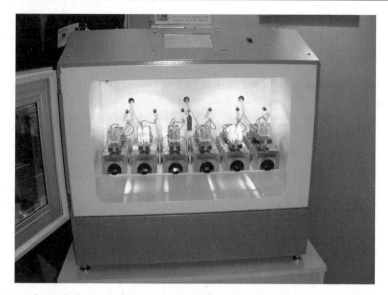

Fig. 6. Six-channel platform which has been developed in the authors' laboratory for testing of multiparametric sensor chips with living cells. It included an automatic fluid perfusion system, a thermostated cabin and electronic interfaces

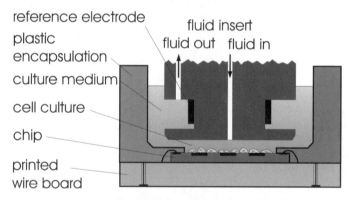

Fig. 7. Schematic cross section of a single chip package with cell culturing volume and fluidic setup

bots and redesigned pipetting tips (Fig. 8). In this way, exchange of culture media, addition of drug solutions and the adjustment of microvolumes necessary for the sensitive determination of cell metabolic rates can be performed on large chip arrays [8].

After sterilization, cells are seeded on the packaged sensor chips with the desired concentration. If necessary, the chips can be pretreated with substances of the extracellular matrix in order to create a uniform surface chemistry for cell

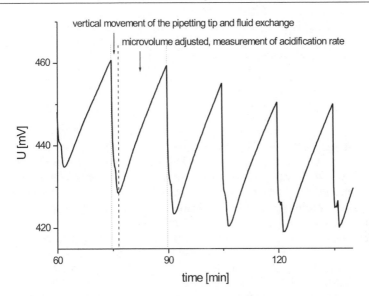

vertical movement of the pipetting tip and fluid exchange

microvolume adjusted, measurement of acidification rate

Fig. 8. Measurement sequence reflecting the extracellular acidification of LS 174 T cells (a cell line derived from human colon adenocarcinoma) growing on an ISFET. For regular fluid exchange, a liquid handling system based on a pipetting principle was used

adhesion. Seeding of cells occurs either while the measurement is running or prior to the start, leaving time for preculturing in the cell culture incubator.

There are efforts to stabilize and to conserve mammalian cells on solid substrates for the preparation of cell-based biosensors with extended shelf lives. Such efforts may become meaningful for applications such as environmental monitoring, where robust and portable devices are most important but further work seems to be necessary [9].

4
Transducers for Cellular Output Signals

Currently, chip-based sensors are capable of detecting four different cellular output signals:
1. Changes in the rate of proton extrusion are detected with pH microsensors, namely ion-sensitive field-effect transistors (ISFETs) or light-activated potentiometric sensors (LAPS).
2. Changes in the rate of oxygen exchange are detected with planar oxygen sensors such as amperometric oxygen sensors or so-called pO_2-FETs (an ISFET–electrode combination, see below).
3. Changes in cell morphology and membrane-associated electrical effects are monitored with impedance sensors such as interdigitated electrode structures (IDES).
4. Rapid changes in the membrane potential of electrically active cells, e.g. neurons or muscle cells, are detected with microelectrode or FET arrays.

In most systems, only one parameter is recorded on a given chip. In the future, however, the development of multiparametric chips combining several complementary sensors is expected to increase [10, 11].

4.1
pH Sensors

For the analysis of proton fluxes across the cell membrane pH-ISFETs with gate regions of about $10 \times 100 \ \mu m^2$ (L×W) are used. In these transistors, a low current passing through a semiconductor channel between two electrodes ("source and drain") is controlled by the voltage of a third non-metallic "gate electrode" above this channel. In pH-ISFETs, the gate electrode binds protons selectively, resulting in a pH-dependent potential. With silicon nitride as gate insulator material, the pH sensitivity is between 40 and 55 mV/pH. Due to the pH sensitivity of the structures the rate of extracellular acidification of cells growing directly on the sensors can be detected [12]. Since the typical noise level of pH recordings by ISFETs is about 0.5 mV, pH variations of about 0.01 can be detected. Most important for a low noise level is a proper liquid junction to the external reference electrode. In a cell culture setting, it proved to be favourable to use the culture medium itself with a constant chloride ion concentration as the electrolyte solution of the Ag/AgCl element.

An alternative chip-based technology to analyse pH is the LAPS, which has been employed successfully for microphysiometry. A description of this transducer technology and of the physical chemistry and cell biology underlying the measurement of extracellular acidification is given in [6, 13]. In a comparative study, LAPS and ISFETs showed similar features with respect to pH sensitivity and drift [14]. Interesting studies have also been done to evaluate the feasibility of an imaging of surface potentials (which could be influenced by the cell metabolic activity) with highly integrated sensors [15].

4.2
Oxygen Sensors

Recordings of cellular oxygen consumption, mostly reflecting the activity of the mitochondrial respiratory chain, are important for many cell biological questions. Usually amperometric Clark-type sensor electrodes are employed. The conditions necessary for these recordings can be calculated from a typical oxygen consumption rate of mammalian cell cultures, which is roughly between 10^{-16} and 10^{-17} mol/cell*s [16–18]. Despite the development of technological solutions for planar miniaturized amperometric oxygen sensors since about 1970 [19] and despite the potential advantages of planar, chip-integrated sensors for the monitoring of cell cultures, only a few published reports exist with an application in cell biology [20, 21]. The reason is probably the poor long-term stability of most of these sensors. Besides, the fabrication technology, which includes the deposition of reference electrode and membrane structures, is not compatible with a standard CMOS process.

Generally, three-electrode systems controlled by a potentiostatic electronic circuit are preferable, since this arrangement prevents a net consumption of oxygen in the vicinity of the cathode. Among the different technological solutions for planar amperometric oxygen sensors tested in our laboratory, a variant without a gas-permeable membrane seems promising. The constant concentration of chloride ions in cell culture media is able to stabilize the reference electrode potential. Even a bare palladium electrode in a cell culture medium provides a reference potential sufficiently stable for an amperometric measurement (pseudo-reference electrode). The applicability of such simplified sensor configurations, however, is limited to defined electrolyte solutions. The geometry of the electrode structure is shown in Fig. 9. A measurement with cells growing on the bare electrode structures is shown in Fig. 10. At present, different operating conditions and the practical use of oxygen sensors without gas-permeable membranes are evaluated. The advantage would clearly be the feasibility of a low-cost, CMOS-compatible fabrication of these sensors.

Another method to analyse cellular oxygen consumption is based on the formation of OH^- ions during oxygen reduction and the corresponding pH shift [22, 23]. If the pH-sensitive gate area of an ISFET is surrounded by an electrode, the ISFET will detect an increase of pH when the palladium (or platinum) electrode catalyses oxygen reduction at a cathodic potential (–750 mV vs. Ag/AgCl). In order to prevent deterioration of electrocatalytic activity, intermittent anodic potentials are applied. At present this method is tested for its practicability in the long-term monitoring of cell respiration.

Fig. 9. Palladium electrode structure consisting of a central working electrode (diameter 25 μm) and surrounding counter and pseudo-reference electrodes. This structure was used for amperometric oxygen measurement in cell culture media

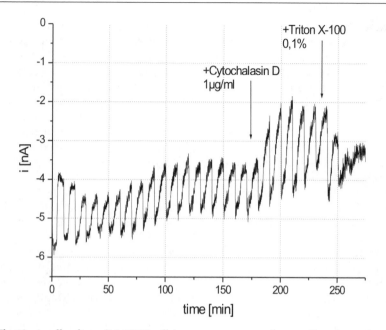

Fig. 10. A cell culture (LS 174 T cells) was grown to confluence directly on the electrode structure shown in Fig. 9. During the stop intervals of the fluid systems, the cellular oxygen consumption results in decreasing current values. Addition of cytochalasin D, which inhibits the cellular Na$^+$/glucose transport, increases the oxidative metabolism. Cellular oxygen consumption is stopped by disrupting cellular membranes with 0.1% Triton X-100. The constant potential of the palladium working electrode (vs. Ag/AgCl) was –0.75 V

4.3
Sensors Directed to Biomembranes and Morphological Patterns

Further critical parameters in cell culture that can be assessed with sensors on chips are the number and the growth of viable cells and changes in cell adhesion or cell morphology. Adherent cells are cultured directly on IDES. The measured impedance values reflect the process of cell spreading/cell adhesion and subtle rearrangements of the cytoskeleton, which is linked to cell–cell and cell–matrix junctions [24, 25]. For example, changes in intracellular calcium ion concentration are known to alter the structure of the cell cytoskeleton, resulting in morphological changes that can be detected by impedance sensors [26]. Figure 11 illustrates the experimental setup while Fig. 12 shows an example experiment with the recordings of two IDES on the same glass chip, detecting the morphological response of a confluent monolayer of Hela cells to stimulation of the histamine (H1) receptor.

The impedance is the ratio of a sinusoidal voltage applied to the pair of electrodes to the sinusoidal component of the current. Unless the system is purely resistive, the impedance is a complex value, because the current and voltage have different phase angles. The impedance is measured at 10 kHz and the am-

Geometry:

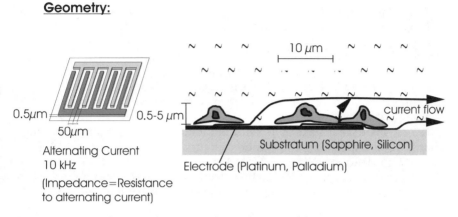

0.5μm ____ 0.5-5 μm

50μm

Alternating Current
10 kHz
(Impedance = Resistance
to alternating current)

10 μm

current flow

Substratum (Sapphire, Silicon)

Electrode (Platinum, Palladium)

Fig. 11. Geometry of IDES and IDES-based impedance measurement on cell cultures

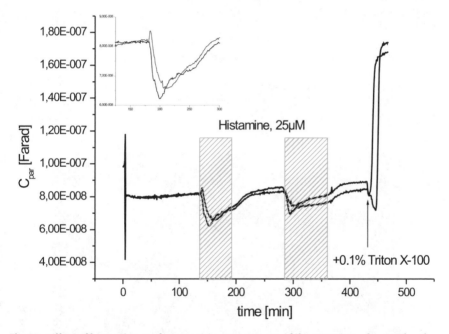

Histamine, 25μM

+0.1% Triton X-100

C_{par} [Farad]

time [min]

Fig. 12. Effect of histamine on the capacitive component of the IDES impedance. The glass chip has two identical IDES; both were grown with a confluent Hela monolayer. Although no distinct morphological alteration can be detected with the microscope, a reversible effect occurs, possibly on cell adhesion. At the end of the experiment the cells were killed with 0.1% Triton X-100

plitude of the applied current is less than 1 μA. The width and the distance of the electrodes is typically 50 μm. Living cells with cytoplasmic membranes are effective insulators for alternating current at the frequency used.

As the most simple equivalent model for the system composed of electrode, cell layer and cell culture medium we selected a circuit with a resistor and a capacitor in parallel. Mostly the capacitive component was used to represent the measured data, since this component is most influenced by cell morphological alterations. Although refined equivalent circuits of the cell–electrode interface can be designed, the value of such models for the interpretation of measured impedance data is limited and depends on additional experimental information [27].

Not only IDES may serve as sensors for monitoring cell morphological alterations. Other impedimetric electrode configurations have also provided good results. Impedance data have been used to analyse a variety of cell behaviours and conditions including cell growth, cell attachment and spreading, cell motility and cell migration, barrier function of confluent layers and cell-substratum spacing [28–30].

Another application of impedance measurement on living cells is the detection and quantification of viable bacteria in aqueous solutions. Prior to the impedance measurement, the electrode structures are used to accumulate the bacteria in their vicinity. The grade of accumulation can be changed by the voltage applied to the sensor electrodes. Figure 13 shows an accumulation of E. coli at an electrode structure at 1 V. Varying the voltage causes a change in impedance, which depends also on the concentration of the bacteria.

A completely different transducer type which has been used for the detection of morphological changes of adherent cells is a piezoelectric quartz crystal microbalance. Perturbations of the cytoskeleton of endothelial cell layers are

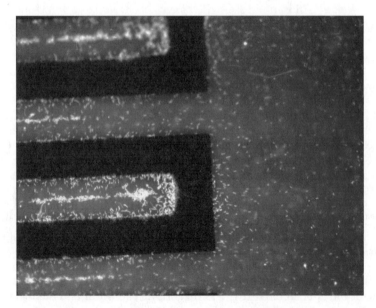

Fig. 13. E. coli bacteria, stained with a fluorescent dye, are attracted to the electrodes by dielectrophoretic forces

reported to cause a shift in resonance frequency and resonance admittance [31, 32].

4.4
Sensors for Electrophysiological Activity

Inspired by the pioneering work of G. Gross [33, 34], several groups started to use multielectrode arrays on planar substrates for the multisite analysis of activity patterns of explanted networks of neuron cells or of muscle cell cultures [35–40]. For the first time it was feasible to address experimentally questions such as how the nature of a neuron's connections with its neighbours affects its ability to generate action potentials. Arrays with different size, spacing and geometry of electrodes are meanwhile offered commercially (even arrays with three-dimensional, protruding electrodes are available). The arrays are produced on glass using varying photolithographic techniques. Cultures of explanted primary cells as well as whole tissue slices can be placed on the arrays to monitor spontaneous – or microelectrically stimulated – electrophysiological activity with remarkable sensitivity. However, the network of connections between neighbouring neurons is typically so complex that it is difficult to assign the inputs of any given cell. In order to overcome this problem, attempts are made to guide neurons on the chip surface by patterning organic compounds attracting or repelling the cells [41]. Starting with the work of Fromherz, who succeeded in recording the electrical activity of a leech neuron by placing it on top of a silicon field-effect transistor [42], the use of ISFETs for electropyhsiological measurements is now emerging [43–45]. The noise level of measurements, however, seems to be consistently higher compared to microelectrodes. Most recently, the fabrication of an array of 16,384 sensors on a single silicon chip was published, enabling records of cellular electrical activity with high spatio-temporal resolution [46].

5
Conclusions, Applications and Future Prospects

Having produced a multiparametric sensor chip, there is now the demand for developing practical analytical tools. For assay and screening applications the sensor chips have to be arranged in large arrays, preferably in multiwell plate formats. Figure 14 shows a six-well plate as a base for further development.

To achieve this aim, substantial technological efforts in chip processing are necessary. One of the most important benefits of CMOS technology is the possibility of on-chip circuitry for signal amplification, data analysis and sensor self-testing. On-chip sensor multiplexing is a precondition for the construction of 96- or 384-multiwell arrays, since the number of necessary electric connections becomes unmanageable. However, the standard CMOS process used has always to be combined with additional thin-film deposition steps.

Any on-chip circuitry and electric connections have to be sealed from the probed liquid compartment in the cell culture vessel. The involved packaging process will contribute most significantly to the overall costs of chips or chip ar-

Fig. 14. Prototype of a multiwell plate for the continuous monitoring of cells with sensor chips, which replace the bottom of the culture wells

rays and therefore deserves particular attention. For mass production, a strategy would be to adapt convenient chip mounting principles such as flip-chip technology to the encountered special material requirements and to combine them with plastic encapsulation for safe liquid handling. The device must be sterilized prior to use in order to be delivered in a ready-to-use condition (e.g. by irradiation) and the used sterilization process has to be in agreement with existing norms. This holds true regardless of whether the device is a one-way product or a reusable one.

Microfluidics must be characterized properly in terms of flow dynamics and physics of the interactions of aqueous samples with the different surfaces. With glass chips, the combination of electrical sensors with optical sensors, e.g. for pH and oxygen, could be a promising approach. Another challenge is the maintenance of a proper "cell culture environment". Groups working with cell-based assays are frequently facing problems of gradual cell culture deterioration in the time course of the studies. Therefore, conditions including the protection of cell culture media from evaporation, adjustment of the gas composition and measures against contamination must be guaranteed either by the fluidic system or by additional equipment.

Finally, the importance of computation of the final results from sensor raw data, statistical design and extraction of information from response libraries will increase with increasing array densities.

The field of application for cell–chip systems is as large as the variety of different cell types to investigate. Obviously, the quality of analysis depends largely on the sourcing and preparation of cells and tissues. While primary cell cultures often provide the greatest approximation to the in vivo phenotype, they also present potential sources of variability. The discovery of different types of stem cells and the possibility of inducing differentiation in vitro presents a new opportunity to obtain a source of cells with fewer functional defects and alterations than found in tumour cell lines commonly used. On the other hand, a range of applications strictly requires cellular material freshly derived from individuals. An example is a predictive, individualized drug screening assay in oncology, performed in a functional way directly on tumour tissue biopsies. Since the amount of the tissue material available for any test is very limited, the miniaturization of multiparametric chips and the possibility of following up directly the response profiles of short-term cultures in the course of the drug treatments makes sensor-chip technology most promising for this type of cell analysis [47–49]. Again, however, the preparation of the tissue (namely, consideration of functional interactions between malignant cells and stromal cells) and the composition of the test medium (e.g. presence of serum components) have to be considered carefully.

For the field of cell-based biosensors, where cellular recognition and signalling capabilities are exploited (e.g. in on-line biomonitoring of waste water effluents), additional demands must be observed [50]. Apart from the critical selection of the appropriate target cell/organism (which may be genetically manipulated) for the specific range of compounds to be detected, there are many issues to consider. These include sample preparation (filtration, bioavailability, sterilization), response time, possible effects of physiological adaptation by subtoxic concentrations of chemical compounds, life-support systems for robust and invariable sensitivities and well-controlled abiotic factors like temperature and light exposure. The question of the minimal detectable concentration is linked not only to the inherent sensitivity of the cellular target, but also to the determination of any sources of noise in the system. The probability of false alarm should be minimized by correlation analysis of parallel sensors.

Although the size of single sensor elements on chips can be as small as about $100 \ \mu m^2$ (e.g. the sensitive gate area of a pH-ISFET), it is generally not intended to analyse single cells on sensor chips. However, single-cell measurements which do not take into account the typical social context of cells have to be assessed critically. Efforts dedicated to metabolic imaging using highly integrated two-dimensional chip arrays are only in their infancy. Generally, the strength of the approach is its non-specificity, enabling the detection of a great variety of cellular responses. This is because cell metabolism and cell morphology itself are closely coupled to the signalling apparatus and can thus be regarded as first-step signal transducers. For the interpretation of results, a multiparametric and kinetic analysis may provide first indications of a drug's mode of action. For interpretations on a molecular level, methods like the comparison of different, transformed cell lines or the use of specifically acting biochemical inhibitors or antibodies have to be applied. Thus, it is likely that cell–chip systems will find their place in the screening of pharmaceutical agents for activity or toxicity, as an additional source of information about cellular behaviour.

The combination of sensors for cell metabolic activity and cell morphology also enables intriguing research in cell biology. As an example, Fig. 15 shows the response of the human colon carcinoma cell line LS 174 T to a treatment

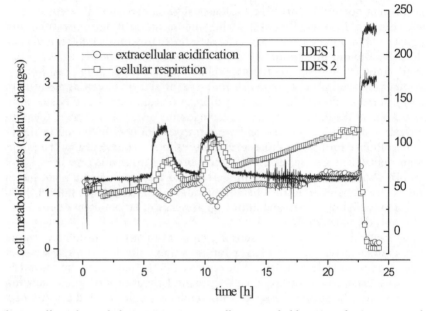

Fig. 15. Effect of cytochalasin D on LS 174 T cells as revealed by microphysiometry. At the end of the experiment, the cells were killed with 0.1% Triton X-100

with cytochalasin D. Upon drug addition (the drug was added twice with a concentration of 1 µg/ml) there was an immediate response in all three parameters. The clearly visible effect on the impedance, represented by the capacitive component, reflects the cell morphological alterations accompanied by the perturbation of the actin cytoskeleton. In the microscope, the visible part of these alterations is a retraction of cytoplasmic protrusions. Cellular proton release was inhibited, probably due to the inhibition of Na^+/glucose transport proteins and a resulting depletion of intracellular glucose. At the same time, in a precisely regulated fashion, oxygen consumption increased, indicating a rapid switch from aerobic glycolysis to oxidative energy providing pathways. The presence of this so-called Crabtree-effect is remarkable for this kind of tumour cell, and tells us that the activity of the respiratory chain can be instantaneously upregulated on demand.

Acknowledgements: The authors are greatly indebted to current and former staff of the group of Prof. Wolf involved in the sensor development and biological application presented in this work, notably Margarethe Remm, Gudrun Teschner, Alfred Michelfelder, Paul Pilawa, Reiner Rampf, Werner Baumann, Ralf Ehret, Mirko Lehmann and Ingo Freund. Much of the work would not have been possible without the support of Micronas GmbH, Freiburg i.Br., the Heinz Nixdorf Foundation, the Verein Deutscher Ingenieure (VDI), the Bundesministerium für Bildung und Forschung (BMBF) and the organization of "Bayern Innovativ".

References

1. Dumont ED, Pécasse F, Maenhaut C (2001) Cell Signal 13:4457
2. Kraus M, Wolf B (1995) Structured biological modelling: a new approach to biophysical cell biology. CRC, Boca Raton
3. Hartwell LH, Hopfield JJ, Leibler S, Murray AW (1999) Nature 402:C47
4. Owicki JC, Parce JW (1992) Biosens Bioelectron 7:255
5. Owicki JC, Bousse LJ, Hafeman DG, Kirk GL, Olson JD, Wada HG, Parce JW (1994) Annu Rev Biophys Biomol Struct 23:87
6. http://www.agilent.de
7. Cooper JM (1999) Trends Biotechnol 17:226
8. Farinas J, Chow AW, Wada HG (2001) A Microfluidic Device for Measuring Cellular Membrane Potential, Anal Biochem 295:138-42
9. Bloom FR, Price P, Lao G, Xia JL, Crowe JH, Battista JR, Helm RF, Slaughter S, Potts M (2001) Biosens Bioelectron 16:603
10. Wolf B, Brischwein M, Otto AM, Grothe H (2002) mstnews 1/02:37
11. Martinoia S, Rosso N, Grattarola M, Lorenzelli L, Margesin B, Zen M (2001) Development of ISFET array. Biosens Bioelectron 16:1043
12. Baumann W, Lehmann M, Schwinde A, Ehret R, Brischwein M, Wolf B (1999) Sens Actuators B 55:77
13. Hafner F (2000) Biosens Bioelectron 15:149
14. Fanigliulo A, Accossato P, Adami M, Lanzi M, Martinoia S, Paddeu S, Parodi MT, Rossi A, Sartore M, Grattarola M, Nicolini C (1996) Sens Actuators B 32:41
15. George M, Parak WJ, Gaub HE (2000) Sens Actuators B 69:266
16. Schlage WK, Bereiter-Hahn J (1983) Microsc Acta 87:19
17. Peterson A, Walum E (1985) In Vitro Cell Dev Biol 21:622
18. Casciari JJ, Sotirchos SV, Sutherland RM (1992) J Cell Physiol 151:386
19. Lambrechts M, Sansen W (1992) Biosensors: microelectrochemical devices. IOP, Bristol
20. Amano Y, Okumura C, Yoshida M, Katayama H, Unten S, Arai J, Tagawa T, Hoshina S, Hashimoto H, Ishikawa H (1998) Hum Cell 12:3
21. Brischwein M, Otto AM, Henning T, Grothe H, Motrescu ER, Wolf B (2002) 2nd European engineering conference, Vienna, 4–8 December 2002. IFMBE Proceedings, vol I, pp 320–321
22. Lehmann M, Baumann W, Brischwein M, Gahle HJ, Freund I, Ehret R, Drechsler S, Palzer H, Kleintges M, Sieben U, Wolf B (2001) Biosens Bioelectron 16:195
23. Sohn B, Kim C (1996) Sens Actuators B 34:435
24. Ehret R, Baumann W, Brischwein M, Schwinde A, Stegbauer K, Wolf B (1997) Biosens Bioelectron 12:29
25. Ehret R, Baumann W, Brischwein M, Schwinde A, Wolf B (1998) Med Biol Eng Comp 36:365
26. Coppolino MG, Dedhar S (2000) Int J Biochem Cell Biol 32:171
27. Wegener J, Zink S, Rösen P, Galla HJ (1999) Eur J Physiol 437:925
28. Giaever I, Keese CR (1993) Nature 366:591
29. Wegener J, Keese CR, Giaever I (2000) Exp Cell Res 259:158
30. Luong JH, Habibi-Rezaei M, Meghrous J, Xiao C, Male KB, Kamen A (2001) Anal Chem 73:1844
31. Marx KA, Zhou T, Montrone A, Schulze H, Braunhut S (2001) Biosens Bioelectron 16:773
32. Wegener J, Seebach J, Janshoff A, Galla HJ (2000) Biophys J 78:2821
33. Gross GW, Wen W, Lin J (1985) J Neurosci Methods 15:243
34. Gross GW, Kowalski JM (1991) In: Antognetti P, Milutinovic V (eds) Neuronal networks: concepts, applications and implementations. Academic, New York, p 47

35. Duport S, Millerin C, Muller D, Corrèges P (1999) Biosens Bioelectron 14:369
36. Hämmerle H, Egert U, Mohr A, Nisch W (1994) Biosens Bioelectron 9:691
37. Blau A, Ziegler C, Heyer M, Endres F, Schwitzgebel G, Matthies T, Stieglitz T, Meyer JU, Göpel W (1997) Biosens Bioelectron 12:883
38. Hofer E, Urban G, Spach MS, Schafferhofer I, Mohr G, Platzer D (1994) Am J Physiol 266: H2136
39. Keefer EW, Gramowski A, Gross GW (2001) J Neurophysiol 86:3030
40. Bove M, Martinoia S, Verreschi G, Guigliano M, Grattarola M (1998) Biosens Bioelectron 15:601
41. Stenger DA, Hickman JJ, Bateman KE, Ravenscroft MS, Ma W, Pancrazio JJ, Shaffer K, Schaffner AE, Cribbs DH, Cotman CW (1998) J Neurosci Methods 82:167
42. Fromherz P, Offenhäuser A, Vetter T, Weis J (1991) Science 252:1290
43. Offenhäuser A, Sprössler C, Matsuzawa M, Knoll W (1997) Biosens Bioelectron 12:819
44. Offenhäuser A, Knoll W (2001) Trends Biotechnol 19:62
45. Besl B, Fromherz P (2002) Eur J Neurosci 15:999
46. http://www.infineon.com/bioscience
47. Henning T, Brischwein M, Baumann W, Ehret R, Freund I, Kammerer R, Lehmann M, Schwinde A, Wolf B (2001) Anticancer Drugs 12:21
48. Metzger R, Deglmann CJ, Hoerrlein S, Zapf S, Hilfrich J (2001) Toxicology 166:97
49. Ekelund S, Nygren P, Larsson R (1998) Anticancer Drugs 9:531
50. Schubnell D, Lehmann M, Baumann W, Rott FG, Wolf B, Beck CF (1999) Biosens Bioelectron 14:465

Bioanalytical Application of Impedance Analysis: Transducing in Polymer-Based Biosensors and Probes for Living Tissues

G. Farace, P. Vadgama

Abstract. Impedance measurements have been a somewhat neglected method of electrochemical transduction. This is despite the fact that these measurements are both extremely sensitive and ideally suited to interfacial phenomena. This review will highlight the work of the authors and others in the development of both biosensors and devices for bacterial and cellular monitoring which incorporate such measurements. The review will pay particular attention to the novel combination of conducting polymers and impedance measurements that has been developed to create label-free or reagentless biosensor instruments. Finally the review will examine some of the new directions of research that may ultimately make impedance measurements more accessible to the general science community.

Keywords: Impedance, Biosensor, Reagentless monitoring, Conducting polymers, Bacteria and cell measurements

1
Introduction

While amperometric and potentiometric measurements have been widely utilized and accepted as key components of the biosensor transduction cascade, impedimetric measurement has yet to achieve this status and its use for transduction in either biosensors or bioanalytical instrumentation is still somewhat limited. This is despite the fact that the technique itself is extremely sensitive and capable of both material, notably film, characterization and quantitation of interfacial phenomena at sensing devices. Moreover, interrogation of both electrochemically active and non-electrochemically active systems is possible. The use of impedance measurements within biosensor devices warrants further exploration and appears to be both logical and advantageous as regards contributing analytical advantages [1]. The higher cost, seemingly complicated nature of the instrumentation and the difficulties in interpreting the raw data certainly are drawbacks and may help explain why impedance measurement methodologies have been under-utilized. However, it is not necessary to be an experienced electrochemist to undertake these measurements using contemporary instruments and for practical sensing, empirical interpretations are often sufficient to acquire meaningful bioanalytical information. The relatively poor understanding of the power of impedimetric measurement has limited its use to certain niche areas within the bioanalytical field; two areas where the technique has been firmly established are label-free or reagentless detection and cell- and bacteria-associated measurements.

Our group is one of an increasing number actively working in this field [2] and it is the intention of the review to highlight both our own strategy and the work of others in these particular areas. As a basis for the review, the background and concept behind impedance and the nature of the impedance experiment will be described.

Electrochemical impedance was first described by Oliver Heaviside in the 1880s and was further developed and advanced in representational terms by the vector diagrams and complex number terminology of Kennelly in 1893 and Steinmetz in the 1920s. Impedance is a more general concept than resistance because it is associated with a time domain, and in classical AC impedance encompasses phase differences in the current–voltage relationship. When a monochromatic voltage signal $v(t)=V_m\sin(\omega t)$ with an associated single frequency of $v\equiv\omega/2\pi$ is applied to a cell, the resulting current $i(t)$ can be expressed as $I_m\sin(\omega t+\theta)$ where θ is the phase angle difference between the voltage and current. The value for θ would be zero in a purely resistive system. Based upon these relations, impedance can be defined as $Z(\omega)\equiv v(t)/i(t)$ and its magnitude or modulus as $|Z(\omega)|=V_m/I_m(\omega)$ and the phase angle $\theta(\omega)$.

Impedance can also be described using the vector quantity $Z(\omega)=Z'+jZ''$, and can be plotted using rectangular or polar coordinates. In this case the rectangular coordinate value will be

$$Re(Z) \equiv Z' = |Z|\cos(\theta) \tag{1}$$

for the real contribution and

$$Im(Z) \equiv Z'' = |Z|\sin(\theta) \tag{2}$$

for the imaginary contribution. The phase angle will therefore be

$$\theta = \tan^{-1}\left(Z''/Z'\right) \tag{3}$$

and the modulus

$$|Z| = \left[\left(Z'\right)^2 + \left(Z''\right)^2\right]^{1/2} \tag{4}$$

As defined in the above equations, Z is frequency dependent and conventional impedance measurements are performed as a function of either v or ω over a wide range of frequencies; it is from the structure of the $Z(\omega)$ *vs* ω plot that valuable information can be derived.

The general approach to an impedance measurement is to apply an electrical stimulus (a known voltage or current) across a resistor through electrodes and to observe the response (the resulting corresponding current or voltage). During electrical stimulation, a multitude of microscopic processes occur across the resistor system and, taken as whole, these form the overall response noted at the recording instrument. The processes occurring can include the transport of electrons through an electronic conductor, the transfer of electrons at the electrode/electrolyte interface, to or from charged and uncharged molecular species that results from some oxidation or reduction reaction within the electrochemical cell, or the flow of charged atoms or atomic agglomerates through the electrolyte. It is the aim of the experiment to determine the dynamics of bound and mobile charged species in the bulk or interfacial regions of any solid or liquid material (*viz* ionic semiconducting, mixed electronic-ionic, insulators/dielectrics) present within the electrochemical cell. The properties that can be investigated may be divided into two main groups; those which are pertinent to the electrode itself (conductivity, dielectric constant, the mobility of charge, the equilibrium concentration of charged species, the formation/recombination rates of bulk species) and those that are associated with the interface between the material and the electrode (adsorption/reaction rate constants, capacitance of the interfacial region, the diffusion coefficient of neutral species within the electrode).

There are three types of electrical stimuli that can be utilized in an impedance experiment. The first occurs in transient measurements where a step function of voltage ($V(t)=V_0$ for $t>0$ and $V(t)=0$ for $t<0$) is applied at $t=0$ to the system and the resulting time-dependent current $(i)\tau$ is determined. The ratio $V_0/(i)t$ (the indicial impedance) describes the impedance due to the step function voltage perturbation of the electrochemical interface. Generally, these time-varying results are Fourier-transformed into the frequency domain, yielding frequency-dependent impedance. However, this transformation is only valid when $|V_0|$ is sufficiently small that the response of the system is linear. The advantages of this approach are that the instrumentation required is easy to realize and the rate of the electrochemical reaction at the interface is dependent on the voltage applied. The main disadvantage is that there is no direct control over the frequency spectrum and so impedance cannot be measured over a defined frequency range.

A second approach is to apply a signal $V(t)$ composed of random (white) noise to the interface and measure the resulting current. As above, the results are Fourier-transformed to obtain the impedance signal. This methodology offers the advantage of fast data collection because only one signal is applied to the interface for a short period of time. The disadvantage is that the technique requires true white noise.

The third, and most common, approach is to measure impedance directly in the frequency domain by applying a single-frequency voltage to the interface and measuring the current phase shift and amplitude, expressed as real and imaginary parts, of the current resulting at that frequency. Readily available commercial instrumentation is able to perform these impedance measurements in frequency ranges from milli- to mega-Hertz. The main advantage of this approach is that the individual frequency windows can be selected and investigated, and so it is possible to monitor only those frequencies that are of direct interest to the investigator.

None of these approaches is mutually exclusive, and it is possible to combine them to create other experimental techniques. One of the most important of these combination techniques is AC polarography, which combines the first and third approaches outlined above through the simultaneous application of a linearly varying unipolar transient signal and a much smaller single-frequency sinusoidal signal [3].

2
Impedance for Transduction in Biosensors

Impedance as a transduction mechanism has been utilized by a number of different groups and has been the focus of our own studies utilizing impedance across planar conducting surfaces. Impedimetric detection has been incorporated into (1) traditional electrochemical biosensors where an electroactive species is generated or destroyed by the biological detection element, so the measurement detects the dynamic change in the concentration of this species, (2) less conventional reagentless systems where direct impedimetric monitoring of a bioaffinity reaction, but without a degradative product, achieves detection of the analyte.

A number of biosensor devices have been reported using impedance-based detection in a traditional transducer. Gallardo Soto et al. [4] utilized a traditional electrochemical biosensor configuration where the formation of ions was monitored at a planar electrode. Their system was a urease-activated sensor based on interdigitated gold electrodes (IDEs) with AC impedance as the transduction mechanism. The hydrolysis of urea by urease is a particularly easy reaction to follow by impedance because the breakdown products are all ionic (ammonium, bicarbonate and hydroxide), though the ionization status of each species is a function of pH. Therefore, changes in impedance within the background electrolyte in the vicinity of the electrode can be used for quantitative determination of the concentration of urea in the sample. They also utilized impedance measurements to evaluate different immobilization strategies, including entrapment within gelatin and Nafion membranes, and crosslinking with glutaraldehyde. Through such measurements they observed that, as with most biosensors, the choice of immobilization method was crucial in terms of sensor response and sensor shelf life.

Alfonta et al. [5] also described a traditional biosensor configuration utilizing an impedimetric assay. They demonstrated their approach for enzyme, antibody–antigen and DNA–DNA detection with a redox process being the basis for transduction. In each instance, the formal detection required horseradish peroxidase as an indicator/label, which was used to catalyse the oxidation of 4-chloro-1-naphthol to the insoluble product 4-chlorohexadienone. They then used impedance to monitor the change in electron-transfer resistance as the insoluble product covered the subject electrode surface. The increase in resistance caused by the insoluble product insulated the surface of the electrode, and was proportional to the amount of enzyme present on the electrode; in turn this was proportional to the amount of labelled target analyte present.

In a study of a monitored polymer phase, Saum et al. [6] reported a different approach whereby detection of collagenase was achieved by monitoring impedance changes at a gelatin film coated onto a gold IDE. Exposure of the protein film to the enzyme in deionized water led to film digestion with dissolution of the protein layer. As the integrity of the film was lost, impedance increased at a rate that could be directly related to enzyme activity. This sensor configuration provides an example of reagentless detection, and should be applicable to other film-degrading systems.

Reagentless detection is an important practical advantage for biosensors and increases the possibility of non-skilled analysis, thus providing a route to a mass market and commercialization. Biosensor devices that have successfully made the transition from the laboratory to the home testing market in general have tended to be either simple qualitative systems, as used for pregnancy testing, or rely on a relatively straightforward reaction chemistry, e.g. that using an oxidase enzyme for glucose measurement. Both types of device meet the criterion of being easy to use with straightforward data interpretation, but neither is truly simplified to the degree that no indicator element is required. Thus a label of some sort is needed for a pregnancy test and an enzymic product is demonstrated for glucose measurement. The ultimate in simplicity is a reagentless system where the transducer directly reports the interaction of the target analyte with the recognition element. The advantage is that in a direct detection mechanism, it becomes possible to monitor affinity reactions such as antibody–antigen interactions or DNA hybridization that undergo no further modification without the need for a secondary label. Direct detection of the event is only possible because the complex formed creates a physical parameter change that can be picked up above the background properties of its local microenvironment. A number of biophysical techniques are capable of directly measuring physical parameters, e.g. the quartz crystal microbalance, which measures gravimetric and viscoelastic changes, or surface plasmon resonance, which detects dielectric changes at a waveguide surface. However, it is currently difficult to envisage either of these techniques becoming easy to operate or reducible to a consumer product. Electrochemistry-based devices, however, have already been developed and marketed for general use (over the counter testing), so the possibility for the development of reagentless sensors based on electrochemistry principles, notably impedance measurement, can at least be contemplated. Our work has been directed to reagentless detection of antibody–antigen binding using impedance. The approach taken has been to entrap the antibody within a conducting polymer layer and to monitor impedance changes in the plane of the films during exposure to the target antigen.

Conducting polymers have been known since the 1860s when Letheby achieved such a structure following anodic oxidation of aniline in dilute sulphuric acid. In the past 25 years they have become the focus for a range of different research areas, latterly including biosensors. Various approaches have been taken within the sensor field, and a number of excellent reviews have been published covering the more widely researched applications; gas sensors and electronic noses [7], enzyme electrodes [8], controllable barrier membranes [9] and whole cell monitoring [1].

Impedance and conductivity measurements at conducting polymers have been reported in the literature over the past 20 years and the two elements have combined to create a variety of biosensor constructs. The most common type of biosensor described is where the sensing element is entrapped within the conducting polymer or within a phase adjacent to the conducting polymer. Coentrapment of biomolecules into polymeric films was demonstrated by Taylor et al. [10] who reported a sensor which incorporated cholinergic receptors within a polypyrrole film with ligand binding monitored directly via changes in film electrical impedance.

Nishizawa et al. [11] described a sensor system where an array of platinum IDEs was coated with a layer of polypyrrole grown electrochemically on the surface of the electrodes. The polymer layer was then coated with a layer of crosslinked penicillinase. The reaction of penicillin with the penicillinase induced a localized pH change. This local change in pH altered the ionization and redox state and, therefore, the conductivity of the underlying polypyrrole film. Here, monitoring the bulk film conductivity was used to follow the pencillin/penicillinase reaction at the surface. Contractor et al. [12] devised a similar electrode, but instead entrapped the enzyme within the conducting polymer, a polyaniline. As with Nishizawa, addition of the enzyme substrate resulted in a localized change in film pH and redox potential. The film reporter appears to be suitable for a number of different substrate/enzyme pairs including glucose/glucose oxidase, urea/urease, neutral lipid/lipase and haemoglobin/pepsin.

Sargent and Sadik [13] described a method for monitoring antibody–antigen reactions within polypyrrole by impedance. Here, films containing antibodies to human serum albumin were generated galvanostatically at planar electrode surfaces. During impedance measurements, after exposure to human serum albumin, a marked reduction in double layer capacitance was observed. This change in capacitance was observed in films held at marginally negative potentials, and was attributed to structural effects at the conducting polymer allowing charge to be carried more easily.

In our studies on antibodies entrapped within polypyrrole films [2], the films were grown using cyclic voltammetry on gold IDEs (15-μm-wide fingers with a 15 μm separation between the fingers, total area 2 mm). Cyclic voltammetric growth was chosen over galvanostatic because the former method produced smoother films with more controlled lateral growth able to bridge the inter-electrode gaps more readily. Also, more evenly coated films across both electrodes and spaces between the electrodes were generated. Antibody-loaded polypyrrole films were created from a 10 ml solution of 0.5 M pyrrole monomer and 0.125 mg/ml antibody. Importantly, the antibody solution was extensively dialysed against deionized water to remove trace buffer salts. The IDEs were part of a three-electrode cell using platinum foil (approximately 1 cm^2) as an auxiliary electrode and a Ag/AgCl reference electrode. Electropolymerization was performed by cycling the potential between −0.9 and +0.9 V for 15 scans at a scan rate of 50 mV.

A number of different polyclonal and monoclonal antibodies were also successfully incorporated into polypyrrole films, including antibodies against luteinizing hormone (LH), hen egg white lysozyme and avidin. The process of antibody capture in the growing films was likely to include an adsorption/entrapment sequence as well as one driven by charge, with the protein contributing a charged counterion to the oxidized (cationic) polypyrrole.

The formed films were characterized using cyclic voltammetry in solutions of sodium chloride (Fig. 1) and in-plane impedance spectroscopy in deionized water (Fig. 2). The cyclic voltammograms demonstrated that with the antibody-loaded films there were two redox processes, whereas films created using sodium chloride as the counterion produced only a single redox peak. The presence of two redox peaks has also been observed with other protein-loaded films. The large peak observed at the more positive potential is due to mobile chloride ions within the film, while the smaller peak is due to the immobile protein entrapped within the polymer film. This differentiation between mobile and immobile counterions was previously noted by Talaie and Wallace [14]. Impedance spectroscopy measurements of the different antibody-loaded polypyrrole films (Fig. 2) resulted in broadly similar Cole–Cole plots. This similarity is likely to reflect the similarity of the IgG antibodies in each instance. With other proteins of different size and shape, variants in the Cole–Cole plots are observed indicating a differential effect on the movement of charge across films.

Exposure of an antibody-loaded film to the target antigen results in no initial change in the Cole–Cole plot. However, following a single potentiodynamic cycle (–0.1 V to -0.9 V to –0.1 V) distinctive and reproducible changes can be observed in the plots. These changes do not occur if the film is exposed to mole-

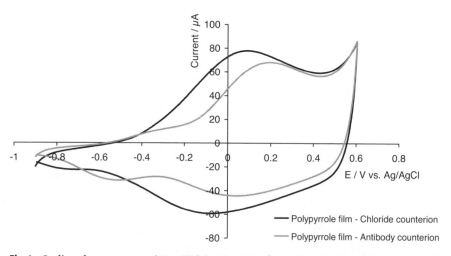

Fig. 1. Cyclic voltammograms (50 mV/s) in 50 mM sodium chloride for polypyrrole-modified 15-μm IDE loaded with chloride (50 mM sodium chloride, 0.5 M pyrrole, 15 cycles) (black line) and antibody to luteinizing hormone (0.25 mg/ml antibody, 0.5 M pyrrole, 15 cycles) (grey line)

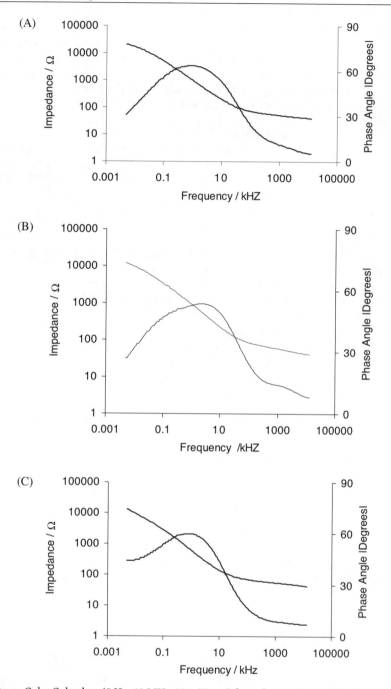

Fig. 2a–c. Cole–Cole plots (5 Hz–13 MHz, 20 mV rms) for polypyrrole-modified 15-μm IDE loaded with antibodies to **a** lysozyme, **b** avidin and **c** luteinizing hormone (0.125 mg/ml antibody, 0.5 M pyrrole, 15 cycles). Measurements performed in phosphate buffer, pH 7.4

cules which are not capable of binding specifically to the antibody, and indicate that the Cole–Cole plots are indeed due to the formation of the affinity complex and not to either independent alterations in films following the potentiodynamic step or to non-specific protein deposition on the film. The fact that the maximum potential reached in the cycle is below that required to over-oxidize the polypyrrole film, and to degrade the polymer, makes it less likely that a chemical change in the polymer film due to cycling is the cause. A plausible explanation is that there is some relaxation (rearrangement) of the polymer chains around the affinity complex as the film is allowed to contract and expand during the cycling step; any polymer conformational change would lead to impedimetric variation via changes in conjugation state.

Polypyrrole exposure to different concentrations of LH analyte resulted in changes to the Cole–Cole plots (Fig. 3) with the predominant change being in

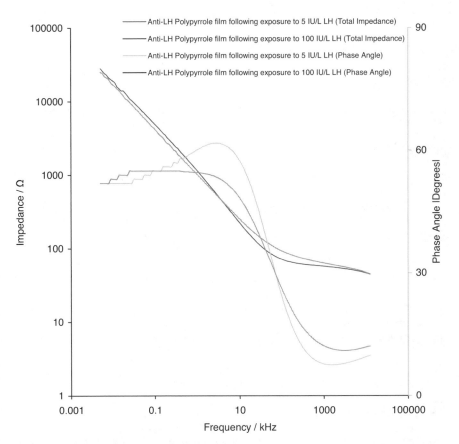

Fig. 3. Cole–Cole plot (5 Hz–13 MHz, 20 mV rms) for monoclonal antibody to luteinizing hormone-loaded polypyrrole films (0.25 mg/ml antibody to luteinizing hormone, 0.5 M pyrrole, 15 cycles) exposed to two concentrations of luteinizing hormone [5 (black line) and 100 (grey line) IU/l] in phosphate buffer, pH 7.4

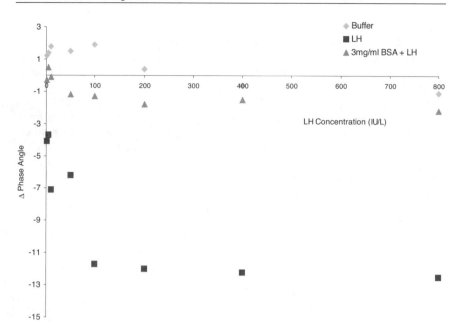

Fig. 4. Concentration-dependent response curve for a polypyrrole film loaded with a monoclonal antibody to luteinizing hormone (0.125 mg/ml antibody, 0.5 M pyrrole, 15 cycles) following exposure to buffer (filled diamond), luteinizing hormone (empty square), and luteinizing hormone in the presence of 3 mg/ml of BSA (filled triangle)

the phase angle peak. Based on this, it was possible to construct a concentration-dependent curve for LH and polypyrrole films containing antibodies to LH (Fig. 4). This demonstrated the quantitative measurement capability of this technique and a truly reagentless immunosensor system.

The effect of solution matrix interferents has yet to be fully explored for impedimetric devices. In our studies, inclusion of bovine serum albumin (BSA) as a surface-active protein may have masked the binding process (Fig. 4). It is possible that pre-sample cleanup, extraction or membrane-based separation is needed for avoiding such matrix effects. The exquisite sensitivity offered by impedance monitoring and its surface dependence are potential drawbacks to practical measurements.

3
Impedance as a Basis for Probing Bacteria and Cells

Monitoring of electrochemical processes within and around bacteria, eukaryotic cells and tissues is not new, indeed the first paper on this subject was presented by G.N. Stewart at the 1898 meeting of the British Medical Association in Edinburgh. This work was later published in the Journal of Experimental Medicine under the title "The changes produced by the growth of bacteria in the molecular concentration and conductivity of culture media". The study described the investigation of the putrefaction of both blood and serum following infection by bacteria, with the electrical response curves presented being remarkably similar to those obtained by modern measurement systems. The time span of these experiments (30 days), however, was considerably longer than the matter of hours required for modern analysis [15]. As with many other biophysical techniques founded in the late nineteenth and early twentieth centuries, impedance microbiology remained a poorly investigated phenomenon until electronic instrumentation evolved to the point where dedicated measurement systems became available. Impedance studies on eukaryotic cells and tissues followed a similar course with sporadic articles appearing in the early literature. Reports considered examination of explanted neural tissue [16], erythrocytes and whole blood [17, 18] and cultured cell suspensions [19].

There has been renewed interest in impedance measurements on bacteria and cells since the 1970s. A steady stream of articles has been published with some pioneering observations made by Ur and Brown [20], Cady [21], Richards et al. [22] and Hause et al. [23]. The body of work produced by these and other groups led to the introduction of the first commercial instruments, the Malthus and the Bactometer. Much of the subsequent literature in this field has concentrated on the application of these instruments The first general area to be tackled was analysis for the food industry, with Gnan and Luedecke [24] being amongst the first, for example, to demonstrate the ability to monitor total bacterial counts in raw milk. Monitoring of total bacterial counts has since been demonstrated for grain products [25], UHT low-acid foods [26], fish [27] and meat products [28]. It is also possible to use the technique to detect yeast cells in wine [29] and fruit juices [30]. With further refinement, particular (potentially dangerous) organisms can be detected, e.g. Gram-negative bacteria in pasteurized milk [31], Escherichia coli in potable water [32] and coliforms in meat [33]. The importance and usefulness of impedance monitoring of bacteria was recognized when it became a recommended methodology for the screening of salmonella in animal feeds under the 1989 Processed Animal Protein Order [34].

The pharmaceutical industry has also adopted impedance microbiology as a quality control tool to monitor raw materials and finished products for bacterial contamination. However, the technique has also moved into the research laboratory where it has been demonstrated as a non-optical tool for the testing of new antibacterial agents [35] and for determining the bactericidal potency of sanitizer compounds [36].

Microbiology impedance measurements offer an indirect measure of bacterial concentration since the technique can quantitatively follow the time taken for a sample to reach a preset conductivity threshold. Prior to performing an analysis of an unknown sample it is, however, necessary to create a standard curve prepared from known concentrations of the target bacterial strain under investigation, with the unknown sample then incubated in specially prepared media. The conductivity of the sample, monitored to a fixed conductivity threshold, is then termed the time detection threshold or TDT. The bacterial concentration coinciding with this is then determined from a standard curve. A variant on this is used for samples with a salt concentration that is either too high or too low to allow impedance measurements to be performed. Here, the culture medium is inoculated with the sample in one chamber and the measurement electrodes in a second chamber are immersed in a solution of strongly alkaline potassium hydroxide. As the bacteria multiply, they produce more CO_2, which then reacts with hydroxide ions to produce carbonate ions and water. The result is a large net change in the molar conductivity of the alkaline solution, detected directly by the electrode. With ionization, the molar conductivity change here is proportional to the amount of CO_2 adsorbed; in turn this is directly related to the microbial biomass present [37]. This method, however, will tend to underestimate the bacterial concentration since the efficiency of CO_2 transfer will not be 100% due to retention of CO_2 by the culture medium, pH changes in the absorbent solution, surface area restriction, volume effects of the culture medium and the absorbent solution and the dead volume headspace of the two chambers.

Impedance microbiology is now a tool for quality control measurements, since not only can data obtained be correlated with plate counts, but also the assay time required is much reduced in comparison to traditional plate counting. The drawback is that the technique is not a direct measure of bacteria in the sample, and requires standardized growing conditions and specialized media in order to complete the electrochemical measurements, conditions which may not suit all types of bacteria.

Much of the modern interest in impedance measurements for eukaryotic cells can be linked to the group of Giaever. They were the first to examine the impedance signatures derived from small populations of cells (20 to 80 cells) cultured on electrodes significantly larger than the cells being studied. In 1984, he and Keese [38] reported a system where standard polystyrene tissue culture dishes were modified to include four working electrodes, each 3×10^{-4} cm^2, together with a single large reference electrode (2 cm^2) (Fig. 5). They cultured hu-

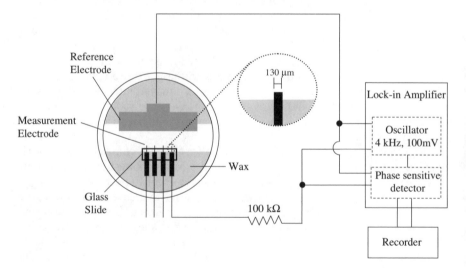

Fig. 5. An impedance measurement system for cells. Cells are cultured in a standard polystyrene tissue culture dish. Impedance measurements are performed on a series of gold microelectrodes. Adapted from [38]

man lung fibroblast cells (WI-38 and WI-38/VA-13) within these modified cell culture dishes and applied an AC voltage through a 100-kΩ resistor to a single electrode. This provided a virtually constant current source and so it was possible to determine the impedance signal through the measurement of the resulting voltage. A lock-in amplifier was then used to observe the effects of cell proliferation (increase in impedance) as well as small microscopic movements and translocations of the cells on the surface (dynamic fluctuation in the observed impedance signal) [38]. This system was the first to demonstrate monitoring of such dynamic processes by an electrochemical technique.

The non-optical nature of the technique allowed Giaever, Keese and co-workers to observe nanometre motions of the cells that were invisible to conventional optical microscopy [39]. From the first study in 1984 they expanded their work to examine the effect of different proteins on cell adhesion, spreading and motility [40] and, moreover, devised a mathematical model to describe the motion of cells [41]. More recently, the effect of parameters such as temperature, CO_2 and glucose concentration on the micromotion of the cells [42, 43] and morphological response to pulsed AC fields [44] were reported. It appears that endothelial cells can be monitored under fluid flow, thus more closely mimicking the in vivo environment [45], and there is no limit to the cell type as indicated by work on macrophages [46], bovine pulmonary microvessel cells, bovine pulmonary artery endothelial cells (BPMVEC and BPAEC) [47] and human umbilical vein endothelial cells (HUMVEC) [48]. A model for Madine–Darby canine kidney epithelial cells (MDCK) allowed a measure of the distance between the basal cell surface and underlying substrate and the determination of the capacitance of the apical, basal and lateral cell membranes [49]; elucidation of these parameters is difficult by any other technique.

In studies of bioactive factors, Smith et al. [50] investigated the effect of prostaglandin E_2 on the morphology of orbital fibroblast cells derived from patients with Graves opthalmopathy, Pei et al. [51] described the effect of pSV2-neo plasmid on the motion of NIH 3T3 cells and Noiri et al. [52] investigated the role of nitric oxide on endothelin-induced endothelial cell migration.

In a modification of the systems developed above, Connolly et al. [53] added a glass ring around the electrode construct to retain cell culture media and, by inserting a permeable cellulose nitrate membrane to divide this ring into two separate halves each with a working electrode, were able to run controls on the same electrode construct as the experiment. This reduced background and noise effects and provided more reliable data. The modified arrangement was used to monitor the attachment, spreading and motion of MDCK and baby hamster fibroblast (BHK) cells. Impedance characteristics of individual BHK cells were measured using reduced-area electrodes. These were fabricated by the selective deposition of gold onto a glass surface to create 10-μm-wide areas together with relevant interconnects and bondpads. Insulating layers of 1-μm-thick silicon nitride or 4-μm-thick polyamide were used as coatings and 8×10-μm areas were etched to create gold working electrodes, with bondpads also etched for electrical connections. Platinization of electrodes reduced electrode impedance. Full coverage of an electrode was possible by a single cell. Measurements on BHK cells continued fluctuations in impedance of 10–20% – a phenomenon not observed on the control electrodes; this difference they attributed to the motion of the cells on the electrode.

Impedance made possible a measurement of interfacial effects of surface roughness, notably on cell adhesion [54]. Three electrode finishes were assessed: a smooth gold surface, rough platinized gold electrodes and gold electrodes roughened following a dry etching process. The roughened gold electrodes produced inconclusive results with BHK cells and, while the rough platinized electrode seemed to promote adhesion, it was not clear if this was due solely to surface roughness. In a study with neurons from Lymnaea stagnalis [55] a correlation was seen between impedance and extracellular recorded action potentials.

A number of electrode variant designs have been suggested for cell impedance measurement. Hagedorn et al. [56] described an electrode construct (Fig. 6) where two vertically aligned electrodes were separated by a silicon membrane containing a single pore. The electrical circuit therefore runs from one electrode to the other via the pore. The small size of the pore (10 μm diameter) relative to the electrodes means that it is the region of the pore that dominates the impedance. This apparatus enabled the cellular and filopodial motion of single 3T3 and L929 mouse fibroblast cells to be distinguished. A silicon membrane containing an array of pores allowed a population of cells to be investigated. In this configuration when cells covered the pores, the observed impedance decreased. The advantages of this configuration are that it is more compatible with current MEMS technology and thus easily integrated into such systems, and that the silicon surface is more biocompatible than previous electrode surfaces.

Wegener et al. [57] utilized a voltage divider technique for the determination of impedance across a range of frequencies (1 Hz to 10 kHz). With cultured porcine brain microvessel cells, epithelial cells from porcine choroids plexus and

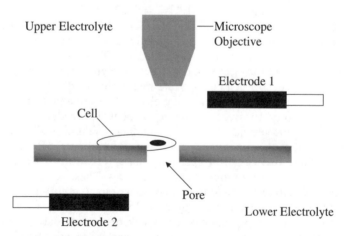

Fig. 6. An impedance measurement chamber for cells. Two electrolyte reservoirs are separated by a porous silicon membrane that is used as substrate for cell attachment and growth. Impedance measurements are performed at two platinum electrodes (15 mm^2). Adapted from [56]

MDCK cells directly onto gold electrodes, transepithelial and transendothelial resistances were followed for events such as cell proliferation.

IDEs within a 5×5-mm area with electrode fingers of 50 μm, separated by 50 μm, have been used to monitor the attachment, spreading and morphology of CV-1 (fibroblast cells from the kidney of an adult male African green monkey) and LS147T (epithelial cells from a human colon adenocarcinoma) cells [58]. Cadmium-induced cell membrane degradation was also detectable.

Gold electrode arrays coated with different proteins to promote cell adhesion to the electrode surface showed that concanavalin A gave the best results [59]. With these arrays the effect of cellular inhibitors (mercury chloride, trinitrotoluene, trinitrobenzene and 2-amino-4,6-dinitrobenzene) on the motility and viability of insect cells could be followed.

The body of work probing cells and cell membranes using impedance measurements thus indicates that a number of different electrode configurations are possible to investigate eukaryote cells. Cell adhesion, spreading and motion can be followed and physical parameters, e.g. cell membrane capacitance, can be determined. Monitoring of the effects of external factors such as culture media, pharmacological compounds and toxins on an individual cell provides a link to future biosensor constructs.

4
Future Directions

Despite the body of work that has been accumulated over the past 30 years regarding impedance measurements for sensors and tissue measurements, there still remain a large number of directions to explore. One promising direction is the creation of imaging sensors using scanning impedance techniques. Zou et al. [60] described a scanning instrument capable of such localized impedance measurements. The instrument consists of a probe of two platinum microelectrodes 10 μm in diameter. The AC current distributions measured using the probe give good agreement with analytical and computer simulations. A version of this instrument is now commercially available through Uniscan Instruments Ltd, UK. Currently, the instrument has been used to map corrosion pits in coatings; however, the micrometre resolution and the ability to scan in solution means that the instrument could be equally utilized within the biological field.

A scanning photoinduced impedance microscope has been developed [61]. Here, photocurrent measurements are made at a field-effect structure. A focused pulsed light beam is scanned across the surface of the sample, producing local photocurrents; this provides information about the distribution of dielectric properties of the sample that can be used to create an image map of the sample. The reported micrometre resolution is more than sufficient to image tissue processes and may lead to a novel class of optical biosensors.

Another major opportunity is in relation to bio-microelectromechanical systems (bio-MEMS) and 'lab-on-a-chip' devices. Thus, it is eminently possible to imagine the applications described in this review being performed at small microchip arrays. Indeed many of the applications described are already being performed on miniature electrode structures and so it does not require a major adaptation to integrate these into chip-based devices.

Given the increase in interest in impedance monitoring over the past few years, it is unlikely that the technique will remain an esoteric niche area of research and it seems possible that 'mainstream' research usage will emerge.

References

1. Wallace GG, Kane-Maguire LAP (2002) Manipulating and monitoring biomolecular interactions with conducting electroactive polymers. Adv Mater 14:953–960
2. Lillie G, Payne P, Vadgama P (2001) Electrochemical impedance spectroscopy as a platform for reagentless bioaffinity sensing. Sens Actuators B 78:249–256
3. Smith DE (1966) AC polarography and related techniques: theory and practice. In: Bard AJ (ed) Electroanalytical chemistry, vol 1. Marcel Dekker, New York, pp 1–155
4. Gallardo Soto AM, Jaffari SA, Bone S (2001) Characterisation and optimisation of AC conductimetric biosensors. Biosens Bioelectron 16:23–29
5. Alfonta L, Bardea A, Khersonsky O, Katz E, Wilner I (2001). Chronopotentiometry and Faradaic impedance spectroscopy as signal transduction methods for the biocatalytic precipitation of an insoluble product on electrode supports: routes for enzyme sensors, immunosensors and DNA sensors. Biosens Bioelectron 16:675–687
6. Saum AGE, Cumming RH, Rowell FJ (1998) Use of substrate-coated electrodes and AC impedance spectroscopy for the detection of enzyme activity. Biosens Bioelectron 13:511–518
7. Albert KJ, Lewis NS, Schauer CL, Sotzing GA, Stitzel SE, Vaid TP, Walt DR (2000) Cross-reactive chemical sensor arrays. Chem Rev 100:2595–2626
8. Lewis TW, Wallace GG, Smyth MR (1999) Electrofunctional polymers: their role in the development of new analytical systems. Analyst 124:213–219
9. Wang LX, Li XG, Yang YL (2001) Preparation, properties and applications of polypyrroles. React Funct Polym 47:125–139
10. Taylor RF, Marenchic IG, Spencer RH (1991) Antibody-based and receptor-based biosensors for detection and process control. Anal Chim Acta 249:67–70
11. Nishizawa M, Matsue T, Uchida I (1992) Penicillin sensor based on a microarray electrode coated with pK-responsive polypyrrole. Anal Chem 64:2642–2644
12. Contractor AQ, Sureshkumar TN, Narayanan R, Sukeerthi S, Lal R, Srinivasa RS (1994) Conducting polymer-based biosensors. Electrochim Acta 39:1321–1324
13. Sargent A, Sadik OA (1999) Monitoring antibody–antigen reactions at conducting polymer-based immunosensors using impedance spectroscopy. Electrochim Acta 44:4667–4675
14. Talaie A, Wallace GG (1994) The effect of the counterion on the electrochemical properties of conducting polymers – a study using resistometry. Synthetic Met 63:83–88
15. Whitley D (1999) Introduction to principles of impedance. Impedance microbiology. Don Whitley Scientific, Shipley. http://www.dwscientific.co.uk/pdf/pw/pw2.pdf. Cited 27 Aug 2002
16. Cole KS, Curtis HJ (1939) Electrical impedance of squid giant axon during activity. J Gen Physiol 22:649–670
17. Fricke H (1926) The electric capacity of suspensions with special reference to blood. J Gen Physiol 9:137–152
18. Fricke H, Curtis HJ (1935) The electric impedance of hemolyzed suspensions of mammalian erythrocytes. J Gen Physiol 18:821–836
19. Schwan HP (1957) Electrical properties of tissues and cell suspensions. In: Lawrence JH, Tobias CA (eds) Advances in biological and medical physics. Academic, New York, vol 5, 147–209

20. Ur A, Brown DFJ (1975) Impedance monitoring of bacterial activity. J Med Microbiol 8:19–27
21. Cady P (1978) Progress in impedance measurements in microbiology. In: Sharpe AN, Clark DS (eds) Mechanising microbiology. Charles C. Thomas, Springfield, pp 199–239
22. Richards JCS, Jason AC, Hobbs G, Gibson DM, Christie RH (1978) Electronic measurement of bacterial growth. J Phys E 11:560–568
23. Hause LL, Komorowski RA, Gayon F (1981) Electrode and electroyte impedance in the detection of bacterial growth. IEEE Trans Biomed Eng 28:403–410
24. Gnan S, Luedecke LO (1982) Impedance measurements in raw milk as an alternative to the standard plate count. J Food Prot 45:4–7
25. Sorrells KM (1981) Rapid detection of bacterial content in cereal grain products by automated impedance measurements. J Food Prot 44:832
26. Coppola K, Firstenberg-Eden R (1988) Impedance-based rapid method for detection of spoilage organisms in UHT low-acid foods. J Food Sci 53:1521
27. Van Spreekens KJA, Stekelenburg FK (1986) Rapid estimation of the bacteriological quality of fresh fish by impedance measurements. Appl Microbiol Biotechnol 24:95–96
28. Firstenberg-Eden R (1983) Rapid estimation of the number of microorganisms in raw meat by impedance measurement. Food Technol 37:64–70
29. Henscke PA, Thomas DS (1988) Detection of wine yeast by electronic methods. J Appl Bacteriol 53:123–133
30. Deak T, Beuchat LR (1993) Comparison of conductimetric and traditional plating techniques for detecting yeasts in fruit juices. J Appl Bacteriol 75:546–550
31. Nieuwenhof FFJ, Hoolwerf JD (1987) Impedance measurement as an alternative to the plate-count method for estimating the total count of bacteria in raw milk. J Food Prot 50:665–668
32. Colquhoun KO, Timms S, Fricker CR (1995) Detection of Escherichia coli in potable water using direct impedance technology. J Appl Bacteriol 79:635–639
33. Firstenberg-Eden R, Klein CS (1983) Evaluation of a rapid impedimetric procedure for the quantitative estimation of coliforms. J Food Sci 48:1307–1311
34. Ministry of Agriculture, Fisheries and Food (1989) The processed animal protein order, 1989. Statutory Instrument 1989 No 661, Schedule II, Part b
35. Gould IM, Jason AC, Milne K (1989) Use of the Malthus microbial-growth analyzer to study the post-antibiotic effect of antibiotics. J Antimicrob Chemother 24:523–531
36. Andrade NJ, Bridgeman TA, Zottola EA (1998) Bacteriocidal activity of sanitizers against Enterococcus faecium attached to stainless steel as determined by plate count and impedance methods. J Food Prot 61:833–838
37. Owens JD, Thomas DS, Thompson PS, Timmerman JW (1989) Indirect conductimetry: a novel approach to the conductimetric measurement of conductivity changes. Lett Appl Microbiol 9:245–249
38. Giaever I, Keese CR (1984) Monitoring fibroblast behaviour in tissue culture with an applied electric field. Proc Natl Acad Sci USA 81:3761–3764
39. Giaever I, Keese CR (1993) A morphological biosensor for mammalian cells. Nature 366:591–592
40. Giaever I, Keese CR (1986) Use of electric fields to monitor the dynamical aspect of cell behaviour in tissue culture. IEEE Trans Biomed Eng 33:242–247
41. Giaever I, Keese CR (1989) Fractal motion of mammalian cells. Physica D 28:128–133
42. Lo CM, Keese CR, Giaever I (1993) Monitoring motion of confluent cells in tissue culture. Exp Cell Res 204:102–109
43. Lo CM, Keese CR, Giaever I (1994) pH changes in pulsed CO_2 incubators cause periodic changes in cell morphology. Exp Cell Res 213:391–397
44. Ghosh PM, Keese CR, Giaever I (1994) Morphological response of mammalian cells to pulsed ac fields. Bioelectrochem Bioenerg 33:121–133
45. DePaola N, Phelps JE, Florez L, Keese CR, Minnear FL, Giaever I, Vincent P (2001) Electrical impedance of cultured endothelium under fluid flow. Ann Biomed Eng 29:648–656

46. Kowolenko M, Keese CR, Lawrence DA, Giaever I (1990) Measurement of macrophage adherence and spreading with weak electric fields. J Immunol Methods 127:71–77
47. Tiruppathi C, Malik AB, Delvecchio PJ, Keese CR, Giaever I (1992) Electrical method for detection of endothelial-cell shape change in real-time assessment of endothelial barrier function. Proc Natl Acad Sci USA 89:7919–7923
48. Moy AB, Van Engelenhoven J, Bodmer J, Kamath J, Keese C, Giaever I, Shasby S, Shasby DM (1996) Histamine and thrombin modulate endothelial focal adhesion through centripetal and centrifugal forces. J Clin Invest 97:1020–1027
49. Lo CM, Keese CR, Giaever I (1995) Impedance analysis of MDCK cells measured by electric cell-substrate impedance sensing. Biophys J 69:2800–2807
50. Smith TJ, Wang HS, Hogg MG, Henrikson RC, Keese CR, Giaever I (1994) Prostaglandin E_2 elicits a morphological change in cultured orbital fibroblasts from patients with Graves ophthalmopathy. Proc Natl Acad Sci USA 91:5094–5098
51. Pei ZD, Keese CR, Giaever I, Kurzawa H, Wilson DE (1994) Effect of the pSV2-neo plasmid on NIH 3T3 cell motion detected electrically. Exp Cell Res 212:225–229
52. Noiri E, Hu Y, Bahou WF, Keese CR, Giaever I, Goligorsky MS (1997) Permissive role of nitric oxide in endothelin-induced migration of endothelial cells. J Biol Chem 272:1747–1752
53. Connolly P, Clark P, Curtis ASG, Dow JAT, Wilkinson CDW (1990) An extracellular microelectrode array for monitoring electrogenic cells in culture. Biosens Bioelectron 5:223–234
54. Lind R, Connolly P, Wilkinson CDW, Breckenridge LJ, Dow JAT (1991) Single cell mobility and adhesion monitoring using extracellular electrodes. Biosens Bioelectron 6:359–367
55. Breckenridge LJ, Wilson RJA, Connolly P, Curtis ASG, Dow JAT, Blackshaw SE, Wilkinson CDW (1995) Advantages of using microfabricated extracellular electrodes for in vitro neuronal recording. J Neurosci Res 42:266–276
56. Hagedorn R, Fuhr G, Lichtwardtzinke K, Richter E, Hornung J, Voigt A (1995) Characterization of cell movement by impedance measurement on fibroblasts grown on perforated Si membranes. BBA-Mol Cell Res 1269:221–232
57. Wegener J, Sieber M, Galla HJ (1996) Impedance analysis of epithelial and endothelial cell monolayers cultured on gold surfaces. J Biochem Biophys Methods 32:151–170
58. Ehert R, Baumann W, Brischwein M, Scwinde A, Stegbauer K, Wolf B (1997) Monitoring of cellular behaviour by impedance measurements on interdigitated electrode structures. Biosens Bioelectron 12:29–41
59. Luong JHT, Habibi-Rezaei M, Meghrous J, Xiao C, Male KB, Kamen A (2001) Monitoring motility, spreading and mortality of adherent insect cells using an impedance sensor. Anal Chem 73:1844–1848
60. Zou F, Thierry D, Isaacs HS (1997) A high-resolution probe for localized electrochemical impedance spectroscopy measurements. J Electrochem Soc 144:1957–1965
61. Krause S, Talabani H, Xu M, Moritz W, Griffiths J (2002) Scanning photoinduced impedance microscopy – an impedance-based imaging technique. Electrochim Acta 47:2143–2148

Chapter 9

Noninvasive Electrical Sensor Devices to Monitor Living Cells Online

Andreas Janshoff, Claudia Steinem, Joachim Wegener

Abstract. Experiments that rely on isolated animal cells in vitro are indispensable in all branches of biomedical research, especially in those areas addressing molecular interactions like drug discovery or molecular medicine. From the analytical viewpoint, these scientific goals not only require assays and devices capable of monitoring the interaction between two molecular species, but also to detect whether any biological functionality is generated. Accordingly, assays based on living cells as the sensing elements are often required and advantageous. The present article describes two transducer devices capable of monitoring changes in the three-dimensional shape of living cells when these are exposed to chemical, biological, or physical stimuli. In both technologies, the cells are cultured on thin conducting gold electrodes. The readout is based on electrochemical impedance spectroscopy or the quartz crystal microbalance technique. Besides a condensed outline about the methodological backgrounds, this article presents some selected examples about their successful application as whole-cell biosensors.

Keywords: Quartz crystal microbalance, QCM, Electric cell–substrate impedance sensing, ECIS, Whole-cell biosensor

1
Introduction

Most physiological processes in living organisms are mediated by molecular recognition. For instance, enzymes specifically recognize their substrates, hormones bind to an individual set of cell surface receptors, and cells communicate and interact with each other only after recognition of certain cell surface markers. Biosensors make use of the specificity of molecular recognition in order to detect one particular species, the analyte, in a complex physiological matrix. One approach to build such a sensor is to isolate the receptor molecules from their natural environment and to integrate them into a transducer device that is capable of indicating analyte binding by alteration of a physical quantity. There are numerous studies about biosensor devices that follow this basic concept. The question of whether these receptors change their biological specificity or affinity due to the artificial environment is difficult to answer but a reasonable concern.

An alternative concept for biosensor development is to leave the specific receptors in their native environment, which was optimized and adapted for efficient molecular recognition during evolution, and to use intact cells as the biologically active element. Biosensors of this kind are referred to as *whole-cell biosensors* [1] or cell-based biosensors [2]. At first sight, it appears that using living cells as the recognition element only introduces complications. However, the kind of information that is accessible makes it worthwhile. Compared to devices based on reconstituted receptors that only return binding of a particular ligand, and hence allow for an accurate determination of concentration, *whole-cell* sensors provide information about functional activity induced by receptor occupancy [1]. Successful binding of a hormone to its cell surface receptor is certainly the best indicator for correct molecular specificity, but whether the hormone activates the receptor to induce the subsequent signal transduction or simply blocks its binding site cannot be inferred from binding assays. Even if a hormone analog binds to the receptor with high affinity, traces of it may still be cytotoxic, for instance, on a receptor-independent pathway. Due to the very unique opportunity to monitor biological functionality, whole-cell biosensors are attracting more and more interest and already replace or reduce animal experiments in the early stages of drug development [3, 4].

Very different technical concepts have been proposed to monitor the functional response of living cells in a sensory system. Many approaches address changes in the cells' energy metabolism when exposed to certain stimuli. Since energy metabolism is ubiquitous in all cells, biosensors of this kind can be realized with a huge variety of different cell types. Changes in the rate of energy metabolism were proven to be very sensitive to a broad range of physiological alterations [5]. Metabolic activities are most often measured by pH-sensitive devices that report on extracellular acidification rates, e.g., based on LAPS (light-addressable potentiometric sensors) or ISFETs (ion-selective field-effect transistors) [6–9]. In a comparative study LAPS and ISFETs showed very similar analytical characteristics like sensitivity, drift, and response to a pH shift [10]. In contrast to the measurement of oxygen consumption, determined for example by classical Clark oxygen electrodes, extracellular acidification is sensitive to both glycolytic metabolism and respiration. The most recent devices that perform oxygen partial pressure measurements combine both sensors – pH and O_2 – within one setup based on ISFET technology [11]. When it comes to electrically excitable cells as sensory elements, extracellular recording of stimulated action potentials and membrane potential measurements serve as the analytical parameter [12–15]. Most recently, Woolley et al. [16] have introduced a new technology by growing adherent cells on the surface of gold-film electrodes and measuring the electrical potential difference between the substrate and a reference electrode. The recorded voltages are dependent on the cell type and vanish when the cells leave the electrode surface.

In this chapter we will focus on two *whole-cell biosensor devices* that are sensitive to cell shape. Changes in cell shape are usually only accessible from microscopic images or time-lapse video microscopy [17]. The techniques described here are, however, electrochemical in nature and require growing the cells on conducting surfaces. From either electrochemical impedance analysis of the attached cells or their interactions with an oscillating quartz plate, we can extract changes in cell–cell or cell–substrate contacts in a quantitative and noninvasive fashion. Since the three-dimensional shape of a cell within a tissue-like cell monolayer is the sum of interactions with their neighboring cells and the substrate, these real-time approaches can be classified as morphological biosensors. For each of the techniques we will introduce the basic theoretical background before we illustrate the capabilities of the devices by selected examples.

2
Electrochemical Impedance Measurements to Monitor Living Cells

Compared to other electrochemical techniques impedance analysis in the linear current-to-voltage regime can be performed in an entirely noninvasive fashion, since it only requires small-amplitude signals, either voltage or current, to be applied to the system of interest [18]. It is imperative to apply only small-amplitude signals to avoid significant perturbations of the membrane potential, since the latter controls a multitude of membrane-associated processes and ranges only between 50 and 100 mV in most mammalian cells.

Application of noninvasive impedance measurements to suspended cells and tissues has a long history and goes back to the pioneering work of Schwan and Fricke [19, 20]. Their major objective was to determine the dielectric properties of the tissue over a broad frequency range and to identify the associated relaxation mechanisms. These early studies clearly paved the way for the introduction of impedance measurements into clinical diagnostics. Nowadays *electrical impedance tomography* (EIT) is considered to be a potential future candidate for new diagnostic imaging devices that are not based on ionizing radiation [21].

In the following section we will describe a whole-cell biosensor device that allows monitoring of various aspects of living cells in vitro and that is based on noninvasive impedance measurements. As will be shown in detail, the impedance approach is sensitive to the morphology of substrate-anchored cells. Since it is generally accepted that the shape of a cell is very sensitive to a myriad of chemical compounds as well as physical or biological stimuli, this approach is very versatile and may fill a gap in the pool of analytical tools in cell research. After a short paragraph about its technical background we will present two published examples showing that the impedance technique may become a very useful and powerful tool in drug development, cytotoxicity screening or questions originating from fundamental cell biology.

2.1
Experimental Background of ECIS

The idea of *electric cell-substrate impedance sensing* (ECIS) was introduced by Giaever and Keese, who were first to grow mammalian cells directly on the surface of gold-film electrodes and to record and analyze the concomitant changes of the electrode's electrical impedance [22, 23]. In the meantime several other groups followed this basic concept but applied other electrode materials and geometries [24]. Figure 1a sketches the basic concept of the ECIS technique.

ECIS uses a two-electrode setup. The two coplanar gold-film electrodes are prepared on cell culture dishes and cells are grown on these surfaces under ordinary cell culture conditions. The electrical connection between the two electrodes is provided by the culture medium, a buffered salt solution of ~15 mS/cm conductivity that contains all the nutrients and growth factors the cells require. The two electrodes differ with respect to their surface area. By making the counter electrode at least 100 times larger than the working electrode, the impedance of the latter dominates the readout of the entire system. Thus, the observed changes in electrical impedance can be clearly assigned to changes that occur at the working electrode. To increase the sensitivity of the measurement the working electrode should be kept small [25]. Most data available in the literature were recorded with circular electrodes of 250 μm diameter. Instead of using a coplanar counter electrode it is also possible to work with a dipping electrode reaching into the culture medium from above. As long as the surface of the dipping electrode is large and the associated interface impedance low, the readout of the system is not affected by this change in electrode geometry.

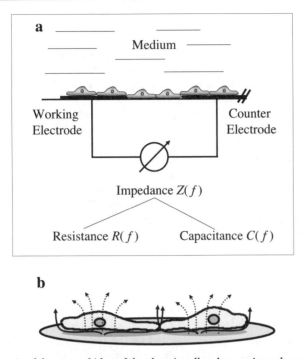

Fig. 1a. Schematic of the general idea of the electric cell–substrate impedance sensor introduced and pioneered by Giaever and Keese. The cells are grown on the surface of two co-planar gold electrodes that are connected to a phase-sensitive impedance analyzer. The frequency-dependent complex impedance is broken down into the resistance and the capacitance of the electrode. The working electrode is at least 100-fold smaller in surface area than the counter electrode, so that the overall impedance signal is dominated by the impedance of the working electrode. b Current pathways across a cell monolayer attached to a conducting surface. Most of the current has to bypass the cell bodies (*solid arrows*) so that it flows along the narrow gap between the lower cell membrane and the electrode surface before it can escape through the intercellular shunt between adjacent cells. At sufficiently high frequencies the current couples through the plasma membrane capacitively. Accordingly, the impedance at various frequencies holds information on cell–substrate and cell–cell contacts as well as the dielectric properties of the plasma membrane

In this experimental approach, impedance data are usually recorded over a frequency range between 1 and 10^6 Hz. For the major fraction of this frequency band (<10 kHz) the cells behave essentially like insulating particles forcing the current to flow around the cell body on paracellular pathways. Current leaving a cell-covered electrode has to flow through the confined and narrow channels between the ventral plasma membrane and the electrode surface before it can escape through the paracellular shunt between adjacent cells into the bulk phase (Fig. 1b). Since the current has to bypass the cell bodies, it picks up impedance contributions in the cell substrate adhesion zone as well as in the adhesion area between neighboring cells. Readings of the total impedance are thus sensitive to changes in cell–cell and cell–substrate contacts or the cell shape in general. For high frequencies, the current can capacitively couple through the cells passing the ventral and the dorsal membrane in the form of a displacement current. Accordingly, high-frequency impedance readings can be used to measure the membrane capacitance and thus the degree of membrane folding. Taken together, frequency-dependent impedance measurements of mammalian cells attached to planar film electrodes provide independent information about changes in the cell–substrate and cell–cell adhesion zones and the morphology of the plasma membrane [25, 26].

Giaever and Keese published a mathematical model that can be applied to frequency-dependent impedance data of cell-covered gold-film electrodes in order to extract and distinguish the various impedance contributions [25]. Herein, the cells are modeled as circular disks with radius r_c hovering at a distance h above the electrode surface. Within this model, the impedance arising in the cell–substrate adhesion zone is accounted for by a parameter α that is defined as (Eq. 1):

$$\alpha = r_c \sqrt{\frac{\rho}{h}},\tag{1}$$

with the specific resistivity underneath the cell ρ. Impedance contributions arising within the intercellular cleft are accounted for by an ohmic resistance R_b whereas the membranes are represented by the capacitance C_m. Thus, time-resolved impedance readings are not only suitable to follow changes in the cellular morphology, but a detailed analysis also provides information on where the changes precisely occur and what their relative contributions are [25]. As yet, time-lapse video microscopy is the only experimental alternative for gathering similar information [17]. However, it has already been proven that ECIS is significantly more sensitive than an optical microscope [25].

2.2
Biomedical Applications

The ECIS technique can be used to monitor various aspects of living cells in culture [23]. Dependent on how the experiment is designed and in which mode the electrical measurements are performed, conditions can be adjusted to be tailor-made for the problem to be solved [27]. In the following paragraphs a set of two very different examples demonstrates the wide applicability of ECIS.

2.2.1
Attachment and Spreading of Mammalian Cells to Adhesive Proteins

The interactions of mammalian cells with in vitro surfaces have attracted more and more interest throughout the last decade, originating from a broad range of biomedical applications that are based on cells immobilized on solid substrates in or ex vivo [28]. In vitro studies of processes like tumor metastasis, wound healing, or cell migration in general focus on the stationary and dynamic interactions of cultured cells with a particular substrate [29, 30]. Even in more technical fields, like the development and modification of biomaterials for anatomical implants [31] or the construction of cell–semiconductor interfaces [14], cell–substrate anchorage is the decisive parameter.

Driven by these scientific endeavors several new techniques have been developed that are capable of measuring the various aspects of cell–substrate interactions optically, electrically or even mechanically. Total internal reflection aqueous fluorescence (TIRAF) [32, 33] and reflection interference contrast microscopy (RICM) [34, 35] have been used to image the adhesion area on solid substrates and to monitor its dynamics. Fluorescence interference contrast (FLIC) microscopy has paved the way for the reliable determination and mapping of cell–substrate separation distances with an accuracy down to a few nanometers [36, 37]. Other techniques impose a mechanical force on the anchored cells and quantify the number of cells resisting this discriminating force and thus remaining attached to the substrate. The forces are applied by means of centrifugal acceleration [38] or exposure of the cells to laminar shear flow [38, 39]. In the most commonly used cell attachment assays, cells are first seeded onto a substrate of interest. After a certain time interval, weakly adhering cells are rinsed off with a stream of buffer, and the remaining cells are quantified either by counting or by indirect colorimetric assays. The latest approaches measure the forces necessary to detach single cells from a given substrate with a force microscope [40] or determine the forces that cells exert on a flexible substrate at oil/water interfaces [41].

ECIS is particularly well suited to monitoring the dynamics of cell attachment and spreading [26]. Attachment and spreading of cells on surfaces is an enormously complex and versatile process. After first molecular contacts between substrate-immobilized proteins and the corresponding cell surface receptors have been established, the cells start to actively spread out (Fig. 2). The cellular cytoskeleton plays a major part during the formation and also maintains the spread cell morphology.

The most sensitive parameter to probe the time course of cell attachment and spreading is the capacitance of the gold-film electrode above a threshold fre-

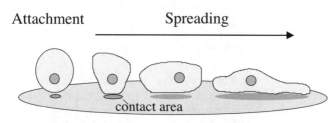

Fig. 2. The process of attachment and spreading of anchorage-dependent cells. In suspension, individual cells with a spherical morphology sediment down to the electrode surface and form initial cell–substrate contacts. With time, they actively increase their contact area with the surface and spread out. This process involves molecular interactions between cell surface receptors and adhesive proteins on the substrate as well as an active contribution of the cytoskeleton. (taken from reference [26])

quency of ~10 kHz. The measured capacitance scales with the surface area of the electrode that is available for capacitive current flow. When cells attach and spread, the cellular bodies block a certain fraction of the available surface area and reduce the electrode capacitance correspondingly. Thus, a linear relationship exists between the fractional surface coverage of the electrode and the decrease in capacitance. However, this very convenient linear relationship only applies as long as the frequency of the AC current is high enough such that its time constant is short compared to the capacitance of the electrode surface covered by the cell body. When these conditions are met, the current does not flow in the narrow clefts underneath the cells and the capacitance readout reflects the degree of cell spreading.

Figure 3a shows the time course of the electrode capacitance at a sampling frequency of 40 kHz when equal numbers of suspended MDCK (Madine–Darby canine kidney)-II cells attach and spread on four ECIS electrodes that had been coated with different protein layers prior to cell seeding. The four traces show that MDCK cells apparently attach and spread much faster to surfaces coated with fibronectin (FN) compared to all other proteins used in this experiment. Two parameters have been extracted to quantitatively compare the dynamics of cell spreading. The parameter s denotes the apparent spreading rate, which is deduced from the slope of the curves at $t=t_{0.5}$ (Fig. 3b). $t_{0.5}$ is the time necessary to achieve half-maximum cell spreading (Fig. 3c). According to both parameters, the kinetics for attachment and spreading to vitronectin (VN) and laminin (LAM) are rather similar. For bovine serum albumin (BSA)-coated electrodes it requires more than 5 h before the cells even start to spread out significantly. The half times $t_{0.5}$ of LAM, VN, and BSA clearly express this huge difference in spreading dynamics. Consistent with these findings BSA is considered to be a nonadhesive protein. The apparent spreading rates s for BSA, however, are surprisingly close to the values for LAM and VN, indicating at first sight that the cells eventually spread on this protein layer with similar kinetics.

The experiment shown in Fig. 4 explains the data in a different way. Here, equal numbers of MDCK cells were inoculated on three electrodes that were all precoated with BSA under identical conditions (first inoculation, Fig. 4a, solid lines). After 20 h the cells were gently removed from the surface and a fresh cell suspension was inoculated on the identical electrodes that had been used in the preceding experiment (second inoculation, Fig. 4a, dotted lines). Now the cells attach and spread much faster and the kinetic parameters of the second inoculation are similar to those that had been determined for a LAM coating (compare Fig. 3).

The explanation for these observations is that the cells that had been first inoculated on the BSA-coated electrodes have synthesized adhesive proteins and secreted them on the electrode surface. Since the biosynthesis of these proteins requires time, it took roughly 5 h before the cells started to spread on their self-made extracellular matrix. When the adherent cells were gently removed from the surface, their adhesive proteins were left behind and the cells that had been inoculated afterwards found a layer of adhesive proteins already on the surface. The spreading characteristics of the second inoculation indicate that the initially seeded cells may have secreted LAM onto the surface. Consistent with this

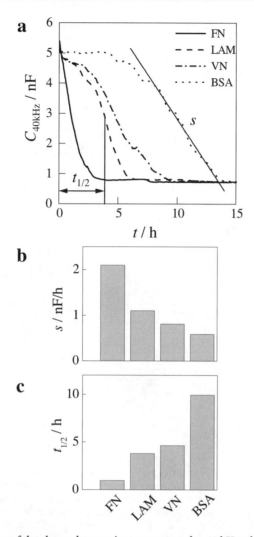

Fig. 3a. Time course of the electrode capacitance measured at 40 kHz when MDCK cells are seeded onto ECIS electrodes that had been coated with different proteins: fibronectin (*FN*), laminin (*LAM*), vitronectin (*VN*), and bovine serum albumin (*BSA*). Two quantities that are useful to describe the kinetics of cell spreading were extracted from each curve: $t_{0.5}$ denotes the time necessary to achieve a half-maximum capacitance drop as indicated in the figure and *s* is equivalent to the apparent spreading rate. **b** Apparent spreading rates *s* as determined from the data shown in **a**. **c** Half-times $t_{0.5}$ as determined from the data shown in **a** (taken from reference [26])

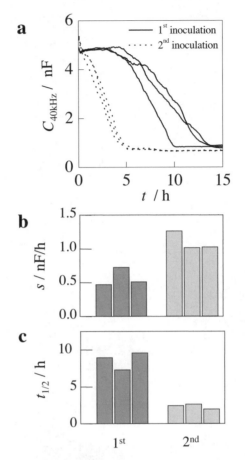

Fig. 4a. Time course of the electrode capacitance measured at a frequency of 40 kHz when MDCK cells are seeded on ECIS electrodes that had been precoated with BSA (*solid lines*). After 24 h the cells were gently removed from the surface and a fresh cell suspension was seeded on the same electrodes (*broken lines*). Serum-free medium was used throughout the entire experiment. **b** Apparent spreading rates s as determined from the data shown in **a**. **c** Half-times $t_{0.5}$ as determined from the data shown in **a** (taken from reference [26])

experiment it has been reported that MDCK cells synthesize and secrete LAM [42].

The outstanding sensitivity of the impedance technique to monitor the substrate anchorage of living cells was nicely demonstrated in experiments, in which adhesion of MDCK cells to FN-coated electrodes was studied in the presence of physiological concentrations of either Ca^{2+} or Mg^{2+}, or both. Binding affinities of the cell surface receptors that are responsible for FN binding, the integrins, are dependent on the presence of divalent cations. Some integrins show a selectivity for Ca^{2+} over Mg^{2+} or vice versa [43]. Figure 5 shows the time course of the electrode capacitance at 40 kHz when initially suspended MDCK cells at-

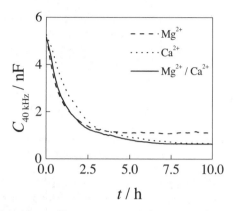

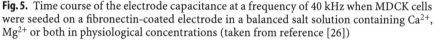

Fig. 5. Time course of the electrode capacitance at a frequency of 40 kHz when MDCK cells were seeded on a fibronectin-coated electrode in a balanced salt solution containing Ca^{2+}, Mg^{2+} or both in physiological concentrations (taken from reference [26])

tach and spread to FN-coated electrodes in the presence of either Ca^{2+}, Mg^{2+} or both.

Apparently, MDCK cells attach and spread slightly faster to a FN coating when Mg^{2+} is present in the culture fluid, either alone or in the presence of Ca^{2+}. When Ca^{2+} is the only available divalent cation, the spreading kinetics are somewhat retarded. It is also apparent that the electrode capacitance drops to smaller values when Ca^{2+} is present in the fluid either alone or in the presence of Ca^{2+}, indicating that the substrate contact is altered compared to the exclusive presence of Mg^{2+}. A more detailed analysis was provided by impedance readings along the full frequency range ($1-10^6$ Hz) immediately after the experiment had been terminated. Subsequent data modeling revealed that the parameter α amounts to $16\ \Omega^{0.5}$ cm in the presence of Ca^{2+} but only $4\ \Omega^{0.5}$ cm in its absence. According to the definition of α this can either be due to significant changes of the radius of the spread cell r_c, the average cell-substrate separation distance h or a significant change in the specific resistivity of the fluid underneath the cells ϱ. Data modeling also revealed significant differences with respect to cell–cell contacts dependent on the divalent cation present in the culture fluid. When Ca^{2+} is present in the medium, the resistance between adjacent cells amounts to ~30 Ω cm^2 compared to 0.5 Ω cm^2 when Ca^{2+} is omitted. This rather drastic difference is caused by the Ca^{2+}-dependency of barrier-forming tight junctions between adjacent MDCK cells that tightly occlude the intercellular space under physiological, normal Ca^{2+} conditions [44]. Tight junctions are well known for their inability to substitute Ca^{2+} by any other divalent cation, which readily explains our findings.

2.2.2
β-Adrenergic Stimulation of Cells Forming the Blood Vessels

In another example of strong physiological and medical relevance, we have addressed the barrier function of the vascular endothelium. All blood and lymphatic vessels are lined by endothelial cells that control the exchange of solutes between the circulating blood stream and the interstitial fluid of the tissue. This barrier function is based on very unique cell–cell contacts, the tight junctions (see above). Tight junctions occlude the intercellular shunt and thereby avoid free and uncontrolled diffusion of solutes along the perimeter of the cells. Dependent on the type of vessel – macrovascular or microvascular – the tightness of the vessel wall differs significantly. Whereas most macrovascular endothelial cells are rather leaky for ions but effectively exclude macromolecules and cells, endothelial cells derived from cerebral capillaries are considered electrically tight, which means that the flow of inorganic ions is impeded. β-adrenoceptor agonists like adrenalin are well known to interfere with the barrier function of endothelial cells in vivo [45, 46]. Many studies have addressed changes in macromolecular permeability of endothelial cells upon stimulation with β-adrenoceptor agonists. These studies consistently showed a significant barrier strengthening associated with a reduced permeability for probes like serum albumin or dextrans [45, 46]. These permeation assays are, however, confined to the application of labeled probes (radiolabels, chromophore labels) and time resolution is rather low. We thus applied the electrochemical impedance technique to study the response of aortic endothelial cells upon stimulation with the synthetic adrenaline analog isoproterenol (ISO) in extended detail, and to check the suitability of the technique to serve as a transducer in drug screening assays [47].

To perform the experiments, we first established confluent monolayers of bovine aortic endothelial cells (BAECs) on gold electrodes. We then recorded impedance data along the entire frequency range in order to select the most sensitive frequency for time-resolved measurements with maximum resolution. Figure 6a shows the impedance spectrum of a circular gold-film electrode (d=2 mm) with and without a confluent monolayer of BAECs. The contribution of the cell layer to the measured impedance values is most pronounced at frequencies close to 10 kHz, which was thus selected as the sampling frequency for further studies.

Figure 6b traces the time course of the impedance magnitude at a frequency of 10 kHz when confluent BAEC monolayers were either challenged with 10 μM ISO or a vehicle control solution at the time indicated by the arrow. The exchange of fluids produces a transient rise of the impedance by 10 to 20 Ω that is not caused by any cellular response but just the reduced fluid height within the measuring chamber. As expected no response of the cells was detected in the control experiment. The cell population exposed to 10 μM of ISO shows a significant increase in electrical impedance that goes through a maximum 10 min after ISO application and then slowly declines. It is important to note that the time resolution in these measurements is ~1 s so even much faster cell responses could be detected. From varying the ISO concentration, we extracted a dose-response relationship (Fig. 6c). Fitting a dose–response transfer function to the recorded data returns a concentration of half-maximum efficiency EC_{50} of 0.3±0.1 μM, which is in close agreement with the binding constant of ISO to β-adrenoceptors on the BAEC surface as determined from binding assays with radiolabeled analogs [45].

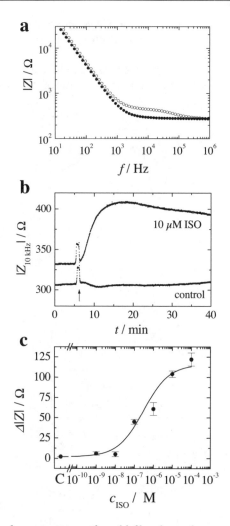

Fig. 6a. Typical impedance spectrum of a gold-film electrode covered with a confluent layer of bovine aortic endothelial cells (BAECs, *open circles*) compared to the same electrode without cells (*filled circles*). **b** Time course of the impedance magnitude $|Z|$ at a frequency of 10 kHz for a confluent monolayer of BAECs when the cells are exposed to 10 μM isoproterenol (*ISO*) or to control conditions. The *arrow* indicates the time of medium exchange. **c** Dose–response relationship for the impedance increase after isoproterenol addition. C given in the abscissa denotes the result of a control experiment. The *solid line* gives the result of fitting a dose–response transfer function to the data yielding a concentration of half-maximum efficiency (EC_{50}) of 0.3±0.1 μM (taken from reference [47])

In order to test the suitability of the electrochemical impedance assay to screen for potent inhibitors of β-adrenoceptors, known as β-blockers, we applied the well-known inhibitor alprenolol (ALP). ALP is a competitive inhibitor of β-adrenoceptors and therefore well suited to block the stimulating activity of ISO. Figure 7a shows the time course of the impedance magnitude at a fre-

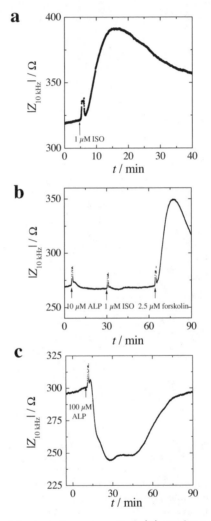

Fig. 7a. Time course of the impedance magnitude $|Z|$ at a frequency of 10 kHz when confluent BAECs were exposed to 1 μM isoproterenol (*ISO*). **b** Time course of the cell response when confluent BAECs were exposed to 1 μm isoproterenol (*ISO*) after 20 min of preincubation with 10 μM alprenolol (*ALP*), a $\beta_1\beta_2$-unselective β-blocker. The typical response of the cells to a 1 μM ISO stimulus as shown in a is abolished by ALP application. **c** Time course of the impedance magnitude $|Z|$ at a frequency of 10 kHz upon treatment of confluent BAECs with 100 μM alprenolol. The decrease in impedance indicates a reversible permeabilization of the cell layer (taken from reference [47])

quency of 10 kHz when BAEC monolayers have been stimulated with 1 μM ISO. Consistent with an EC_{50} of 0.3 μM the cells responded very similar to this 1 μM ISO challenge as they did to the challenge shown for 10 μM ISO in Fig. 6b. When the cell layers were incubated with 10 μM ALP prior to the addition of 1 μM ISO, the cells did not show any ISO response indicating that a tenfold increase of ALP was sufficient to block activation of the receptors (Fig. 7b). To make sure that the cells were not compromised in any way during these experiments we stimulated the same signal transduction cascade via a receptor-independent method at the end of each experiment. This can be easily done by application of forskolin, a membrane-permeable drug that intracellularly activates the same enzyme that is triggered by ISO binding to the receptor. Forskolin stimulation of those cells that had been blocked with ALP earlier in the experiment induces a strong increase of electrical impedance, indicating that the intracellular transduction pathways are functional (Fig. 7b). Accordingly, the inhibitory effect of ALP is due to receptor blockade.

The dosage of ALP and many of its fellow β-blockers has to be adjusted with great care since these lipophilic compounds are known to integrate nonspecifically into the plasma membrane. As is apparent from Fig. 7b, application of 10 μM ALP does not show any measurable side effects. Using ALP in concentrations of 100 μM induces a transient but very pronounced reduction of the electrical impedance (Fig. 7c). This decrease in impedance may be the result of the interaction of ALP with the plasma membrane or an induced contraction of the cell bodies. Either way, impedance measurements as performed here are highly suitable for monitoring unwanted side effects of drugs and revealing hidden cytotoxicity.

The analytical capabilities of the ECIS technique have been demonstrated by two selected examples. The interested reader is referred to references [48–55] to find out more. The enormous future potential of the devices comes from two directions: (1) from the technical side, it is possible not only to develop new measuring modes and to improve existing ones, but also to find additional ways of data analysis; and (2) from the biological side, there is an enormous variety of cells that may be selected dependent on the requirements of the analytical question. In drug development, it is possible to use those cells that are derived from the target tissue of a particular drug.

3
Piezoelectric Approaches to Monitor Living Cells

Acoustic sensors have been widely used in a number of configurations for sensor applications ranging from chemical to biological sensors [56–58]. The quartz crystal microbalance (QCM) technique is one of the most prominent representatives of acoustic sensors, which allow transduction between electrical and mechanical energies [59, 60]. The main component of the device is a thin piezoelectric quartz disk cut from α-quartz sandwiched between two evaporated metal electrodes and commonly referred to as a thickness shear mode (TSM) resonator. An oscillating potential difference between the surface electrodes leads to a shear displacement of the quartz disk. This mechanical oscillation responds very sensitively to any changes that occur at the crystal surface.

Sauerbrey was the first to show that the resonance frequency of such a quartz resonator decreases linearly when a rigid mass is deposited on the quartz surface in air or vacuum [61]. Hence, changes in the resonance frequency can be attributed to mass deposition on the quartz surface in the subnanogram regime, which gave the technique the name *microbalance* [61–64]. For instance, the mass sensitivity of a typical 5-MHz crystal is ~0.057 Hz cm^2 ng^{-1}. The QCM technique is a well-established technique in air and vacuum. However, for the majority of bioanalytical applications it is necessary to follow adsorption processes under physiological conditions, i.e., in a liquid environment. Only in the 1980s were high-gain oscillator circuits that can overcome the viscous damping of the shear oscillation under liquid loading developed, eventually paving the way for the QCM technique to be applied to biological problems [65–67]. These days, the QCM is used to follow a multitude of biomolecular recognition processes like, for instance, antigen-antibody binding or ligand-receptor interactions. From readings of the characteristic shear wave parameters such as the resonance frequency, it is then possible to extract binding constants and binding kinetics of proteins to surface-confined receptors with outstanding time resolution [56, 57, 68]. More recently, the QCM has also been applied to problems arising in cell biology, since the electromechanical analysis of the signal offers the possibility of extracting a number of relevant parameters in dynamical mechanics [57, 58, 69–71]. Thin films are extremely difficult to study with respect to their mechanical properties such as viscoelasticity using standard techniques of dynamical mechanical analysis [72–76]. The analysis of the resonance behavior of a quartz resonator offers a new way to access such parameters. The following section will deal with the physical basis of TSM resonators followed by examples from applications of the QCM as transducer for whole-cell biosensors.

3.1
Physical Background of the QCM Technique

3.1.1
Thickness Shear Mode Resonators

The converse piezoelectric effect describing a mechanical deformation of a material induced by an external electric field is the fundamental basis of the QCM methodology, as an oscillating potential difference induces the piezoelectric crystal to oscillate mechanically. Due to the electromechanical coupling within quartz crystals, alterations of the mechanical oscillation can then be conveniently inferred from electrical measurements. Although a large number of crystals exhibit piezoelectricity, only quartz and a few artificial materials such as $GaPO_4$ provide the unique combination of mechanical, electrical, chemical, and thermal properties which have led to its commercial significance [63, 77, 78].

For QCM devices, thickness shear mode resonators, which are also known as bulk acoustic wave (BAW) resonators, are applied. In most cases, so-called AT-cut crystals are used, which are prepared by slicing a quartz wafer with an angle of 35.25° to the optical z-axis of the quartz mother crystal. AT-cut quartz crystals show an outstanding frequency stability of $\Delta f/f \approx 10^{-8}$ and a temperature coefficient close to zero at temperatures between 0 and 50 °C [63, 78, 79]. Typical fundamental eigenfrequencies range from 4 to 30 MHz. We use 5-MHz quartz resonators with diameters of 14 mm, which are 330 μm thick. When an alternating potential difference is applied to the evaporated gold electrodes of the quartz plate, shear waves of opposite polarity are generated at the electrodes on either side of the crystal with the shear displacement in plane with the crystal surface. Both waves traverse the quartz thickness and are reflected at the opposing crystal face (phase shift π) and then returned to their origin (Fig. 8a). Constructive interference of incident and return waves occurs if the acoustic wavelength is an odd multiple of twice the crystal thickness ($2d_q$) and standing wave condi-

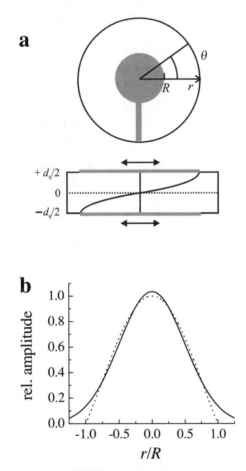

Fig. 8a. Circular quartz resonator of thickness d_q coated with evaporated electrodes of radius R shown in *gray*. **b** Radial distribution of the relative oscillation amplitude at the quartz surface according to a Bessel distribution (*dotted line*) and a Gaussian distribution (*solid line*) with $a=2$. The Bessel function drops to zero at the electrode edges ($r/R=1$), whereas the Gaussian function includes motion near the electrode edges

tions are then established. Particle displacement of the standing acoustic wave is maximum at the crystals' surfaces ($z=\pm d_q/2$) and a displacement node exists in the center of the resonator ($z=0$). When a foreign mass layer is deposited rigidly and uniformly on the crystal surface, the additional mass film increases the resonator's thickness and the eigenfrequency of the crystal is diminished. Thus, the deposited mass is accessible from measurements of the eigenfrequency of the crystal and the device operates like a microbalance.

Sauerbrey has provided the first mathematical treatment of mass deposition on the quartz sensor [61]. By solving the one-dimensional equation of motion he showed that a thin, rigid and uniformly deposited mass on the quartz plate results in a frequency decrease Δf that is proportional to the deposited mass Δm when operated in air or vacuum. If the density of the mass layer is equal to that of the quartz crystal, the following relation applies (Eq. 2):

$$\Delta f = -\frac{2nf_0^2}{A\sqrt{\bar{c}_{66}\rho_q}} = \Delta m = -S_f \Delta m \tag{2}$$

A denotes the piezoelectric active area, c_{66} the piezoelectric stiffened elastic modulus, ρ_q the density of the quartz, and f_0 the fundamental frequency. The integral mass sensitivity or Sauerbrey constant S_f depends on the square of the fundamental frequency but increases only proportionally to the overtone number n.

A more detailed theoretical treatment of the propagating acoustic wave reveals that the shear amplitude along the crystal surface is not uniform but radially symmetric, which in turn means that the quartz resonator is not uniformly sensitive to an adsorbed material [57, 80]. The amplitude is at a maximum in the center of the evaporated electrode ($r=0$) and decreases monotonically with increasing distance from the center, vanishing at the electrode edges ($r=R$). Martin and Hager were the first to show that the amplitude of oscillation is nonzero beyond the electrode edges ($r=R$) due to field fringing, which is enhanced in an environment of high dielectric permittivity such as water [80]. The radial distribution of the shear amplitude can empirically be described by a Gaussian function as shown in Fig. 8b.

3.1.2
Electromechanical Coupling in TSM Resonators

TSM sensors are capable of providing information about a variety of different material-specific properties [81]. The device can be used to extract mechanical properties of thin films like viscosities of liquids or shear modulus of viscoelastic materials [64, 72, 73, 75, 77, 82, 83]. Since the *mechanical* interactions between the resonator and the material in contact with its surface are measured *electrically*, it is necessary to understand the conversion between mechanical and electrical parameters in order to fully recognize the capabilities of the QCM.

Mechanical models can be transformed into electrical equivalent circuits, which permit a complete description of the oscillation in the presence of an adsorbent. The first general one-dimensional acoustic wideband model was suggested by Mason and provides a basis for the theoretical description of complex composite resonators as they occur in life science [59, 60]. Under certain assumptions, like low load and in close vicinity to the resonance, the wideband model can be transformed into an equivalent circuit with lumped elements – the so-termed Butterworth-van Dyke (BVD) circuit (Fig. 9a) [82].

The BVD circuit combines a parallel and series resonance circuit (motional branch). The motional branch consists of L_q, C_q, and R_q. Additionally, a parallel capacitance C_0 occurs due to the presence of the dielectric quartz material between the two surface electrodes. Hence, in an impedance spectrum obtained in the frequency range close to the resonance frequencies of the quartz plate, the quartz resonator shows two resonance frequencies corresponding to a phase shift $\phi=0$ at minimum and maximum magnitude of the impedance $|Z|$, respectively, if damping is negligible ($R_q\rightarrow0$). The resonant frequencies are referred to

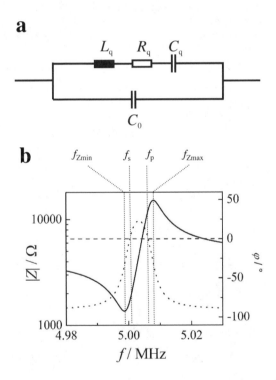

Fig. 9a. Butterworth-van Dyke (BVD) equivalent circuit. Near resonance the three-port Mason model can be transformed into an equivalent circuit composed of discrete impedance elements. The capacitance C_q represents the mechanical elasticity of the quartz, the inductance L_q the initial mass, and the resistance R_q the energy losses due to viscous effects, internal friction, and damping induced by the crystal holder. The static capacitance C_0 determines the admittance away from resonance, while the motional components dominate near resonance. **b** Simulated impedance spectrum from the BVD equivalent circuit with marked resonance frequencies f_{Zmin}, f_s, f_p, and f_{Zmax} for the case of $R_q>0$

as resonant frequency f_R and antiresonant frequency f_A. If $R_q>0$, four resonance frequencies can be assigned according to the general definitions of resonance frequencies. Two of them are defined at the minimum and maximum magnitude of the impedance $|Z|$ (f_{max}, f_{min}), differing from the two resonance frequencies at $\phi=0$ (f_s, f_p) (Fig. 9b) [78, 82, 84, 85].

Equivalent circuit modeling with the BVD circuit allows the extraction of characteristic properties of the quartz resonator itself as well as the contacting material. To fully describe the composite loading situations as common in life science and biosensor applications, the theoretical framework developed by Bandey et al. [82] is most suited. The corresponding equivalent circuits for different load situations are depicted in Fig. 10. For three different load situations which are particularly interesting in biological applications [81], the results of the theoretical treatment are given in Table 1.

Table 1. Expressions for the inductance and resistance for different load situations. K is the electroacoustic coupling factor, m the additional mass, ϱ_L the density, and η_L the viscosity of the liquid. G is the complex shear modulus, in which $Re(G)$ is the storage modulus G' and $Im(G)$ the loss modulus G''. ϱ_F is the density of the viscoelastic film.

Load	L	R
Rigid mass	$L_2 = \dfrac{\pi m}{4K^2 A\omega C_0\sqrt{\bar{c}_{66}\rho_q}} = S_f\dfrac{m}{A}$	0
Newtonian liquid	$L_3 = \dfrac{\pi}{4K^2\omega C_0\sqrt{\bar{c}_{66}\rho_q}}\sqrt{\dfrac{\omega\rho_L\eta_L}{2\omega}}$	$R_3 = \dfrac{\pi}{4K^2\omega C_0\sqrt{\bar{c}_{66}\rho_q}}\sqrt{\dfrac{\omega\rho_L\eta_L}{2}}$
Viscoelastic body	$L_L = \dfrac{\pi}{4K^2\omega^2 C_0\sqrt{\bar{c}_{66}\rho_q}}\sqrt{\dfrac{\rho_F\left(\left|G'\right|-G''\right)}{2}}$	$R_L = \dfrac{\pi}{4K^2\omega C_0\sqrt{\bar{c}_{66}\rho_q}}\sqrt{\dfrac{\rho_F\left(\left|G'\right|-G''\right)}{2}}$

a

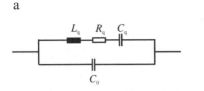

Unperturbed resonator

b

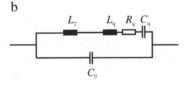

Rigid mass

c

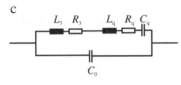

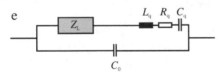

Newtonian liquid

d

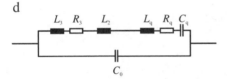

Combination of rigid mass
and Newtonian liquid

e

Viscoelastic body

f

Thin viscoelastic body
and Newtonian liquid

Fig. 10a–f. Extended equivalent circuits derived from the BVD circuit for different load situations. **a** Unperturbed quartz plate, **b** rigid mass, **c** Newtonian liquid, **d** rigid mass and Newtonian liquid, **e** thick viscoelastic layer, and **f** thin viscoelastic body and Newtonian liquid (adapted from reference [58])

3.1.3
Devices, Setups, and Measuring Principles

3.1.3.1
Active Mode

In the active mode, the quartz plate is the frequency-determining element. Once excited to resonance an appropriate feedback circuit keeps the oscillation stable. The active mode is a very sensitive and inexpensive way to follow changes at the crystal surface like mass deposition or changes in the viscosity of the adjacent solution [65, 66, 86]. However, frequency measurements alone are not sufficient to investigate the viscoelasticity of thin films as provided by biological material, such as cell monolayers grown on a quartz disk. To obtain a second parameter characteristic for viscoelastic properties of a film, damping of the resonator can be recorded by using an amplitude-controlled oscillator circuit that monitors the amplitude apart from the resonance frequency [87].

A typical experimental setup used to study the adhesion of cells by the active mode is shown in Fig. 11. An AT-cut quartz disk with a fundamental frequency

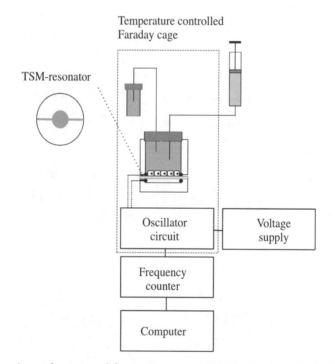

Fig. 11. Experimental setup used for QCM measurements in the active mode. The quartz resonator is mounted in a crystal holder, which itself is placed in a temperature-controlled Faraday cage. The quartz resonator is connected as the frequency-controlling element to an electronic oscillator circuit that compensates damping losses and thereby drives the shear oscillation. Exchange of solution is performed using a syringe (taken from reference [58])

of 5 MHz (d=14 mm) is used, which is a good compromise between practicability and sensitivity. Although quartz crystals with higher fundamental frequency are much more sensitive, as the mass-sensitivity increases with the square of the fundamental resonance frequency, the quartz plates become more susceptible to high acoustic load as provided by cells growing on the substrate. Inert gold electrodes are evaporated on both sides of the quartz crystal (thickness ~100 nm, d=6 mm), which enable the electromechanical transduction and serve as growth substrate for the cells. Glass cylinders (d=10 mm) are mounted on the quartz disk with a noncytotoxic silicon glue to form a culture-dish-like chamber. An oscillator circuit capable of exciting the quartz crystal to its resonance frequency in a liquid environment has been developed in our laboratory and is based on a TTL IC SN74LS124 N from Texas Instruments. The crystal holder is designed to minimize parasitic damping due to mounting losses and to prevent the occurrence of longitudinal waves [86]. Crystal holder and oscillator circuit are placed in a temperature-controlled Faraday cage, while the resonance frequency is recorded with a frequency counter outside the cage. A personal computer is used to control the measurement and record the change in frequency. A time resolution of practically 0.1–1 s can be readily achieved with a frequency resolution of 0.1–0.5 Hz for a 5-MHz resonator.

3.1.3.2
Passive Mode

Although the active mode is an elegant and inexpensive way to follow, for example, the time course of cell attachment at the quartz surface, a detailed analysis of the acoustic physics of a composite layer, such as cell monolayers requires more sophisticated methods. AC impedance analysis in the vicinity of the resonance frequency provides a more comprehensive analysis of the shear oscillation as detailed in the preceding section. Changes in energy dissipation can be readily distinguished from changes in mechanically stored energy.

In the passive mode an impedance or network analyzer is used to excite the crystal to a forced vibration near resonance while monitoring the complex system response as a function of the applied frequency. In AC impedance spectroscopy, an impedance analyzer applies a sinusoidal voltage over a small band of frequencies to a system and measures the corresponding current through the system in a phase-sensitive manner. The ratio of voltage and current amplitudes – the magnitude of impedance $|Z|$ – and the phase angle ϕ between both quantities are then computed for the various frequencies and subsequently analyzed. The transfer function of the BVD equivalent circuit is, as a first approximation, fitted to the measured impedance spectra by means of least-square methods. This procedure provides a set of parameters (L, C, R, C_0). By using the aforementioned formalism, it is possible to extract mechanical parameters of the system of interest from the parameters of the BVD network. A prerequisite for the detailed modeling of the QCM response is to recognize the parts of the cellular body and the interface between the ventral membrane and the extracellular matrix that contribute to the change in the impedance spectrum.

Figure 12a schematically shows a general narrowband model for epithelial cells on a quartz resonator. Three major contributions arising from the extracellular matrix (ECM), the cell body, and a water layer determine the system response.

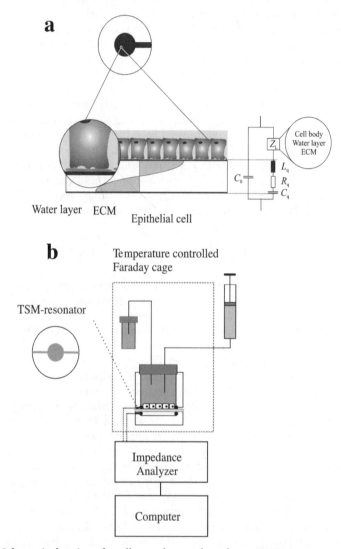

Fig. 12a. Schematic drawing of a cell monolayer cultured on a TSM resonator and the corresponding equivalent circuit including the extracellular matrix (*ECM*) and a water layer. Z_L represents the combined viscoelastic properties of adherent cells and ECM and an adjacent water layer between cell and surface. **b** Experimental setup used for QCM measurements in the passive mode. The quartz resonator is connected to an impedance analyzer SI 1260 (Solartron Instruments, Farnborough, UK) controlled by a PC (taken from references [57, 58])

Alternatively, a network analyzer might be employed in the reflectance mode to measure the complex impedance in which the reflectance r is a complex value and Z_0 the input impedance (Eq. 3):

$$Z = Z_0 \frac{1+r}{1-r} \tag{3}$$

The setup used in our group is depicted in Fig. 12b. It comprises a crystal holder that mechanically fixes the quartz dish and contacts the gold electrodes on either side of the quartz plate. The impedance analyzer is connected via shielded cables. Impedance analysis is typically carried out in a frequency range of ±30 kHz around the resonance frequencies of the quartz plate applying a sinusoidal excitation voltage of 150 mV amplitude (rms).

3.2
Cell Adhesion Monitoring

The idea to investigate and quantify the attachment and spreading of mammalian cells on the quartz resonator becomes clear considering the fact that the QCM is a highly sensitive tool for studying the interface between the solid surface of the crystal and the adjacent nonpiezoelectric material. Attachment and spreading of cells on surfaces is an enormously complex and versatile process and results in a growing contact area. With time the state of adhesion to the substrate alters due to changes in the number and type of binding proteins, the strength of adhesion, and alteration of the cytoskeleton.

3.2.1
Cell Adhesion Monitored in the Active Mode

To explore the suitability of the QCM technique to monitor the attachment and spreading of mammalian cells in real time, it is instructive to analyze the time course of the resonance frequency when initially suspended cells are seeded in a QCM chamber and subsequently attached to the resonator surface.

Despite the reasonable potential for applications in cell biology there are only very few reports in the literature dealing with adhesion monitoring of mammalian cells using a QCM [88–90]. Redepenning and coworkers showed that the attachment and spreading of osteoblasts to such an oscillating quartz crystal results in a frequency shift that is nearly linearly dependent on the fractional surface coverage as determined by scanning electron microscopic inspection of gradually covered resonators [90]. Gryte et al. [89] detected the attachment of African green monkey kidney (Vero) cells using this technique. Matsuda et al. [88] followed the adhesion of platelets to a QCM surface. More recently, Marxs et al. [91] presented a study in which they used the QCM to detect microtubule alterations induced by nocodazole in endothelial cells as the transduction element. Cans et al. [71] measured the dynamics of exocytosis and vesicle retrieval at cell populations using the QCM-D technique. The cell lines NG 108-15 and PC 12 were grown on the TSM resonator to confluence and examined separately with respect to exocytosis and vesicle retrieval. Loss of mass was detected when the cells were stimulated by increased potassium concentrations, indicating exocytosis. In the absence of Ca^{2+} the effect was completely suppressed. The measurements show how complex dynamic processes may be followed using the QCM technique [71].

We studied the adhesion characteristics of three different cell lines – strains I and II of the epithelial cell line MDCK and Swiss-3T3 fibroblasts – as a function of seeding density [69]. Figure 13a shows a set of curves that were recorded when increasing numbers of MDCK-II cells were supplied to the quartz dish. The data reveal that the frequency response of the QCM device is governed by the number of cells that spread on the resonator surface. When no cells are seeded into the quartz dish (curve a), the frequency shift is negligible indicating that the adsorption of medium ingredients does not significantly contribute to the QCM response. Tracing the maximum frequency shift versus the applied seeding density, as shown in Fig. 13b for MDCK-II cells, leads to a typical saturation behavior. Remarkably, saturation of the resonance frequency was observed for cell numbers that exceeded the number of cells necessary to form a complete monolayer. This observation indicates that the QCM is only capable of monitoring cells that are in close contact with the surface, and does not detect cells

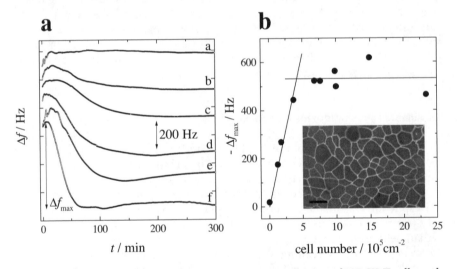

Fig. 13a. Time courses of the resonance frequency upon adhesion of MDCK-II cells seeded in increasing densities on quartz resonators: *a* without cells, *b* $1.3 \cdot 10^5$ cells/cm^2, *c* $1.8 \cdot 10^5$ cells/cm^2, *d* $3.7 \cdot 10^5$ cells/cm^2, *e* $7.7 \cdot 10^5$ cells/cm^2, *f* $1.5 \cdot 10^6$ cells/cm^2. **b** Maximum frequency shifts as deduced from the time courses in **a** as a function of the applied seeding density. The *solid lines* are the results of fitting a straight line in the range of $0-5 \cdot 10^5$ cells/cm^2 and a horizontal line at cell numbers higher than $5 \cdot 10^5$ cells/cm^2. The slope of the straight line is $(1.2 \pm 0.1) \ 10^{-3}$ Hz cm^2/cells. The *horizontal line* depicts the maximum frequency shift associated with the establishment of a confluent MDCK-II cell monolayer and amounts to 530 ± 25 Hz. The *insert* shows a fluorescence micrograph of a confluent, ZO-1 stained MDCK-II cell monolayer. Cell boundaries are clearly visualized and the geometric surface area of an individual cell can be determined. The *scale bar* represents 20 μm (taken from reference [69])

loosely attached atop an already established first cell layer. Similar results were obtained in an earlier work by Redepenning et al. [90]. They concluded that the shear wave does not propagate beyond an established cell layer; hence, the device cannot detect cells on top of the first layer.

Figure 13b shows a simple two-case approach to analyze the data, each of which is represented by a straight line. The straight line with the positive slope approximates the region of the curve in which the resonator is gradually covered by the seeded cells. The horizontal line represents conditions under which the resonator surface is completely covered by a cell monolayer and the surplus of cells not contributing to the monolayer, and not affecting the resonance frequency. The intersection between the ascending and the horizontal lines provides the minimum cell number necessary to establish a confluent monolayer. The average surface area of an individual MDCK-II cell within a confluent monolayer was obtained from fluorescence micrographs of MDCK-II cells (insert) stained for the tight junction associated protein ZO-1. A surface area of $180\pm10\ \mu m^2$ corresponds to a cell number of $(5.5\pm0.3)\cdot10^5$ cells/cm^2, which is in good agreement with the value of $(4.3\pm0.5)\cdot10^5$ cells/cm^2 extracted from QCM data. The analysis of other cell lines used in this study is listed in Table 2. Interestingly, the maximum frequency shifts obtained for confluent cell layers strongly depend on the seeded cell type and are typically in the range of 200–600 Hz. Applying Sauerbrey's equation, implying that cells behave as an additional rigid mass, a cell monolayer on a 5-MHz quartz resonator is supposed to induce frequency shifts on the order of several thousand Hertz, which is at least a factor of 10 larger than the maximum frequency shifts experimentally observed for the three cell lines. In agreement with our results Redepenning et al. [90] and Gryte et al. [89] also found that experimentally observed frequency shifts are far smaller than expected from the Sauerbrey equation. Thus, the simple model of an additional mass attached to the resonator does not apply for attached cells. In the forthcoming sections we will describe efforts to find a more appropriate model to describe this particular load situation. The encouraging

Table 2. QCM-based adhesion assay of three mammalian cell lines, MDCK-I, MDCK-II, and 3T3 fibroblasts in the active mode. Δf is the frequency shift associated with the formation of a confluent monolayer of the respective cell type. Values were obtained from the horizontal regression line of the adhesion isotherm at high cell numbers (Fig. 13Bb). $\Delta f_{max}/\Delta N$ represents the slope of the ascending regression lines for low cell numbers. The intersection of the two lines indicates the minimum cell number necessary to form a complete monolayer. The geometric surface area of an individual cell (A_{cell}) and the cell number per unit area (N_A) are deduced from microscopic images. $T=37$ °C

Cell species	Δf Hz	$\Delta f_{max}/\Delta N$ Hz/cell cm^{-2}	Intersection cell cm^{-2}	A_{cell} μm^2	N_A cell cm^{-2}
MDCK-II	530±25	$(1.2\pm0.1)\cdot10^{-3}$	$(4.3\pm0.5)\cdot10^5$	180±10	$(5.5\pm0.3)\cdot10^5$
MDCK-I	320±20	$(1.0\pm0.1)\cdot10^{-3}$	$(3.1\pm0.4)\cdot10^5$	270±15	$(3.7\pm0.2)\cdot10^5$
3T3	240±15	$(1.3\pm0.3)\cdot10^{-3}$	$(1.9\pm0.5)\cdot10^5$	745±30	$(1.3\pm0.1)\cdot10^5$

finding that different frequency shifts are obtained from the attachment of different cell lines suggests that it might be possible to infer individual properties of individual cell species from QCM measurements.

In order to gather further information about the signal sources that lead to individual readings for different cell lines, we addressed the question of how far the attachment of the cells to the resonator surface governs the frequency response. The closest contacts between mammalian cells and artificial surfaces are mediated by so-called focal contacts in which membrane-bound integrins and extracellular matrix proteins such as fibronectin, displaying high-affinity peptide sequences to the receptor family of integrins, are playing the major role. It is well known that fibronectin is recognized by integrins due to a sequence of four amino acids, namely RGDS (Arg-Gly-Asp-Ser) located within the FN-III$_{10}$ subunit. The sequence was also shown to be present in other extracellular matrix proteins like laminin, vitronectin, collagen, and thrombospondin. Integrins are transmembrane proteins connecting the intracellular cytoskeleton to the ECM, which provides mechanical stability to cell–substrate adhesion.

The impact of focal contact formation on the QCM response was studied by means of adhesion experiments similar to the ones shown in Fig. 13 but with varying concentrations of antiadhesive peptides (Fig. 14). Adding water-soluble peptides with a sequence including RGDS to the medium abolishes the formation of strong noncovalent bonds between the integrins and the ECM components. When integrin-mediated adhesion of MDCK-II cells was completely inhibited due to the presence of 1 mM GRGDS in the medium, the resonance frequency of the crystal remained fairly constant. Since the frequency does not alter in the presence of this antiadhesive peptide, we conclude that sedimenta-

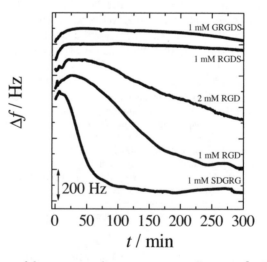

Fig. 14. Time courses of the resonance frequency upon seeding ~8·10^5 cells/cm^2 MDCK-II cells in the quartz dishes in the presence of antiadhesive peptides varying in their amino acid sequence. The peptides were added in different concentrations to the cell suspension before seeding them (taken from reference [69])

tion of cellular bodies followed by loose contact with the resonator surface due to nonspecific interactions does not influence the response of the quartz crystal. In order to ensure the specificity of the GRGDS effect, we performed a similar experiment but added SDGRG, a pentapeptide with reverse order of amino acids to the medium. Addition of 1 mM SDGRG leads to a typical adhesion curve indistinguishable from those recorded without any peptide added. Adhesion of MDCK-II cells is apparently not affected by the SDGRG peptide. Thus, competitive inhibition of the integrin binding site for many ECM proteins by the soluble GRGDS pentapeptide interferes strongly with the ability of MDCK-II cells to firmly attach to the surface and to induce a considerable shift in resonance frequency.

We extended the study to check for the number of relevant amino acids that would abolish formation of focal contacts detectable by QCM. The tetrapeptide RGDS – missing the first arginine residue – shows a similar inhibitory effect on cell adhesion as GRGDS when applied in the same concentration (Fig. 14). However, the peptide RGD, which is missing the first arginine and the last serine residue, is significantly less effective than GRGDS; adhesion of MDCK-II cells is only partly inhibited in the presence of 1 mM RGD, though inhibition can be gradually enforced by adding higher RGD concentrations [69].

In conclusion, the quartz crystal microbalance is indeed a powerful technique to monitor quantitatively the attachment and spreading of mammalian cells to in vitro surfaces in real time without the use of microscopic techniques. However, the origin of the frequency response cannot be attributed to pure mass loading. As will be discussed below, the rather complex material composition of an attached cell layer, with extracellular matrix proteins and water enclosed beneath the cells, the cellular plasma membrane, the cytoskeleton, and the cytosol, has to be taken into account.

3.2.2
Cell Adhesion Monitored in the Passive Mode

When starting an impedance analysis of cell monolayers grown on a quartz disk it is instructive to have a first glance at the individual impacts of different mammalian cell types on the Butterworth–van Dyke (BVD) parameters of the shear oscillation [92]. Therefore cells were grown to confluence on a resonator surface and subjected to impedance analysis. The cell layer was then mechanically removed from the surface and the cell-free resonator was measured again. Typical impedance spectra of a quartz resonator before and after removing MDCK-II cells are depicted in Fig. 15.

It is apparent that the presence of firmly attached cell bodies damps the quartz oscillation considerably, as concluded from the diminishing impedance maximum (Fig. 15a) and the decreasing phase maximum (Fig. 15b). The attached cell layer increases the energy dissipation of the shear wave, indicating that a cell monolayer has to be treated like a lossy overlayer rather than a rigid mass. This observation was later independently confirmed by Fredriksson et al. [93, 94], who have simultaneously monitored the resonance frequency and the dissipation factor D of quartz resonators using the QCM-D technique. They ob-

a   **b**

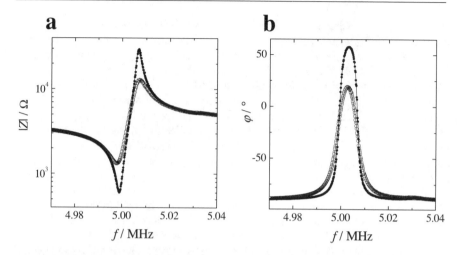

Fig. 15a,b. Influence of a confluent MDCK-II cell monolayer on the impedance spectrum of a 5-MHz quartz resonator ($|Z|$ (**a**) and ϕ (**b**)) before (*open circles*) and after (*filled circles*) scrapping the cell layer from the surface. The *solid lines* are the results of curve fitting using the BVD equivalent circuit (taken from reference [92])

served a decrease in resonance frequency and an increase in the dissipation factor. The latter finding indicates that damping of the shear vibration occurs upon cell adhesion. By fitting the parameters of the BVD-equivalent circuit to the recorded impedance spectra, we obtained characteristic values of the motional resistance R and inductance L for the different cell species. As it is impossible to directly separate the load-specific contributions to R and L from those of the unperturbed resonator, we either describe our experimental data in terms of the total resistance R and the total inductance L or in terms of changes of these two parameters (ΔR, ΔL) with respect to the values of the same resonator without cells but covered with culture medium. The obtained results are summarized in Table 3. The largest shift in resonance frequency f was determined for MDCK-II cells and this corresponds to the most pronounced changes in R and L as deduced from impedance analysis. MDCK-I cells generally display smaller shifts for f, R, and L, while 3T3 fibroblasts have the smallest impact on all three parameters. It is well known that epithelial cells tend to attach very firmly to artificial surfaces with most of their lower (basal) plasma membrane in close contact with the substrate. Fibroblastic cells like the Swiss 3T3 cells are known to touch the substrate only in small areas, while the rest of the basal plasma membrane is further away. These facts, together with the QCM results, support our hypothesis that the number of substrate contacts and the resulting contact area are decisive for the QCM response.

Adhesion-promoting proteins immobilized on the solid substrate are required to allow mammalian cells to anchor to the surface by means of specific molecular bonds. We therefore investigated the influence of this protein film on the QCM response. ECM proteins on the growth surface can originate from spontaneous adsorption of serum proteins or secretion of endogenous proteins

Table 3. Summary of experimental data recorded by passive mode QCM for five different mammalian cell lines: MDCK-I, MDCK-II, epithelial cells from porcine choroid plexus, bovine aortic endothelial cells (BAEC), and 3T3 fibroblasts. ΔR and ΔL are the differences in motional resistances and inductances associated with the presence of a confluent cell layer of the specified species with respect to the same resonator covered merely with cell culture medium. $T=23\,°C$

Cell species	$\Delta R/(\Omega)$	$\Delta L/(\mu H)$
MDCK-II	990±40	10.5±0.8
MDCK-I	755±35	6.7±0.7
Choroid plexus	800±45	16±1.5
BAEC	55±12	3.1±0.5
3T3	280±20	2.5±0.5

Table 4. Changes in motional resistance ΔR_1 and inductance ΔL_1 due to the presence of complete cell monolayers with respect to the same resonator after cleaning. ΔL_2 and ΔR_2 denote the differences in the motional branch that are associated with the presence of the extracellular matrix proteins immobilized on the substrate underneath the attached cell monolayers with respect to the same resonator after cleaning. These data were obtained by removing established cell layers from their growth substrate using a protocol that leaves the extracellular matrix proteins behind. In either case the resonators were characterized by impedance spectroscopy. The last row shows the influence of an adsorbed collagen film on the impedance elements. $T=23\,°C$

Cell species	$\Delta L_1/\,(\mu H)$	$\Delta L_2/\,(\mu H)$	$\Delta R_1/\,(\Omega)$	$\Delta R_2/\,(\Omega)$
MDCK-I	11.6±0.5	9.7±0.2	685±40	20±10
MDCK-II	4.8±0.3	2.6±0.4	485±40	35±10
3T3	2.0±0.1	1.9±0.3	260±20	10±5
Collagen	–	7.4±0.6	–	43±15

by the cells themselves. Therefore the amount and composition of ECM proteins on the surface may vary from cell type to cell type. Coverage of a quartz resonator with a collagen layer alters the motional inductance L by 7.4±0.6 μH and the motional resistance R by 43±15 Ω (Table 4). Compared with the changes in R recorded for confluent cell layers, the influence of a collagen film on this parameter is rather small. However, the inductance change is considerable and in the same range as that observed for the entire cell layer including the ECM. Removing adherent cells from their culture substrate but leaving the ECM protein layer behind provides the necessary evidence that the ECM actually exists underneath a confluent cell layer. QCM parameters obtained from this procedure are summarized in Table 4. The data indicate that the contribution of the ECM to the total change in L observed for intact cell layers is indeed rather large while R is barely influenced. Consistently, the observed inductances for the actually existing ECM underneath the cells are in the same range as those measured for a collagen-coated quartz crystal. In conclusion, interpretation of QCM data in terms of mechanical properties of adherent cells might end up in a misleading hypothesis since changes in L may be at least partly attributed to the protein layer beneath the cell.

3.3
Manipulating the Actin Cytoskeleton

A two-dimensional actin network underlies the plasma membrane of mammalian cells and regulates its mechanical properties (cell cortex). Filamentous actin also forms thick bundles, so-called stress fibers, which interconnect different sites of adhesion to the substrate [95]. These fibers provide cell–substrate contacts with a remarkable mechanical stability. In order to investigate how far disruption of the actin cytoskeleton alters the QCM parameters R and L of adherent cells, confluent MDCK-II cell layers were treated with cytochalasin D that is known to inhibit the polymerization of actin monomers. Since the depolymerization of filaments is not affected, treatment with cytochalasin D results in a net contraction of the filaments and eventually in their disappearance. Figure 16a shows that immediately after addition of 5 μM cytochalasin D to the equilibrated cell layer, the motional resistance R drops from its baseline value by 150 Ω within 75 min but starts to recover subsequently.

The inductance, however, decreases monotonically after adding cytochalasin D with only small changes toward the end of the experiment. To correlate the time courses of R and L with the actual disruption of this part of the cytoskeleton, the filamentous actin was visualized by fluorescence microscopy (Fig. 16b) during cytochalasin D exposure using fluorescence-labeled *phalloidin*. While at time zero actin bundles are clearly visible, lying almost parallel to the substrate plane and along the cell perimeter, 25 min after cytochalasin exposure most of the stress fibers are already degraded to point-like structures. After 75 min of cytochalasin exposure the filamentous actin completely disappears. This coincides with the minimum of the motional resistance R and underlines its correlation. There are two major conclusions from these experiments:
1. The QCM device is suitable and sensitive to monitor changes in the actin cytoskeleton of adherent cells. This observation renders this technique an interesting tool for many physiological processes that involve cytoskeletal changes.
2. The mechanical properties of the cell cortex apparently contribute to the overall response of the QCM to adherent cells and have to be included in a detailed model of cells attached to acoustic devices.

3.4
Double-Mode Impedance Analysis of Cell-Substrate and Cell-Cell Contacts

The quartz crystal microbalance is an extraordinary tool to monitor noninvasively the establishment and modulation of cell–substrate interactions in situ. However, the QCM does not provide information about cell–cell interactions

A

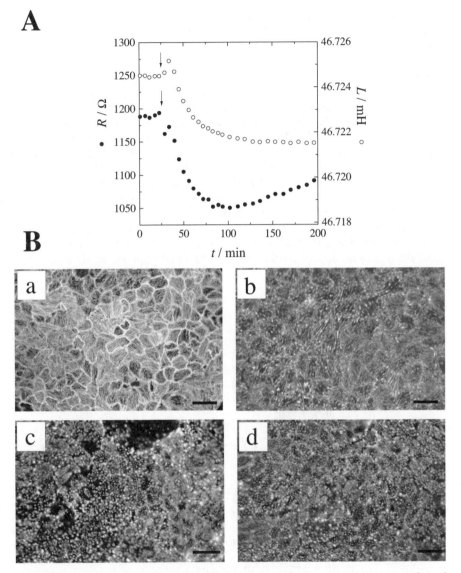

B

Fig. 16A. Time course of resistance R (*filled circles*) and inductance L (*empty circles*) of a confluent monolayer of MDCK-II cells after adding 5 μM cytochalasin D to the medium. The injection of cytochalasin is marked by an *arrow*. **B** Fluorescence micrographs of MDCK-II cell layers stained with fluorescently labeled phalloidin, which binds to the actin cytoskeleton. Cells were exposed to cytochalasin D for *a* 0 min, *b* 25 min, *c* 75 min, and *d* 100 min. A considerable alteration of the actin network is discernible. The *scale bars* correspond to 25 μm (taken from reference [92])

affecting morphology and differentiation of cells. Therefore it is straightforward to combine both the ECIS and the QCM in the passive mode (double-mode (DM) impedance analysis) in one experiment using the experimental set-up depicted in Fig. 17.

As pointed out in the previous section on ECIS, the frequency-dependent electrical impedance for such a system is dependent on the impedance of the cell-free electrode, the resistance between adjacent cells R_b, the capacitance of the cell membranes C_m, and an additional impedance contribution that is associated with the current flow between the basal plasma membrane and the substrate. The resistance R_b is dependent on the width of the intercellular cleft between adjacent cells and the establishment of tight junctions, and is therefore a sensitive parameter for cell–cell contacts. Thus, impedance analysis of the shear displacement – referred to as the quartz mode (Fig. 18a) – and impedance analysis of the cell-covered surface electrode – the cell mode (Fig. 18b) – provide independent quantitative information about cell–substrate- and cell–cell-contacts. The core component of the setup is a relay used to connect either the two gold electrodes on opposite sides of the quartz plate (quartz mode in the pas-

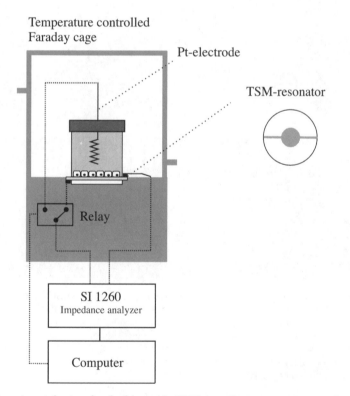

Fig. 17. Experimental setup for double-mode (DM) impedance spectroscopy of confluent cell monolayers. The core component is the relay that allows switching between the cell and the quartz modes (taken from reference [58])

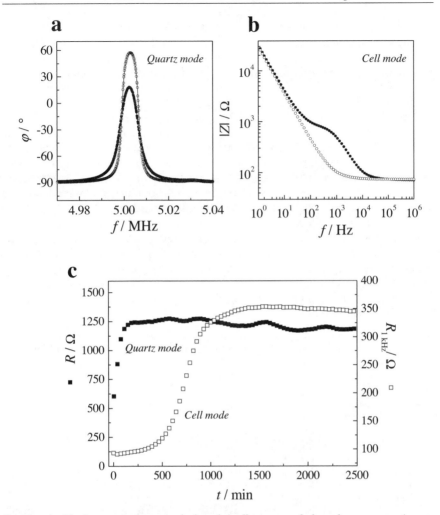

Fig. 18. **a** and **b** show typical spectra before the cells were seeded on the resonator (*empty circles*) and after the confluent cell layer had been established (*filled circles*) (taken from reference [58]). **c** Time course of the motional resistance R (*filled squares*) as deduced from quartz mode measurements and the resistance R at 1 kHz (*empty squares*) as deduced from cell mode measurements during adhesion and differentiation of MDCK-II cells

sive mode) or the cell-covered gold electrode together with the dipping electrode (cell mode) to the impedance analyzer. Quartz mode data were recorded as described above, while cell mode data were taken in the low frequency range between 1 and 10^6 Hz. For the time course experiments, we show the electrical resistance at a frequency of 1 kHz as it was returned by the impedance analyzer, since the equivalent circuit modeling is only valid for complete cell monolayers. The 1 kHz resistance directly mirrors the establishment of cell–cell contacts between epithelial MDCK-II cells that were examined here.

The power and versatility of the DM impedance technique is best displayed when the task is to follow the establishment and differentiation of epithelial cell layers. Figure 18c shows DM impedance results of MDCK-II cells seeded on the resonator surface at time zero. According to the quartz mode data, cell spreading is completed within 200 to 250 min after seeding, as inferred from changes in the motional resistance R. Strikingly, formation of tight junctions occurs 500 min *after* seeding according to cell mode measurements. The results provide clear evidence that the cells need a certain time after they have anchored on the substrate to synthesize and sort the various cell adhesion molecules and eventually the molecular components of intercellular tight junctions. The DM impedance technique is so far the most versatile electrochemical method to scrutinize complex biochemical processes occurring in cell monolayers in great detail and with good time resolution that may help to understand them.

In conclusion, ECIS and QCM turned out to be well suited as transducers capable of monitoring changes in the three-dimensional shape of living cells when these are exposed to chemical, biological or physical stimuli. A detailed understanding of the available parameters of these transducers in combination with attached cell bodies as the biologically active component will pave the way to whole-cell biosensors providing invaluable information about the functional activity of various analytes.

Acknowledgements: JW is supported by a grant provided by the Deutsche Forschungsgemeinschaft (WE-2711-1/1).

References

1. Bousse L (1996) Sens Actuators B 34:270
2. Ziegler C (2000) Fresenius J Anal Chem 366:552
3. Ehret R, Baumann W, Brischwein M, Lehmann M, Henning T, Freund I, Drechsler S, Friedrich U, Hubert ML, Motrescu E, Kob A, Palzer H, Grothe H, Wolf B (2001) Fresenius J Anal Chem 369:30
4. Stenger DA, Gross GW, Keefer EW, Shaffer KM, Andreadis JD, Ma W, Pancrazio JJ (2001) Trends Biotechnol 19:304
5. Owicki JC, Parce JW, Kercso KM, Sigal GB, Muir VC, Venter JC, Fraser CM, McConnell HM (1990) Proc Natl Acad Sci USA 87:4007
6. McConnell HM, Owicki JC, Parce JW, Miller DL, Baxter GT, Wada HG, Pitchford S (1992) Science 257:1906
7. Owicki JC, Bousse LJ, Hafeman DG, Kirk GL, Olson JD, Wada HG, Parce JW (1994) Annu Rev Biophys Biomol Struct 23:87
8. Parce JW, Owicki JC, Kercso KM, Sigal GB, Wada HG, Muir VC, Bousse LJ, Ross KL, Sikic BI, McConnell HM (1989) Science 246:243
9. Hafner F (2000) Biosens Bioelectron 15:149
10. Fanigliulo A, Accossato P, Adami M, Lanzi M, Martinoia S, Paddeu S, Parodi MT, Rossi M, Sartore M, Grattarola M, Nicolini C (1996) Sens Actuators B 32:41
11. Lehmann M, Baumann W, Brischwein M, Gahle H, Freund I, Ehret R, Drechsler S, Palzer H, Kleintges M, Sieben U, Wolf B (2001) Biosens Bioelectron 16:195
12. Gross GW, Rhoades BK, Azzazy HM, Wu MC (1995) Biosens Bioelectron 10:553
13. Fromherz P, Offenhausser A, Vetter T, Weis J (1991) Science 252:1290
14. Fromherz P, Stett A (1995) Phys Rev Lett 75:1670
15. Fromherz P (1999) Eur Biophys J 28:254

16. Woolley DE, Tetlow LC, Adlam DJ, Gearey D, Eden RD (2002) Biotechnol Bioeng 77:725
17. Dieterich P, Odenthal-Schnittler M, Mrowietz C, Kramer M, Sasse L, Oberleithner H, Schnittler HJ (2000) Biophys J 79:1285
18. MacDonald JR (ed) (1987) Impedance spectroscopy: emphasizing solid materials and systems. Wiley, New York
19. Fricke H (1953) Nature 172:731
20. Schwan HP (1957) In: Lawrence JH, Tobias CA (eds) Advances in biological and medical physics, vol V. Academic, New York, p 147
21. Grimnes S, Martinsen OG (2000) Bioimpedance and bioelectricity basics. Academic, New York
22. Giaever I, Keese CR (1984) Proc Natl Acad Sci USA 81:3761
23. Giaever I, Keese CR (1993) Nature 366:591
24. Ehret R, Baumann W, Brischwein M, Schwinde A, Wolf B (1998) Med Biol Eng Comput 36:365
25. Giaever I, Keese CR (1991) Proc Natl Acad Sci USA 88:7896
26. Wegener J, Keese CR, Giaever I (2000) Exp Cell Res 259:158
27. Keese CR, Giaever I (1994) IEEE Eng Med Biol 402
28. Kasemo B (2002) Surf Sci 500:656
29. Horwitz AR, Parsons JT (1999) Science 286:1102
30. Lauffenburger DA, Horwitz AF (1996) Cell 84:359
31. Klee D, Höcker H (1999) Adv Polym Sci 149:1
32. Todd I, Mellor JS, Gingell D (1988) J Cell Sci 89:107
33. Gingell D, Heavens O (1996) J Microsc 182:141
34. Marchi-Artzner V, Lorz B, Hellerer U, Kantlehner M, Kessler H, Sackmann E (2001) Chemistry 7:1095
35. Parak WJ, Domke J, George M, Kardinal A, Radmacher M, Gaub HE, de Roos AD, Theuvenet AP, Wiegand G, Sackmann E, Behrends JC (1999) Biophys J 76:1659
36. Iwanaga Y, Braun D, Fromherz P (2001) Eur Biophys J 30:17
37. Braun D, Fromherz P (1997) Appl Phys A 65:341
38. Channavajjala LS, Eidsath A, Saxinger WC (1997) J Cell Sci 110:249
39. Xiao Y, Truskey GA (1996) Biophys J 71:2869
40. Sagvolden G, Giaever I, Pettersen EO, Feder J (1999) Proc Natl Acad Sci USA 96:471
41. Keese CR, Giaever I (1991) Exp Cell Res 195:528
42. Salas PJ, Vega-Salas DE, Rodriguez-Boulan E (1987) J Membr Biol 98:223
43. Kirchhofer D, Grzesiak J, Pierschbacher MD (1991) J Biol Chem 266:4471
44. Gonzalez-Mariscal L, Contreras RG, Bolivar JJ, Ponce A, Chavez De Ramirez B, Cereijido M (1990) Am J Physiol 259:C978
45. Zink S, Rosen P, Sackmann B, Lemoine H (1993) Biochim Biophys Acta 1178:286
46. Zink S, Rosen P, Lemoine H (1995) Am J Physiol 269:C1209
47. Wegener J, Zink S, Rosen P, Galla H (1999) Pflügers Arch 437:925
48. Wegener J, Keese CR, Giaever I (2002) Biotechniques 33:348
49. Keese CR, Bhawe K, Wegener J, Giaever I (2002) Biotechniques 33:842
50. Wegener J, Sieber M, Galla HJ (1996) J Biochem Biophys Methods 32:151
51. Wegener J, Hakvoort A, Galla HJ (2000) Brain Res 853:115
52. Lo C-M, Keese CR, Giaever I (1993) Exp Cell Res 204:102
53. Lo CM, Keese CR, Giaever I (1994) Exp Cell Res 213:391
54. Keese CR, Karra N, Dillon B, Goldberg AM, Giaever I (1998) In Vitr Mol Toxicol 11:183
55. Giaever I, Keese CR (1989) Physica D 38:128
56. Kaspar M, Stadler H, Weiss T, Ziegler C (2000) Fresenius J Anal Chem 366:602
57. Janshoff A, Galla H-J, Steinem C (2000) Angew Chem Int Ed 39:4004
58. Wegener J, Janshoff A, Steinem C (2001) Cell Biochem Biophys 34:121
59. Rosenbaum JF (1988) Bulk acoustic wave theory and devices. Artechhouse, Boston
60. Mason WP (1965) Physical acoustics. Academics, New York
61. Sauerbrey G (1959) Z Phys 155:206

62. Schumacher R (1990) Angew Chem 102:347
63. Buttry DA, Ward MD (1992) Chem Rev 92:1355
64. Benes E, Gröschl M, Burger W, Schmid M (1995) Sens Actuators A 48:1
65. Barnes C (1991) Sens Actuators A 29:59
66. Bruckenstein S, Shay M (1985) Electrochim Acta 30:1295
67. Nomura T, Okuhara M (1982) Anal Chim Acta 142:281
68. Höök F, Rodahl M, Brzezinski P, Kasemo B (1998) Langmuir 14:729
69. Wegener J, Janshoff A, Galla H-J (1998) Eur Biophys J 28:26
70. Janshoff A, Wegener J, Sieber M, Galla H-J (1996) Eur Biophys J 25:93
71. Cans A-S, Höök F, Shupliakov O, Ewing AG, Eriksson PS, Brodin L, Orwar O (2001) Anal Chem 73:5805
72. Lucklum R, Behling C, Cernosek RW, Martin SJ (1997) J Phys D Appl Phys 30:346
73. Lucklum R, Hauptmann P (2000) Sens Actuators B 70:30
74. Granstaff VE, Martin SJ (1994) J Appl Phys 75:1319
75. Johannsmann D (1999) Macromol Chem Phys 200:501
76. Martin SJ, Bandey HL, Cernosek RW (2000) Anal Chem 72:141
77. Ballantine DS, White RM, Martin SJ, Ricco AJ, Zellers ET, Freye GC, Wohltjen H (1997) Acoustic wave sensors. Academic, San Diego
78. Bottom VE (1982) Introduction to quartz crystal unit design. van Nostrand Reinhold, New York
79. Martin BA, Hager HE (1989) J Appl Phys 65:2630
80. Martin BA, Hager HE (1989) J Appl Phys 65:2627
81. Yang M, Thompson M (1993) Anal Chem 65:1158
82. Bandey HL, Martin SJ, Cernosek R (1999) Anal Chem 71:2205
83. Reed CE, Kanazawa KK, Kaufman JH (1990) J Appl Phys 68:1993
84. Nwankwo E, Durning CJ (1999) Sens Actuators A 72:99
85. Nwankwo E, Durning CJ (1999) Sens Actuators A 72:195
86. Lin Z, Ward MD (1995) Anal Chem 67:685
87. Zhang C, Feng G (1996) IEEE Ultrasonics Symp 43:942
88. Matsuda T, Kishida A, Ebato H, Okahata Y (1992) ASAIO J 38:M171
89. Gryte DM, Ward MD, Hu WS (1993) Biotechnol Prog 9:105
90. Redepenning J, Schlesinger TK, Mechalke EJ, Puleo DA, Bizios R (1993) Anal Chem 65:3378
91. Marxs KA, Zhou T, Montrone A, Schulze H, Braunhut SJ (2001) Biosens Bioelectron 16:773
92. Wegener J, Seebach J, Janshoff A, Galla HJ (2000) Biophys J 78:2821
93. Fredriksson C, Kihlman S, Rodahl M, Kasemo B (1998) Langmuir 14:248
94. Fredriksson C, Kihlman S, Kasemo B (1998) J Mater Sci Mater Med 9:785
95. Bray D (1992) Cell movements. Garland, New York

Part IV
Lipid Membranes as Biosensors

Chapter 10

Functional Tethered Bilayer Lipid Membranes

Wolfgang Knoll, Kenichi Morigaki, Renate Naumann,
Barbara Saccà, Stefan Schiller, Eva-Kathrin Sinner

Abstract. Tethered bilayer lipid membranes (tBLMs) constitute a novel experimental concept with very promising features for fundamental biophysical investigations of the general correlation between structural properties and functional processes in model membrane systems. Moreover, these architectures offer a robust platform for membrane-based biosensing principles that could eventually result in the design and fabrication of stable and cheap membrane chips. We first discuss a few synthetic strategies that lead to various types of architectures with unprecedented electrical properties: some of these tBLMs show an electrical resistance that exceeds even that of the best model membrane known so far, i.e., the bimolecular lipid membrane (BLM). The reconstitution of a synthetic ionophore, i.e., the carrier valinomycin, allows one to study the K^+-selective and reversible increase of the membrane conductivity by more than 4 orders of magnitude. Next, the incorporation of various types of integrin receptors into the tethered bilayer membranes will be briefly discussed as an example of the versatility of this model system in membrane binding assays.

And finally, the photopolymerization of polymerizable lipids will be introduced as a way to generate patterned bilayers that can serve as the basic structure for the construction of membrane chips for massive paralleled detection of membrane processes.

Keywords: Tethered bilayer lipid membrane, Polymerizable lipids, Surface plasmon spectroscopy, Impedance spectroscopy, Valinomycin, Integrin receptor

1
Introduction

The set of available model membrane systems – Langmuir monolayers, vesicles, bimolecular lipid membranes (BLMs), or solid-supported bilayers – has recently been complemented by a novel concept: the tethered bilayer lipid membrane [1]. This construct consists of (1) a solid substrate, (2) the so-called tethering layer composed of peptides, proteins, oligosaccharides, oligonucleotides, synthetic organic oligomers or polymers, and (3) the lipid bilayer [2]. The (partial) covalent attachment of the tethers to the substrate, as well as to some anchoring lipids within the proximal monolayer of the membrane, give the system a stability and robustness that is missing in merely supported bilayers. On the other hand, the length and flexibility of the tethers guarantee a sufficient separation of the membrane from the substrate surface such that enough "headspace" for the lipids and – more importantly – for any incorporated protein can be generated. Moreover, despite being closely coupled to the substrate the lipid membrane can

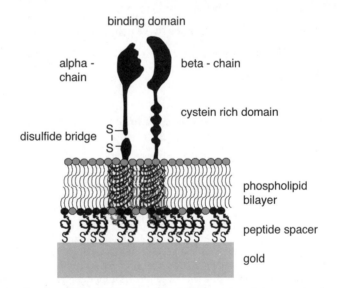

Fig. 1. Schematic drawing of the interfacial architectures prepared for protein reconstitution. Onto a bare gold surface, a mediator layer of a laminin fragment is deposited by a self-assembling process. A lipid (DMPE) monolayer is coupled to the peptide layer by amino coupling. Spreading of integrin-functionalized vesicles results in the final architecture: the integrin-functionalized planar lipid membrane (cf. also Fig. 4)

exist in a fluid, liquid-crystalline state, which is a prerequisite for many membrane-mimicking properties and processes one might be interested in. Figure 1 gives a schematic representation of the desired architecture. This example refers to a strategy for the coupling of a bilayer to Au (electrode) surfaces – hence the thiol-based tethering concept [3].

In this short summary report we will first describe a few essential details for the design and construction of tethered bilayers [4] and then refer to the preparation of patterned membranes [5] as a key step toward the membrane chip. Next we give examples of some of their important performance parameters, i.e., their fluidity and their electrical barrier properties [6]. The last section then deals with the incorporation and functional characterization of integral membrane proteins [7].

2
Designing Tethered Bilayer Lipid Membranes

Tethered bilayer lipid membranes (tBLMs) mimic the biological cytoplasma membrane and are promising platforms for basic research and biosensor applications [4]. They comprise a lipid group linked to a solid substrate, e.g., to gold via a thiol-modified tether as spacer unit between the lipid and the substrate surface. The tether/spacer molecules provide a hydrophilic layer with functional properties resembling those of the cytosol/cytoskeleton. They establish a water-containing submembrane space which reduces the interaction of the membrane with the substrate, thus enabling the functional incorporation of membrane proteins [8].

Moreover, the combination of such ultrathin tBLMs with the robustness of a solid support in a planar configuration allows for the application of surface-sensitive techniques, including surface plasmon resonance spectroscopy (SPR), measurements with the quartz crystal microbalance (QCM), surface-enhanced IR, and electrochemical techniques such as electrochemical impedance spectroscopy (EIS). In the case of a metal substrate investigations under a defined electric field are also possible, thus extending the range of the experimental characterization techniques toward the study of field-sensitive processes, for example ion transport via receptors, channel proteins or carriers such as valinomycin. In the following we will briefly describe the buildup of tethered bilayer membranes by two slightly different synthetic strategies.

2.1
A novel Archaea Analog Lipid as Self-Assembling Telechelic

The first strategy is based on a single molecule, the novel archaea analog 2,3-di-O-phytanyl-sn-glycero-1-tetraethylene glycol-DL-α-lipoic acid ester lipid 1 (DPTL), Fig. 2. Isoprenoid phytanyl chains were chosen as hydrophobic tails due to their low phase transition temperature and their influence on the density and stability of biological membranes. The transition temperature of 2,3-di-O-phytanyl-sn-glycero-1-tetraethylene glycol was determined to be < -80 °C by differential scanning calorimetry (DSC). Furthermore, the 2,3-di-O-phytanyl-sn-glycerol exclusively contains ether linkages in order to prevent hydrolytic cleavage. This moiety is well known from extremophiles or archaea to form sta-

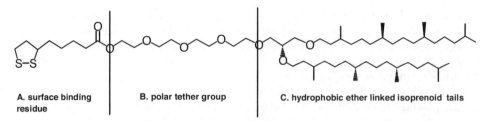

A. surface binding residue   **B. polar tether group**   **C. hydrophobic ether linked isoprenoid tails**

Fig. 2. The novel archaea analog 2,3-di-O-phytanyl-sn-glycero-1-tetraethylene glycol-DL-α-lipoic acid ester lipid 1 (DPTL)

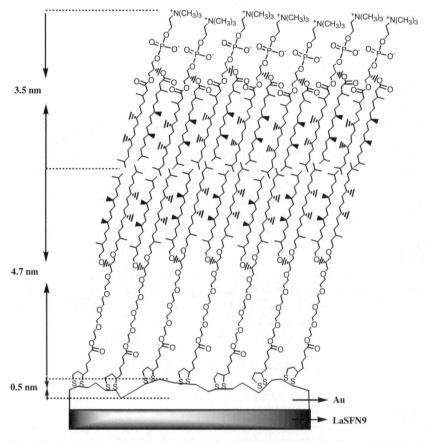

Fig. 3. Schematic view of the arrangement of the DPTL molecules to form the tBLM. Molecular dimensions are in proportion to the ultraflat gold surface to guarantee the perfect alignment of the phytanyl chains forming a dense monolayer. The tetraethylene glycol chains are assumed to be stretched

ble membranes under extreme conditions such as high temperatures. Regarding the tetraethylene glycol spacer, an elongated hydrophilic chain (cf. Fig. 3) can be assumed according to FTIR spectroscopy studies, providing the maximum decoupling distance of the lipid membrane from the substrate.

In order to account for another key feature of the preparation of self-assembled monolayers (SAMs) and tethered membranes, i.e., the structure and morphology of the gold substrate, the smooth surface of template-stripped gold (TSG) [9] was introduced to form tBLMs with good sealing properties. The roughness of TSG was shown by AFM to be 0.5 nm over areas of square microns, one order of magnitude smaller than the molecular dimension of the DPTL molecule itself as illustrated in Fig. 3. This ensures good supramolecular alignment of the functional units resulting in a stable arrangement of the monolayer. Highly resistive tBLMs can then be prepared starting from a SAM of 1 on the TSG surface, by the fusion of liposomes prepared from diphytanoylphosphatidylcholine (DPhyPC) giving rise to the formation of well-defined lipid bilayers as shown by SPR and EIS measurements [10].

2.2
Sequential Deposition of Structural Elements

A different approach, the sequential layer-by-layer deposition of the different structural/functional elements of a tethered lipid membrane, is illustrated in Fig. 4. Shown is the preparation protocol of the different adsorption/coupling

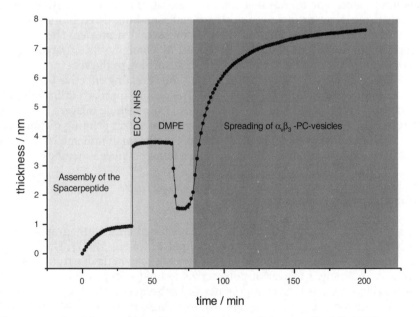

Fig. 4. Example of the assembly of a peptide-supported functionalized lipid bilayer membrane via a layer-by-layer adsorption. For details, see text

steps as they were monitored online by surface plasmon spectroscopic kinetic measurements.

For this experiment a 19-mer spacer peptide CSRARKQAASIKVAVSABR derived from the protein α-laminin was adsorbed onto the gold surface in 10 to 20 min, forming a layer of about 1 nm thickness. Excess of unbound peptide was rinsed off. The activation process was carried out for 10 min by addition of the EDC/NHS solution to the cuvette. (The instantaneous increase in "thickness" is an artifact originating from the change in refractive index of the buffer upon addition of the EDC/NHS solution). Next, detergent-solubilized dimyristoylphosphatidylethanolamine (DMPE) was added and coupled to the activated carboxyl terminus of the surface-attached peptide for 30 min. After rinsing, an additional layer of lipid bound to the peptide of ca. 0.5 nm was found. This is certainly much less than a complete monolayer; however, it is enough to have those lipopeptides acting as anchor lipids in the final tethered membrane architecture [11]. The preparation was completed by adding phosphatidylcholine (soybean) vesicles that contained reconstituted cell adhesion receptors, in this case the $\alpha_v\beta_3$ integrin (as an important therapeutic target). Spreading and fusion of these proteoliposomes was completed within ca. 2 h, resulting in a final peptide-tethered protein/lipid bilayer membrane of ca. 8 nm.

3
Patterning of Tethered Bilayer Lipid Membranes

Micropatterning of tBLMs allows for the creation of a whole microarray of functionalized membrane corrals that will facilitate various new applications, such as high-throughput drug screening for membrane-bound receptors [12, 13]. Patterning has been attempted in the past by several strategies. The first approach is the modification of the substrates. By changing its surface properties one can either prevent the bilayers from adsorbing to the substrate or lower the lateral fluidity of the adsorbed bilayer, thus creating effective barriers to the lateral diffusion of lipid molecules [14]. The second approach utilizes the so-called soft lithography. Thiol-modified lipids [15] or whole bilayers [16] were brought into contact with the surface by the microcontact printing method using a poly(dimethylsiloxane) (PDMS) stamp. Finally, small unilamellar vesicles fused on the substrate surface, thus forming a tethered bilayer membrane.

3.1
Micropatterning by Photopolymerization

Our approach for micropatterning of tBLMs is based on the lithographic photopolymerization of a diacetylene phospholipid 2 (see Fig. 5) [17]. The fabrication process is basically composed of four steps (a schematic illustration is given in Fig. 6). Firstly, a bilayer of polymerizable lipid is deposited onto a solid support (Fig. 6A). The bilayer is polymerized subsequently by UV light irradiation. A designed pattern can be imposed by using a mask (Fig. 6B). The bilayer has to be kept in an aqueous solution up to this step so that the integrity of the membrane is preserved. Once the bilayer is polymerized, it becomes insol-

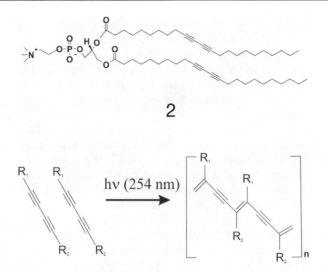

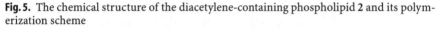

Fig. 5. The chemical structure of the diacetylene-containing phospholipid **2** and its polymerization scheme

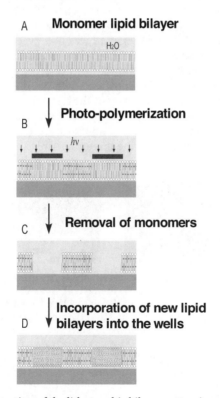

A **Monomer lipid bilayer**

Photo-polymerization

B

Removal of monomers

C

Incorporation of new lipid bilayers into the wells

D

Fig. 6. Schematic illustration of the lithographic bilayer patterning process

uble in organic solvents or detergent solutions. By treating the sample with an organic solvent one removes the monomeric lipids selectively, whereas the polymerized part remains on the substrate (Fig. 6C). The lipid-free areas (corrals) thus created can be filled with another biologically relevant lipid bilayer by vesicle fusion (Fig. 6D). In the case of hydrophilic surfaces such as oxidized silicon, a lipid bilayer adsorbs spontaneously from small unilamellar lipid vesicles by a fusion and reorganization process [18].

We have used the diacetylene-containing phosphatidylcholine 2 as the polymerizable lipid (Fig. 5). Its polymerization has been characterized in detail in liposomes [19]. The polymerization occurs only if the molecules are in a highly ordered state (topochemical polymerization) and the two alkyl chains react only with the same alkyl chain of adjacent molecules (α to α and β to β) [17]. Therefore, cross-linking of two alkyl chains within a single molecule is avoided. Having two polymerizable chains, 2 forms a cross-linked network upon polymerization and becomes insoluble in organic solvents [20].

Bilayers of 2 were deposited on glass substrates from the air/water interface by the Langmuir–Blodgett (LB) and Langmuir–Schaefer (LS) methods. Prior to the photopolymerization, oxygen was removed from the aqueous solution by purging the solution with argon gas. The formation of a highly conjugated polymer backbone induces a strong absorption band in the UV/vis spectral range. The absorption maximum is sensitive to the effective conjugation length that depends on the degree of polymerization and local order of the backbone. In the case of polymerized 2, the absorption band was observed at around $\lambda=470$ nm. Moreover, the polymerized bilayers emitted fluorescence if excited in the wavelength range $\lambda=350$–550 nm. This allowed for the observation of polymerized bilayer areas by fluorescence microscopy, as shown in Fig. 7. In Fig. 7a the bright areas (grid) are the polymerized domains of the bilayer. The dark areas (squares) are either bilayers of monomeric 2 or empty wells after the removal of the monomers, which cannot be distinguished by fluorescence. Only atomic force microscope observation could verify that the monomers were indeed removed.

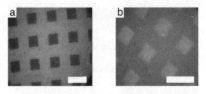

Fig. 7a,b. Fluorescence micrographs of patterned bilayers on glass. **a** The bilayer after lithographic exposure to UV light. The brightly fluorescent area was polymerized by UV irradiation. **b** After removal of the monomeric bilayer of 2 by ethanol, new bilayers containing egg-PC and NBD-PE (1%) were incorporated into the wells by vesicle fusion. The square corrals are brightly fluorescent due to NBD-PE molecules. The scale bars correspond to 100 µm

3.2
Back-Filling the Empty Corrals by Vesicle Fusion

In order to prove that new lipid bilayers can be introduced into the wells, we fused small unilamellar vesicles (extruded through a filter with pores of diameter ~50 nm) of phosphatidylcholine from egg yolk (egg-PC) doped with 1 mol% of the fluorescently labeled lipid NBD-PE (N-(7-nitrobenz-2-oxa-1,3-diazol-4-yl)-1,2-dihexadecanoyl-sn-glycero-3-phosphoethanolamine). Observation by fluorescence microscopy revealed that the corrals between the polymerized domains were intensely fluorescent due to the incorporated lipid bilayers (Fig. 7b). The bilayers were fluid, as proven by fluorescence recovery after photobleaching (FRAP) experiments on a small spot (diameter ca. 4 μm). Furthermore, the polymerized bilayers acted as an effective barrier that confined the lipid molecules, as shown by FRAP experiments of individual corrals or by application of an electric field along the bilayer membrane that resulted in the migration of NBD-PE molecules. If treated with ethanol, the egg-PC bilayers were completely removed, leaving the bare grid structure of polymerized bilayer. The grid structure could be used repeatedly as a 2D master for the incorporation of new lipid bilayers. The bilayers were again fluid and confined in each single corral.

Compared with other approaches that have been demonstrated for the creation of patterned fluid lipid bilayers, the lithographic polymerization of bilayers has a unique feature in that the pattern is created within the bilayer membrane, i.e., the structure is truly two dimensional and independent of the substrate. This fact gives flexibility in the design of synthetic bilayer membranes. And, of course, one can construct patterned bilayers that are separated from the substrates by a thin tethering layer or a soft polymer cushion.

4
Electrical Properties of Tethered Membranes

In addition to their structural integrity as seen, e.g., by optical or X-ray reflectometry measurements, tethered bilayers have to show certain electrical properties in order to be useful as model membranes. Only a lipid bilayer that has a low defect density can act as a barrier for the translocation of ions, and hence shows the required high resistivity in the absence of any synthetic ionophore or reconstituted biological ion transport protein.

4.1
Impedance Spectroscopy

To this end the DPTL/DPhyPC membrane described above was characterized by electrochemical impedance spectroscopy. From these spectra the capacitance and the resistance of the lipid layers were computed by fitting the data to an equivalent circuit [21, 22], consisting of a RC mesh representing the lipid film, the resistance of the electrolyte solution, and the capacitance of the diffuse double layer adjacent to the gold. Typical electrical data are given in Table 1. Data measured for the tBLM varied from 0.45 to 0.8 μF cm^{-2} for the ca-

Table 1. Data obtained by simulations from EIS (capacitance and resistance) and SPR (thickness) measurements. Thicknesses obtained by modelling calculations with CS Chem 3D Pro, are added for comparison

	$C/(\mu F\ cm^{-2})$ Experimental	$R/(M\Omega\ cm^2)$ Experimental	$d/(nm)$ Experimental	$d/(nm)$ Calculated
DPTL, monolayer before vesicle spreading	0.55	1.75	4.7	4.7
Lipid bilayer after vesicle spreading	0.49	4.35	8.5	8
Lipid bilayer + valinomycin	0.64	0.0022		

pacitance and from 2 to 12 $M\Omega\ cm^2$ for the resistance. Thus, they compare quite well with those of BLMs. The resistance, in particular, is sufficiently high so as to allow for investigations of the ion transfer across the bilayer mediated by, e.g., an ion carrier.

4.2
Functionalization by Valinomycin

The effect of valinomycin on the resistance of the tBLM in the presence of potassium ions is shown in Fig. 8. Valinomycin is an antibiotic, produced by *Streptomyces fulvissimus*, complexing specifically potassium (and rubidium) ions to its hydrophilic interior, with a hydrophobic exterior enabling transmembrane ion transport. This gives rise to the so-called inflection point in the EIS spectrum, when the resistance drops to the kilo-ohm range (Table 1) and the capacitance of the diffuse double layer evolves to approximately 5 $\mu F\ cm^{-2}$.

This result is particularly remarkable as it demonstrates that tBLMs, indeed, can be prepared with electrical properties (capacities and resistivities) that match those of the best model membranes, i.e., BLMs. Moreover, the addition of valinomycin in the presence of K^+ ions resulting in an ion-specific increase of the conductivity by more than 3 orders of magnitude demonstrates the potential of these model membranes for transport studies of incorporated ionophores and proteins. In the case of valinomycin it also proves the fluidity of the tethered membrane because this ionophore acts via a shuttle mechanism, i.e., has to diffuse across the hydrophobic barrier of the bilayer, which is possible only if the lipids are in a liquid-crystalline state.

5
Binding Studies with Tethered Membranes

Integrin molecules are indispensable in the process of cell adhesion [23]. They are transmembrane heterodimeric glycoproteins composed of noncovalently associated α and β subunits and involved in the cell adhesion process as anchoring molecules with an intrinsic signaling function [24, 25]. The synthesis of

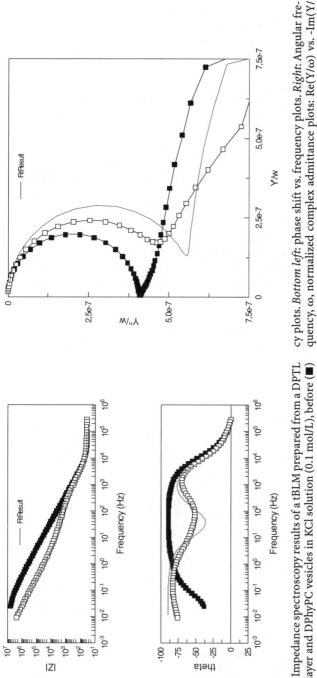

Fig. 8. Impedance spectroscopy results of a tBLM prepared from a DPTL monolayer and DPhyPC vesicles in KCl solution (0.1 mol/L), before (■) and after (□) adding valinomycin (0.1 μM). *Top left*: Re(Z) vs. frequency plots. *Bottom left*: phase shift vs. frequency plots. *Right*: Angular frequency, ω, normalized complex admittance plots: Re(Y/ω) vs. -Im(Y/ω), *solid lines* represent fitted data

anchoring proteins, tightening the cells to surfaces, is initiated by integrin molecules whenever binding epitopes from extracellular matrix proteins, such as collagen, vitronectin or fibronectin, are bound to these receptors [24–26]. Integrin–ligand binding is in the focus of cancer research, because blocking the binding site of tumor-specific integrins is a promising concept to inhibit tumor metastasis and angiogenesis. The $\alpha_v\beta_3$ receptor is an example of an important integrin species and therefore of general interest in the context of protein-functionalized artificial tethered membranes. The binding features of receptor molecules, embedded in a spacer-peptide-supported lipid film, were measured at high resolution by surface plasmon field-enhanced fluorescence spectroscopy (SPFS) [27].

In vivo binding assay using whole biological cells is an experimentally difficult task, because various species are abundant on cell surfaces, interfering in binding specificities. This fact is reflected in high levels of unspecific binding in conventional binding assays. Hence, it was impossible to solve complex molecular interactions using heterogeneous systems, such as cells.

The incorporation of purified integrin receptors in tethered bilayers is an experimental approach aimed at screening for receptor/ligand interactions in a molecularly defined in vitro system [27]. Two examples are shown here, proving the functional incorporation of different integrin species in a tethered membrane system by binding assays with known ligands. In Fig. 4, the preparation of an $\alpha_v\beta_3$-functionalized surface is shown as a stepwise layer assembly. Binding of Cy5-labeled vitronectin (VN) to the immobilized $\alpha_v\beta_3$ receptors is depicted in Fig. 9. The SPR angular scans indicate binding of VN, shifting the minimum

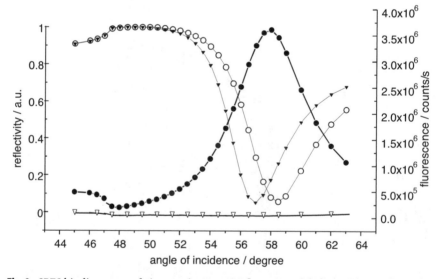

Fig. 9. SPFS binding assay of vitronectin, upon its fluorescent labeling with Cy5 dye, to human $\alpha_v\beta_3$ integrin embedded into a DMPE/PC bilayer coated on the gold surface via a hydrophilic laminin peptide layer. Fluorescence signal upon binding of vitronectin (●) to integrin compared to the background fluorescence signal (▽). The ▼ line indicates the plasmon scan before and after (○) addition of the layer system

of the reflected light to higher angles. Correspondingly, the fluorescence intensity emitted from the surface-bound ligands shows the typical angular dependence following the excitation profile of surface plasmons. The peak intensity is many orders of magnitude above the flat background.

This binding experiment is shown here as an example of monitoring binding processes with high sensitivity. The influence of physiological parameters controlling the integrin/ligand binding, such as cation selectivity, is shown in Fig. 10. An isolated collagen fragment of human collagen type IV was fluorescently labeled and applied to an $\alpha_1\beta_1$-integrin-functionalized tethered membrane. The surface density of the CB3 fragment (ca. 80–100 kDa) binding to the $\alpha_1\beta_1$ receptors on the sensor is not sufficient for SPS detection (data not shown). However, the fluorescence signal indicates binding of collagen fragments with negligible unspecific binding of protein/dye. Removal of bivalent cations over time by EDTA treatment of the surface results in a gradual decrease of the fluorescence signal.

This observation indicates the dissociation of the integrin/collagen complex, whereby the unbound collagen molecules are removed from the aqueous phase by rinsing. The complete loss of the fluorescence signal after ca. 10 h incubation

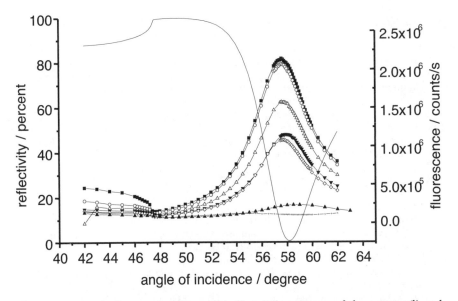

Fig. 10. Decrease in fluorescence due to the dissociation process of the receptor/ligand complex by incubation with EDTA. The background fluorescence scan is indicated by the + line. The ■ symbol indicates the fluorescence scan after addition of Cy5-labeled collagen to an $\alpha_1\beta_1$-functionalized membrane. After incubation in 0.5 M EDTA for 10 min, the scan indicated by (○) was observed. After incubation for 20 min in EDTA, the fluorescence decrease is observed (△). Incubation for 60 min resulted in further decrease in the fluorescence signal (▼); incubation for 600 min resulted in the almost complete loss of the fluorescence signal due to dissociation of the receptor/ligand complex (▲)

with EDTA underlines the physiological behavior of the incorporated integrin receptors. The binding capacity of the surface can be regenerated for further ligand screening. The selective and specific integrin binding assay, using tethered membranes, enables long-term monitoring (up to 3 days) of the binding of collagen and vitronectin, examples of small and multimeric ligands, respectively.

Acknowledgement
Financial support came partially from the National Science Foundation through the Center on Polymer Interfaces and Macromolecular Assemblies (CPIMA).

References

1. Sinner EK, Knoll W (2001) Functional tethered membranes. Curr Opin Chem Biol 5:705–711
2. Knoll W, Frank CW, Heibel C, Naumann R, Offenhäusser A, Rühe J, Schmidt EK, Shen WW, Sinner EK (2000) Functional tethered lipid bilayers. Rev Mol Biotechnol 74:137–158
3. Spinke J, Yang J, Wolf H, Liley M, Knoll W (1992) Polymer-supported bilayer on a solid substrate. Biophys J 63:1667–1671
4. Sackmann E (1996) Science 271:43
5. Morigaki K, Baumgart T, Offenhäusser A, Knoll W (2001) Angew Chem Int Ed Engl 40:172–174
6. Schiller S, Naumann R, Lovejoy K, Kunz H, Knoll W (2003) Archaea analogue thiolipids for tethered bilayer lipid membranes on ultrasmooth gold surfaces. Angew Chem Int Ed Engl 42:208–211
7. Sinner EK, Reuning U, Saccà B, Moroder L, Knoll W, Oesterhelt D (2004) Incorporation of cell adhesion receptors in artificial planar lipid membranes – characterization by plasmon-enhanced fluorescence spectroscopy. Anal Biochem 00:000–000
8. Naumann R, Schmidt EK, Jonczyk A, Fendler K, Kadenbach B, Liebermann T, Offenhäusser A, Knoll W (1999) Biosens Bioelectron 14:651–662
9. Wagner P, Hegner M, Güntherodt HJ, Semenza G (1995) Langmuir 11:3867
10. Bunjes N, Schmidt EK, Jonczyk A, Rippmann F, Beyer D, Ringsdorf H, Gräber P, Knoll W, Naumann R (1997) Langmuir 13:6188
11. Schmidt EK, Liebermann T, Kreiter M, Jonczyk A, Naumann R, Offenhäusser A, Neumann E, Kukol A, Maelicke A, Knoll W (1998) Biosens Bioelectron 13:585
12. Groves JT, Ulman N, Boxer SG (1997) Science 275:651–653
13. Fang Y, Frutos AG, Lahiri J (2002) J Am Chem Soc 124:2394–2395
14. Groves JT, Ulman N, Cremer PS, Boxer SG (1998) Langmuir 14:3347–3350
15. Jenkins ATA, Boden N, Bushby RJ, Evans SD, Knowles PF, Miles RE, Ogier SD, Schönherr H, Vancso GJ (1999) J Am Chem Soc 121:5274–5280
16. Hovis JS, Boxer SG (2000) Langmuir 16:894–897
17. Morigaki K, Baumgart T, Jonas U, Offenhäusser A, Knoll W (2002) Langmuir 18:4082–4089
18. Reviakine I, Brisson A (2000) Langmuir 16:1806–1811
19. Freeman FJ, Chapman D (1988) In: Liposomes as drug carrier. Wiley, New York, pp 821–839
20. Leaver J, Alonso A, Durrani AA, Chapman D (1983) Biochim Biophys Acta 732:210–218
21. Raguse B, Braach-Maksvytis V, Cornell BA, King LG, Osman PDD, Pace RJ, Wieczorek L (1998) Langmuir 14:648
22. Krishna G, Schulte J, Cornell BA, Pace R, Wieczorek L, Osman PD (2001) Langmuir 17:4858

23. Hynes RO (1987) Integrins: a family of cell surface receptors. Cell 48:549–554
24. Tuckwell DS, Humphries MJ (1993) Molecular and cellular biology of integrins. Crit Rev Oncol Hematol 15:149–171
25. Kühn K, Eble J (1994) The stuctural bases of integrin–ligand interactions. Trends Cell Biol 4:256–261
26. Ruoslahti E (1996) RGD and other recognition sequences for integrins (review). Annu Rev Cell Dev Biol 12:697–715
27. Saccà B, Sinner E-K, Fiori S, Kaiser J, Lübken C, Eble J, Moroder L (2002) Identification of the stuctural epitope for cell adhesion to collagen type IV. Chembiochem 9:904–907

Electrostatic Potentials of Bilayer Lipid Membranes: Basic Principles and Analytical Applications

Valeri Sokolov, Vladimir Mirsky

Abstract. The potential distribution on planar lipid bilayers or modification of this distribution can be analyzed by electrical methods based on electrophoretic mobility, compensation of intramembrane field, transport of hydrophobic ions, and nonstationary potential or current measurements. These methods give information on changes of the electric potential averaged in planes that are located at different positions relative to the membrane/water interface. The numerous analytical applications of this approach include quantitative analysis of substance-membrane interactions and development of biosensors.

Keywords: Bilayer lipid membrane, Electrostatic potential, BLM applications, Membrane electrostatics, Drug–membrane interaction, Secondary screening, Biosensor

1
Introduction

Great progress in the investigation of the structure and function of biological membranes was achieved through developing artificial bilayer lipid membranes (BLM[1]) [1]. This simple self-assembled system was immediately recognized as a perfect model of lipid bilayers of biological membranes and has been used intensively over the last 40 years. About 12,000 papers and 120 patents are associated with BLM (Fig. 1). It is interesting that, while saturation in production of scientific publications has been observed since about 1995, the patent activity is still increasing. This trend might signify a shift from the applications of BLM in basic research toward technological applications.

The works performed with BLM can be divided into the following four groups differing in the investigation aims. The first group includes investigation of the physicochemical properties of the BLM itself or interaction of nonmodified BLM with other membranes and membrane structures (electrical breakdown [2–3], membrane fusion [4–8]). These works have already resulted in clinical (tumor therapy by electroporation [9–11] or electropermeabilization [12–14]) and biotechnological (electrofusion of living cells) applications.

The second group includes studies of passive ion transport. Most of the information on mechanisms of action of uncouplers (for example, dinitrophenol, pentachlorophenol, CCCP), ionophores (for example, valinomycin, nonactin, nigericin), and channel formers (gramicidin A, alamethicin, amphotericin and others) has been obtained by using BLM. This knowledge is now not only an essential part of any student textbook on biophysics, biochemistry or electrochemistry, but also a scientific base for numerous applications of these substances in medicine (for example, nystatin and amphotericin B as fungicides), in

---

[1] In some publications, the terms "bimolecular lipid membranes" or "planar lipid bilayers" are used instead of "bilayer lipid membranes".

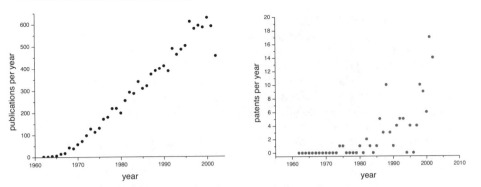

Fig. 1. Annual number of publications (*left*) and patent applications (*right*) on planar lipid bilayers

development and production of ion-selective electrodes and ISFETs (valinomycin, nonactin, crown-ethers), or as a routine instrument in biological and physiological investigations.

The third group includes reconstruction of biological systems on planar lipid bilayers [15–18]. This approach includes incorporation or adsorption of integral or peripheral proteins, for example ion channels [15, 19], bacteriorhodopsin [20–26], and H^+-ATPases [25, 27, 28]. Even after development of the patch-clamp technique, providing a simpler way for functional investigation of ion channels, the reconstruction is still important as a functional test of biochemical preparations. A simple variation of this technique includes the formation of sandwich-like structures by adsorption of proteoliposomes or membrane fragments with biological materials [24, 25, 27, 29, 30][2].

The fourth group includes investigation of the membrane effects of different biologically active substances. An application of BLM in this field started as early as the mid-1960s. The effects of a large number of different substances on the conductance of planar lipid bilayers have been studied; a bibliography or review on many of these publications can be found in [31–33]. The approach allowed the identification of substances which can penetrate lipid bilayers in ionized form. It was then extended to the investigation of the electroneutral transport of such compounds, which can be protonated or deprotonated [34, 35]. Additional possibilities provide measurements of photodynamic [36–38], chemical [39], or biochemical (phospholipase) [40–42] damage of lipid bilayers. Investigation of membrane electrostatics can be considered a comparatively new approach in this field. It allows the study of the electrical structure of the interface between the membrane and aqueous electrolyte. Electrostatic potential can change due to adsorption of biologically active substances on the membrane surface, or due to modification of the lipid structure. In combination with other techniques, the approach based on measuring of electrostatic potential provides a detailed

---

2 See also the review by K. Fendler et al. *Transport proteins on solid-supported membranes: from basic research to drug discovery* in this book (ed).

characterization of the interaction of compounds of interest with the lipid bilayer, including not only their transport through the membrane and membrane toxic effects, but also the influence of these compounds on the profile of electrostatic potential across the membrane, localization of the adsorption plane, and determination of the binding constants of these compounds. This information allows the prediction of electrostatically coupled biological effects on membrane proteins and on some biological processes. A description of experimental approaches to analyze membrane electrostatics is the main topic of this paper.

Besides basic research, investigations of membrane electrostatics has important analytical aspects. Intensive development of pharmaceutical technology and of the drug discovery process, based on combinatorial organic synthesis and subsequent (ultra)high-throughput screening, resulted in a "traffic jump" at the second step of this process, namely in the physicochemical characterization of the promising targets. Application of BLM-based devices is a possible solution. Another application is the monitoring of the modification of potential profiles as a transducer for biosensors. Examples of these applications will also be described.

1.1
Structure of Lipids and Lipid Bilayers

The structure of lipid bilayers has been studied over the last 30 years and these studies were summarized in many reviews and books. Here we overview only the most important information which is necessary for the following description of electrical parameters, and especially for the discussion of the membrane electrostatics.

The most important property of the lipids which is crucial for formation of the bilayer, is that the lipid molecule has amphiphilic structure. It contains hydrophilic polar heads and hydrophobic hydrocarbon tails. In the lipid bilayers the hydrocarbon tails form a hydrophobic region inside the membrane, characterized by low dielectric permeability (about 2), while hydrophilic heads form the outer polar layers. Presumably, the dielectric permeability of the outer layers gradually increases along the normal to the membrane, from the low value close to that of oil inside the bilayers, to the high value of the dielectric permeability of water outside them [43].

The thickness of the bilayer depends on the length of the hydrocarbon tails of the lipids and on the inclusions of the solvent used for membrane preparation [44]. The last factor can result in a much greater thickness of artificial membranes compared to biological membranes. For example, decane, which is the most popular solvent for BLM preparation, increases the membrane thickness approximately twice, which makes questionable the adequacy of this model system in some applications.

The electrical conductance of lipid bilayers is very low. It can be explained by the high electrostatic energy of ions inside the hydrophobic region of the membrane (for example, [45, 46]). Measurements of the conductance of pure lipid bilayers are strongly influenced by impurities in the lipid bilayer. Due to degradation of lipids or poor purification of them after biochemical isolation or organic synthesis, the measured conductance is caused in many cases by these impurities rather than by the lipid bilayer properties themselves. The low conductance of lipid bilayers provides a possibility of using BLM to study induced ion transport, when additives introduced into the bilayers increase the bilayer permeability to certain ions. Such molecules include hydrophobic ions penetrating through BLM, ion carriers, or different kinds of substances forming ion conductive channels. These studies are described in many reviews and monographs [46–48].

1.2
Electrostatic Boundary Potentials

The insulating function of the lipid bilayer is provided by the hydrocarbon chains of lipids forming the hydrophobic core inside the membrane. The polar heads of the lipids, together with oriented water and polar groups of proteins and other membrane-embedded molecules, form local electrostatic fields on the membrane/water interface. These fields play an important role in the electrophysiological function of the membrane. Electrostatic fields in the vicinity of the membrane/water interface can be characterized by interfacial potential. In electrochemistry, the interfacial potential is the difference of the electrical potentials between two points, placed in different phases far enough from the interface to ensure that the medium in the vicinity of these points is electroneutral. However, this definition cannot be applied directly to the interface between the lipid bilayer and water, because one of the phases – the lipid bilayer – is very thin. We could define the interfacial potential as the difference between the potential in water at considerable distance from the membrane and the average potential at the middle median plane of the membrane. However, in practice we will be interested not in the absolute value of the interface potential, but in its change due to external events: adsorption of molecules from water solution, structure rearrangement of lipids etc. These events disturb the electrostatic fields in thin layers near the membrane surface, and it is this change of the interfacial potential that can be measured. One can measure only the difference between the electrical potential of the bulk water solution and the averaged potential at some plane near the membrane boundary. The position of this plane may be different in different methods. Therefore, the comparison of the results obtained by the methods which can measure the potentials averaged at different planes can give some information about the distribution of electric field near the membrane/water interface, and we will discuss later some examples of application of such an approach.

The electrostatic potential distribution near the boundary of the membrane can be divided into two regions, which differ in the origin of potential change and in the choice of appropriate methods to measure it. The first region is the water solution near the membrane. The potential drop in this region arises due to the net charge of the membrane surface (the charge of phospholipids is usually negative); as a result, the electrolyte layer adjacent to the surface is not electroneutral. This layer is referred to as the diffuse layer, and the potential change in this layer as surface potential. According to conventional terminology [49], the surface potential is the potential difference between the bulk water solution and some plane S in the solution immediately at the membrane surface. The important feature of the surface potential is that it is the potential difference between the points placed in the same phase. This potential can be measured, and the appropriate methods will be discussed below.

The second region is the interface between the water and the membrane. The important difference of this region from the first one is that here we deal with the potential change between two different phases: water near the membrane and lipids inside the membrane. The potential distribution in the second region may be complex, since it is influenced by many molecules inside or outside the membrane. The potential drop in this region is usually considered to be caused by dipoles of water molecules, which are oriented on the interface; a contribution of lipid molecules and other molecules at the interface can also be essential. The value of this potential drop is estimated in several hundreds millivolts, the higher potential being in the hydrophobic phase. The potential distribution in this region is often presented as a linear change from some plane outside the membrane to some plane inside it. Such a presentation is very arbitrary; the real distribution of the electric field here is unknown. Moreover, because this is a potential difference between two different phases (the so-called Galvani potential), it cannot be measured in principle [50, 51]; only its variation due to external events can be determined.

1.3
Distribution of the Electric Field Across the Membrane

Taking into account the contribution of the two regions to the boundary electrostatic potential, we can draw a potential distribution across the membrane consisting of two boundary potential drops in the oppositely oriented interfaces and approximately linear potential change inside the hydrophobic region of the membrane (Fig. 2a). One of the important issues is the influence of the applied external voltage on this distribution. To address this issue, the outer regions of the membrane, responsible for the boundary potentials, and the inner hydrophobic part can be represented by three capacitors connected in series (Fig. 2b). The external voltage is divided between these capacitors. Due to greater thickness and lower dielectric permeability, the capacitance of the inner (hydrophobic) layer of the membrane is much smaller than the capacitances of the outer (polar) layers. Therefore, most of the voltage drops inside the membrane, and only a minor part of it (typically less than 0.1%) drops in the outer polar layers. It means that the electrostatic potentials on both boundaries of the membrane are almost unaffected by the applied voltage.

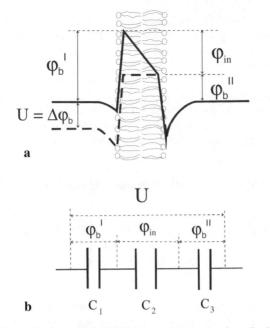

Fig. 2a,b. Potential profile through the lipid bilayer (**a**) can be described by three serially connected capacitors (**b**). The capacitor C_2 corresponds to the capacitance of the inner hydrophobic part of the membrane; C_1 and C_3 correspond to the capacitances of the left and right outer membrane layers including ion double layers, respectively

2
Methods of Measurement of Boundary Potentials

2.1
Mobility of Vesicles in the Electric Field

Although this method can be applied only to suspensions or colloid solutions, it is important as a classical reference for comparison with techniques used with planar bilayers. The method is well described in many student books. The measurement of the mobility of the vesicles, u, in an electric field allows the determination of the ζ-potential by the equation of Helmholtz–Smoluchowski [52]:

$$\xi = \frac{u\eta}{\varepsilon\varepsilon_0} \tag{1}$$

The ζ-potential is the potential at the hydrodynamic shear plane. This plane does not coincide with the surface of the membrane. That is why the ζ-potential is lower than the surface potential calculated theoretically or obtained from other methods. The disparity depends on the ionic strength of the solution and becomes more significant as the ionic strength increases. The surface potential can be calculated from the ζ-potential by the correction of the potential decay with distance from the membrane surface, according to the theory of the diffuse double layer. The best agreement of the corrected surface potential with theoretical predictions, and with the results of other methods, was achieved if the shear plane was assumed to be about 0.2 nm from the membrane surface [53, 54]. The fact that both reference planes used for the potential measurements by this technique (shear plane and aqueous phase far from the membrane) are placed in the same phase is a reason why this method gives information on the absolute value of the surface potential. All other methods reviewed below measure only the change of boundary potential rather than its absolute value.

2.2
Nonstationary Membrane Potentials on Planar Lipid Membranes

If changes of the boundary potentials are fast, one can use direct potentiometry in open circuit conditions to monitor them. This is possible because establishing the equilibrium or steady-state membrane potential takes a definite time depending on the membrane permeability to ions. Note that the equilibrium (or stationary) potential difference between two water solutions divided by the membrane has no relation to the boundary potentials on the membrane/water interfaces; it is determined only by the membrane selectivity to the ions and by their concentrations in the solutions. However, if there is a fast process leading to a change of electrostatic potential on one of the membrane boundaries, the charge distribution across other regions of the membrane remains unaffected. An electrometer can measure this change as transient followed by the relaxation of the potential to the equilibrium or steady-state value. Obviously, an accurate measurement of the boundary potential change by this method is possible only if this change is much faster than the time constant of the membrane discharge. This constant being evaluated as a product of the membrane capacitance and resistance, is about 100 s for low-conductive membranes (10^{11} $\Omega \times 10^{-9}$ F). Consequently, this method can be applied to measure the changes of boundary potential within the timescale of seconds or faster. Examples of its application include investigations of the boundary potential changes due to photoinduced charge transfer [55], flexoeffects (polarization of the membrane due to its fast bending) [56], and changes of surface potential due to fast change of ionic strength [57]. Unfortunately, adsorption processes are usually much slower, and the attempts to apply this method to measure the changes of boundary potentials caused by adsorption gave only qualitative results [58, 59].

2.3
Short Circuit Currents During Modification of Electrical Double Layers

Change of the boundary potentials under short circuit conditions due to adsorption of charged compounds leads to redistribution of the countercharge between the two sides of the membrane; this can be detected as a transient electric current through the external circuit. The process can be formally explained by the model of three capacitors (Fig. 2). The electrical event corresponding to adsorption of a charged substance on the membrane surface (for example, an

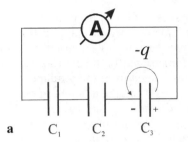

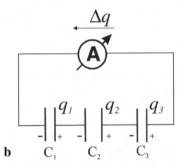

Fig. 3. Changes of boundary potentials caused by an external event (a) lead to a transient current through external circuit (b). For details see the text

anionic substance on the right side of the membrane) can be considered as a transfer of the charge $-q$ through the right-side phase on the membrane surface (Fig. 3). In the short circuit conditions, the countercharge Δq will pass through the external circuit to compensate the potential change on the capacitor C_3:

$$\Delta q = q \frac{C}{C_3},$$ (a)

where C is the total electrical capacitance of the membrane including the capacitance of the electrical double layers:

$$\frac{1}{C} = \frac{1}{C_1} + \frac{1}{C_2} + \frac{1}{C_3}.$$ (b)

This charge can be measured experimentally by integrating the transient short circuit current. It would appear that the value Δq can be used to estimate the surface charge of adsorbed substances. However, the capacitance of the electrical double layer C_3 cannot be measured directly. Its determination requires a time-consuming analysis of ionic strength effects. Additionally, a modification of surface charge results in changes of the capacitance of electrical double layers. The charge can be redistributed not only through the external circuit, but also immediately through the membrane due to leakages, which complicates the analysis. Because of these reasons, determination of the surface charge of the adsorbed substances by integrating the current passing through the external circuit did not find any analytical application. However, direct measurements of these transient currents as quantitative indications of recharging of double layers are used intensively [60]. The relaxation time for the charge distribution, which can be estimated from the product of the resistance of the external circuit (typically <1 MΩ) and the membrane capacitance (typically <10 nF), is much less than the adsorption time. Therefore, the current through the external circuit is determined by the adsorption rate and charge of the adsorbing species and, according to Fick's law, should be proportional to their concentrations and diffusion constants.

A similar consideration can be applied to the adsorption of dipolar compounds; in this case the current through the external circuit is also proportional to the concentration of adsorbing species.

2.4
Determination of Boundary Potential Changes from Measurements of Membrane Conductance

This widely used method is based on the influence of boundary potential on BLM conductance in the presence of ion carriers or hydrophobic ions. Consider a symmetric membrane with equal boundary potentials on both sides in the presence of hydrophobic ions. According to the simplest model, the transport consists of two steps: exchange of hydrophobic ions between water solutions and the membrane, and their jumps through the membrane interior (Fig. 4). We demonstrate here that these two steps are influenced by different parts of the boundary potential. This provides powerful experimental tools for analysis of variations of the potential distribution across the membrane due to adsorption of membrane-active molecules, modification of membrane lipids, and other external events.

The ions adsorb inside the membrane in some layers near the membrane surface, corresponding to the minimum of their potential energy. At low ion concentrations in the membrane phase one can neglect their mutual interactions and saturation effects. The equilibrium amount of ions in the adsorption layer, n, is related to its concentration in water, c_w, and the potential difference between water and the adsorption layer φ_n (Fig. 4) through the Boltzmann distribution:

$$n = \gamma c_w \exp\left(-\frac{zF\varphi_n}{RT}\right),\tag{2}$$

where γ is a coefficient defined by the nonelectrical interaction of the hydrophobic ions with the membrane and water phases (it can be considered as a binding constant).

Hydrophobic ions or ion carriers are usually added in aqueous solution. Because of the very small volume of the membrane phase relative to the aqueous one, the concentration c_w can be considered constant. As a consequence, change of the potential difference between the adsorption layer and bulk water $\Delta\varphi_n$

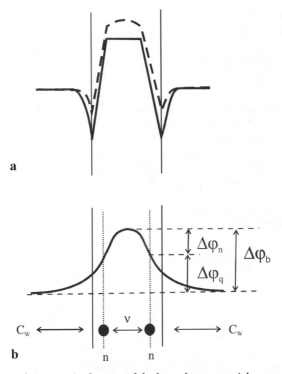

Fig. 4a,b. Influence of symmetric changes of the boundary potentials on membrane transport of hydrophobic ions. The initial and modified potential profiles (**a**) are indicated by *solid* and *dashed lines* respectively; the difference of the profiles is shown in (**b**). For details see the text

leads to a change of the amount of hydrophobic ions inside the membrane from n_0 to n according to the equation

$$\frac{n}{n_0} = \exp\left(-\frac{zF\Delta\varphi_n}{RT}\right). \tag{3}$$

The kinetics of ion transport through the membrane can be described either by the Nernst-Planck equations based on consideration of the membrane as a continuous phase, or by the absolute rate theory considering the ion movement across the membrane as a jump through an energy barrier [45, 46]. In the first approach, the best fitting of experimental current-voltage characteristics was obtained for a trapezoidal energetic barrier [61], whereas in the second approach a sharp barrier shape is assumed. Symmetric changes of two boundary potentials on opposite sides of the membrane due to external events modify the height of the barrier (Fig. 4). The height is affected by that part of the boundary potential, φ_T, which drops between the adsorption plane and the middle of the membrane. It was shown in both approaches that the rate constant of the charge transport through the membrane depends on the height of the barrier according to an Arrhenius-type equation; therefore, regardless of the model selected, a change of the potential $\Delta\varphi_\tau$ leads to a change of the rate constant from v_0 to v:

$$\frac{v}{v_0} = \exp\left(-\frac{zF\Delta\varphi_\tau}{RT}\right) \tag{4}$$

If the membrane conductance is proportional to the product of the concentration of penetrating ions inside the membrane n and the rate constant of their transport v, the influence of the boundary potential on the membrane conductance can be calculated by multiplying Eqs. 3 and 4. As a result, the change of the conductance from g_0 to g caused by the total change of the boundary potential $\Delta\varphi_b = \Delta\varphi_n + \Delta\varphi_\tau$ is described by the following equation:

$$\frac{g}{g_0} = \exp\left(-\frac{zF\Delta\varphi_b}{RT}\right) \tag{5}$$

Equations 3 and 5 demonstrate that the total change in conductance is sensitive to the potential difference between the bulk solution and the plane in the middle of the membrane φ_b, while the distribution coefficient is sensitive to the average potential in the plane coinciding with the plane of adsorption of hydrophobic ions φ_n (Fig. 4). Determination of the boundary potential from membrane conductance, first reported by Lesslauer et al. [62] with iodide, was then used intensively by McLaughlin [63, 49] with the potassium-nonactin complex.

Note that Eq. 5 is valid only if the conductance is proportional to the product of the concentration of penetrating ions inside the membrane and the rate constant of their transport. The BLM conductance to DC current is determined by the rate-limiting step of the ion transport. In this case, Eq. 5 is valid only if the limiting step is the translocation of charged particles across the membrane. This is true for many ion carriers, such as potassium complexes of nonactin and valinomycin, or positively charged hydrophobic ions like tetraphenylarsonium [46, 47, 64]. However, there are also many examples of ion transport systems where it is not the case. For such systems, the assessment of the boundary potentials can be performed from kinetic methods that can determine the rate of ion transport through the inner part of the membrane [65, 66]. The latter approach also gives additional information, allowing one to determine not only changes of the total boundary potential, but also of its parts influencing either the distribution coefficient (Eq. 3) or the rate constant (Eq. 4). Such investigations were performed with rubidium-valinomycin complexes [67, 68], or with hydrophobic anions such as tetraphenylborate [41, 58, 69, 70–72] and dipicrylamine [73].

The main disadvantage of the methods based on conductance measurements is that the conductance change may be caused not only by the modification of boundary potential, but also by other properties of lipid bilayers, such as thickness, dielectric permeability, mobility of charged complexes inside the membrane, or number of binding sites for these complexes in the membrane. That is why a pair of oppositely charged ion translocating complexes is used in many investigations. It is clear from Eqs. 3–5 that the change either of total conductance or of separate parameters of ion transport should be in opposite directions for oppositely charged particles penetrating the membrane. On the other hand, any nonelectrostatic factors influence the ion conductance independently of the sign of the charge of the permeating particles. Comparison of the changes of conductance induced by the pair of systems transporting oppositely charged ions allows one to distinguish electrostatic and nonelectrostatic factors. For example, a comparison of the effects of cholesterol on the conductance of the membrane for negatively charged iodide and for positively charged potassium-nonactin complexes indicated that cholesterol modifies not only dipole potential, but also nonelectrostatic parameters of the lipid bilayer [74].

A pair of hydrophobic ions – tetraphenylboron and tetraphenylarsonium – was also used to estimate the absolute value of boundary potential inside the membrane [75]. The idea was that these anions have similar structure but opposite charges. One could therefore expect that the electrostatic potential is the main factor responsible for a difference in BLM conductance in the presence of each of these two ions. These experiments showed that the boundary potential is positive inside the membrane and its value is about 200 mV. However, exact measurement of the absolute value of boundary potential is impossible in principle, because it is the potential difference between different phases. These results can be regarded therefore only as estimates, because they did not take into account, for example, the difference of hydration energies for oppositely charged ions.

2.5
Methods of Compensation of Intramembrane Field

The measurement of boundary potential difference between two sides of a membrane using the intramembrane field compensation method (further referred to as the IFC method) is based on the following idea. Consider a potential distribution across an asymmetric lipid bilayer (Fig. 2). It can be seen that the field inside the membrane results from the algebraic sum of the potential difference between the aqueous solutions U and the difference of boundary potentials on both sides of the membrane $\Delta\varphi_b = \varphi'_b - \varphi''_b$. The intramembrane field can be represented by the potential drop inside the hydrophobic region of the membrane ϕ_{in}. Therefore,

$$\varphi_{in} = U + \Delta\varphi_b \qquad (6)$$

If the difference of electric potentials between water solutions is zero (short circuit conditions), the asymmetry of the boundary potentials on two sides of the membrane results in an intramembrane electric field. This field disappears when the voltage applied between the water solutions is equal to the difference of boundary potentials (Fig. 2, dashed line). This is the basic idea for measurements of this boundary potential difference. An experimental application of this principle requires a sensitive method for detection of the electric field inside the membrane. For example, one can use the effect of the intramembrane field on the opening of incorporated ion channels [76], on the shape of current–voltage characteristics in the presence of ion carriers [77–79], or on the electric breakdown of the membrane [3]. However, the most convenient methods are certainly based on the influence of the intramembrane field on the electrical capacitance of nonmodified membranes.

The dependence of the membrane capacitance on the applied potential was found as early as 1966 [80]. The mechanism of this phenomenon was the subject of intensive discussions for many years. The main models used for its explanation are based on elastic compression of the lipid bilayer [81][3], on solvent movement from the lipid bilayer into the border or microlenses [82], or on voltage-dependent thermal capillary waves on the bilayer surface [83, 84]. However, application of this phenomenon for measurement of the boundary potential difference does not require a physical explanation of this dependence; the membrane capacitance is only required to be influenced by the *intramembrane* electric field and to reach its minimum at zero intramembrane field. The validity of this assumption follows from the point discussed above, that the applied voltage drops inside the hydrophobic region of the membrane. It was also confirmed by coincidence of the results obtained by the methods using this phenomenon and by other methods.

The idea to measure boundary potentials by capacitance minimization was first mentioned by Shoch and Sargent in 1976 [85]. Later, several groups reported different technical realizations of this principle. The most direct way to find the capacitance minimum was published by Alvarez and Latorre [86], who directly measured the membrane capacitance as a function of the applied voltage. This dependence has a minimum at zero voltage for symmetric membranes; the minimum shifts from zero voltage for asymmetric membranes formed from two monolayers of different lipids. Later, this technique was used by other authors [79, 87]. The technique reported by Abidor et al. [88] employed the fact that the capacitance dependence on the applied voltage should be symmetric relative to the point of minimum. This symmetry follows from the independence of the mechanical force compressing the membrane from the polarity of the electric field.

A more convenient technique, allowing one to perform automated continuous measurements, was developed by Sokolov and Kuz'min [89]. These authors used the second harmonic of the capacitive current generated by nonlinear membrane capacitance under application of sine wave voltage as an indication of the intramembrane field. The generation of higher harmonics on BLM [90, 91] results from the nonlinearity of the membrane capacitance. Any dependence of capacitance C on voltage U near the point of minimum can be approximated by a parabolic function corresponding to the first term of the power series:

$$C = C_0 \left[1 + \alpha \left(U - \Delta\varphi_b \right)^2 \right],$$ (7)

---

[3] See also T. Hianik *Electrostriction of supported lipid membranes and their application in biosensing* in this book (ed).

where C_0 is the capacitance at the point of minimum, and α is the so-called electrostriction coefficient depending on the structure of the lipid bilayer [81, 92]. If the voltage U applied to the membrane is a sum of sine voltage and DC bias

$$U = U_0 + V\cos(\omega t), \qquad (8)$$

the current through the membrane contains the second and the third harmonics

$$I = V\omega C_0 \left[1 + 3\alpha \left(U_0 - \Delta\varphi_b \right)^2 + \frac{\alpha V^2}{2} \right] \sin\left(\omega t\right) -$$

$$-3\alpha C_0 \omega V^2 \left(U_0 - \Delta\varphi_b \right)\sin\left(2\omega t\right) - \frac{3}{4}\alpha C_0 \omega V^3 \sin\left(3\omega t\right) \qquad (9)$$

The magnitude of the second harmonic is proportional to the difference of DC bias U_0 applied to the membrane together with the sine wave voltage and the voltage $\Delta\varphi_b$, corresponding to the point of minimum

$$I_{2\omega} = 3\alpha C_0 \omega V^2 \left(U_0 - \Delta\varphi_b \right) \qquad (10)$$

It is clear that the second harmonic is zero at the point of minimal capacitance, where the intramembrane field is compensated by the external voltage $U_0 = \Delta\varphi_b$. This method of measuring $\Delta\varphi_b$ has essential advantages. Low second harmonic signals can be amplified by highly sensitive selective amplifiers, thus providing an accurate detection of compensation conditions and appropriate measurements of $\Delta\varphi_b$. Besides, the compensation of the intramembrane field can be automated through a negative feedback loop. This provides measurements of slow kinetics of changes in the boundary potentials. The realization of this approach was described by Sokolov and Kuz'min [89]. The second harmonic signal was rectified by the lock-in amplifier and applied to the membrane as DC voltage; due to this feedback, the system spontaneously fixed the applied potential at a level corresponding to the disappearance of the second harmonic of the membrane current, i.e., to compensation of the intramembrane field.

The accuracy of the measurement of the boundary potential difference depends on the magnitude of the second harmonic signal. According to Eq. 10, this magnitude is proportional to the coefficient α representing the "elastic" property of the membrane and to the square of the amplitude of the alternating voltage V applied to the membrane. The smaller the coefficient α, the higher the alternating voltage required to get the same second harmonic signal. In practice, the value V is limited by the membrane stability, and the accuracy of the method depends mainly on α, which is a function of the lipid, and on the solvent composition [81, 92]. The best accuracy (up to 0.1 mV) is observed for bilayer lipid membranes prepared from lipid in decane solution. For "solventless" bilayer lipid membranes, the value of α, and correspondingly the accuracy, are less.

3
Effects of Different Compounds on Electrostatic Potentials of Lipid Bilayers

3.1
Effects of Inorganic Ions

Experiments with planar lipid bilayers are usually performed in aqueous solutions containing inorganic ions. This is a reason for intensive investigation of the effects of inorganic ions on the physical properties of membranes, including electrostatic effects (see, for example, reviews [43, 49, 93]). In most cases, the electrostatic potentials in the presence of inorganic ions can be described by the model of the diffuse double layer. The simplest theory, developed independently by Gouy and Chapman in 1910–1913, is based on a combination of the Poisson equation describing potential distribution, and the Boltzmann equation describing distribution of ions. The model gives a relation between surface charge σ, concentrations of ions c_i, and surface potential φ_s (see the definitions above):

$$\sigma^2 = 2\varepsilon\varepsilon_0 RT \sum_i c_i \left[\exp\left(-\frac{Fz_i\varphi_s}{RT} \right) - 1 \right] \tag{11}$$

For a 1:1 electrolyte, the main equation of the Gouy-Chapman theory has a more simple form:

$$\frac{\sigma}{\sqrt{8\varepsilon\varepsilon_0 RT\, c_1}} = \mathrm{sh}\left(-\frac{zF\varphi_s}{2RT} \right) \tag{12}$$

In the case of lipid bilayers, the electrical double layer is formed by charges of polar heads of lipid molecules, and by the distributed charge formed by ions in the adjacent aqueous solution. The Gouy–Chapman theory is very rough; many investigators criticized it and suggested more advanced theories taking into account, for example, finite sizes of ions, or interaction of the dipolar molecules of water [93–96]. Surprisingly, even this oversimplified theory adequately describes most of the experimental data on BLM electrostatics [49].

According to these equations, the surface potential can be modified either due to change of ion concentration in the solution or due to change of the surface charge. Inorganic ions can influence the surface potentials in both ways. The first is the so-called screening effect of the electrolyte; it is caused by ion redistribution near the charged surface and leads to a decrease of the surface potential at increased ionic strength of the solution. The second effect is adsorption of ions onto the membrane, leading to the corresponding modification of the surface charge, σ. The binding is specific to ions and to the lipid composition. Since the surface charge is usually negative, it is not surprising that the lipid membranes bind cations preferentially. The binding constants determined for the series of different monovalent and divalent cations [97–101] were summarized in several reviews or monographs [43, 96].

A transversal distribution of electrostatic field created by adsorbed ions depends on the position of adsorption sites (adsorption plane) inside the membrane. Information about this distribution can help in identifying the depth at which the adsorption plane is located, as well as in detecting possible structural rearrangements in the membrane initiated by ion adsorption.

Comparison of ζ-potential and potentials obtained by other methods averaged in a different "sensitivity plane" inside the membrane gives the fractions of the total change of boundary potential dropping outside and inside this plane. Application of such an approach to inorganic ions demonstrates that in most cases the measurement of ζ-potential yields the same result as other methods. In other words, the potential created by the adsorption of inorganic ions is located in the diffuse layer of water solution outside the membrane. The coincidence of ζ-potential with boundary potentials determined from the change of conductivity in the presence of nonactin, in the case of adsorption of divalent cations Ca^{2+}, Mg^{2+}, Ni^{2+}, and Cd^{2+}, was shown by McLaughlin and coworkers [49]. The same conclusion was made from the comparison of ζ-potential with the potential measured by the IFC method [54]. An example of such coincidence with Mg^{2+} and Be^{2+} is presented in Fig. 5. However, in some cases of adsorption of ions with high affinity to the lipids (Be^{2+}, Gd^{3+}), the boundary potential measured by the IFC method was significantly higher than the ζ-potential [54, 102]. Therefore, these ions induce potential changes not only in the diffuse double layer outside the membrane (similar to other inorganic ions), but also in the polar region behind the membrane surface. The first component is measured by electrophoretic measurements (ζ-potential), the second, by methods sensitive to the potential change inside the membrane (for example, the IFC method).

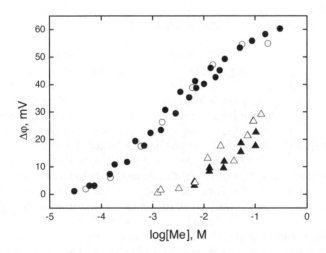

Fig. 5. Changes of boundary potential (*solid symbols*) and surface potential (*open symbols*) due to adsorption of beryllium (*circles*) and magnesium (*triangles*) ions on lipid membranes from diphytanoylphosphatidylcholine. The background electrolyte contained 0.1 M KCl and 20 mM imidazole (pH 6.4). The values of surface potential were calculated from the results of microelectrophoretic measurements considering the slip plane situated at a distance of 2 Å from the surface of liposomes. The data are reproduced from [54]

Probably, this potential drop represents a change of the dipole potential due to structural rearrangement of the membrane initiated by these ions.

3.2
Effects of Amphiphilic Ions

Amphiphilic compounds, combining both hydrophobic and hydrophilic groups in one molecule, display strong affinity to biological membranes and their models. Therefore, it is not surprising that most of these molecules are biologically active. This group includes many biologically active compounds and drugs (for example, local anesthetics, neuroleptica), mediators and components of biological signal transfer (diacylglycerides, phosphatidylinositol, fatty acids, prostaglandins, leucotrienes, thromboxanes, steroid hormones), membrane-fluorescent and ESR probes used in biochemical and biophysical experiments, and detergents.

Amphiphilic molecules are able to adsorb on the membrane/water interface and to influence the electric field distribution there. If organic ions are neutralized by protonation or deprotonation, such neutral molecules can often penetrate through the membrane. Investigation of their adsorption is important for understanding the biological effects of these substances and from an analytical point of view. Modification of the boundary potential is caused mainly by polar groups of amphiphilic molecules, which are either charged (organic ions) or possess significant dipole moment (neutral molecules). The effects of neutral molecules are discussed in a separate section.

3.2.1
Theoretical Models for Adsorption of Amphiphilic Ions

The first theoretical model used to describe adsorption of organic ions assumed that they modify potential drop only in the diffuse double layer, and that its effect on the potential distribution can be described by the Gouy–Chapman theory. However, the experimentally observed dependencies of boundary potential changes on ionic strength for many hydrophobic [103, 104] and amphiphilic [59, 105–110] ions were considerably weaker compared to the predictions of this simplest model. Several explanations have been suggested for this fact. One of them assumed that hydrophobic ions create a very high electric field at the membrane/water interface, leading to the dielectric saturation of the adjacent layer of water. This decrease of the dielectric permeability leads to a higher increase of the electrostatic potential than the Gouy-Chapman theory predicts [104].

Another explanation suggests that the boundary potential caused by the adsorption of these ions consists of two parts. The first one is a potential drop in the diffuse double layer (surface potential) and can be screened by ions of the electrolyte solution. The second component cannot be screened. The origins of the unscreenable potential can be different. If amphiphilic ions have dipole moments, the unscreenable potential can be created by these dipoles [59]. However, the fact that this potential was found for many organic ions of different structures, and its sign always coincides with the sign of charge of these ions, is a strong argument in favor of another model suggested by McLaughlin et al. [49, 103] describing the electrostatic effects of hydrophobic ions, which was extended [111] for amphiphilic ions. According to the authors [103], the ions are adsorbed at some depth δ within the membrane, thus creating a potential drop φ_δ in the membrane surface layer. The total boundary potential φ_b is the sum of the surface potential φ_s in the diffuse layer and the potential drop φ_δ in this layer:

$$\varphi_b = \varphi_s + \varphi_\delta \tag{13}$$

The first term, φ_s, can be calculated from the theory of the diffuse double layer. This potential can be screened by electrolyte ions. Usually, the solubility of organic ions in aqueous electrolytes is low while their affinity to lipid membranes is very high; therefore they are present at very low concentrations and do not influence the total ionic strength of the solution. For a 1:1 electrolyte, the dependence of φ_s on surface charge σ and ionic concentration c_1 can be described by Eq. 12. The second term, called "unscreenable potential" according to [108], (in contrast to [103], where it was called "boundary potential"; in the present review, as well as in our other publications, the name boundary potential denotes the total change of potential φ_b) can be calculated by the simple equation of a capacitor:

$$\varphi_\delta = \sigma / C_\delta \tag{14}$$

It does not depend on the ionic strength. To have a complete set of equations, one needs an additional equation describing the adsorption of the ions on the membrane. If the membrane is formed from neutral lipids, its initial surface charge is zero, and the charge σ is determined by the adsorption of ions. The adsorption of organic ions on the membrane is usually far from saturation, and can be therefore described by the linear equation (Henry isotherm):

$$\sigma = zKC_a \exp\left(-\frac{zF\varphi_b}{RT}\right) \tag{15}$$

It follows from these equations that the effect of ionic strength on the total boundary potential φ_b depends on the relative contribution of the unscreenable potential φ_δ to its value. In the absence of the unscreenable potential (if $\varphi_\delta=0$), the limiting dependence of the potential on the concentrations of electrolyte c_1 and the ions c_a with valence z adsorbing on the membrane can be derived by introducing Eq. 15 into the equation of Gouy-Chapman (Eq. 12) and by approximating the hyperbolic sinus with an exponential (that is valid if the potentials are much higher than RT/F):

$$\varphi \approx \frac{2RT}{(2z+1)F}\ln(c_a) - \frac{RT}{(2z+1)F}\ln(c_1) + \frac{2RT}{(2z+1)F}\ln\left(\frac{zK}{\sqrt{2\varepsilon\varepsilon_0 RT}}\right) \tag{16}$$

This equation shows that the potential induced by the adsorption of monovalent ions (z=1) decreases with increasing ionic strength c_1 with the slope equal to 1/3*2.303 RT/F, or about 20 mV per decade.

In the second limiting case, when the unscreenable potential is dominant, the dependence of boundary potential on ionic strength vanishes. The parameter influencing the contribution of unscreenable potential is the capacitance C_δ in Eq. 14. The physical meaning of this capacitance is clear from the equation of a planar capacitor:

$$C_\delta = \frac{\varepsilon \varepsilon_0}{\delta}$$

According to the model of Andersen et al. [103], C_δ is the capacitance of the layer of the membrane between its interface with water and adsorption plane of the ions. Its thickness δ corresponds to the depth of immersion of the adsorption plane inside the membrane, while ε is the mean dielectric permeability of this layer. A decrease of the capacitance C_δ leads to an increased contribution of the unscreenable component of the boundary potential. Therefore, the deeper the plane of ion adsorption is placed inside the membrane and the lower the dielectric permeability in this region, the weaker is the dependence of the boundary potential on the ionic strength.

It is interesting also to consider the slope of the dependence of boundary potential on the concentration of ions in the solution $\lambda_E = \dfrac{\partial \varphi}{\partial \log C_a}$ (Esin–Markov coefficient). It can be shown that the limiting value of this coefficient in the absence of unscreenable potential (Eq. 16) is equal to $2/3 * 2.303 * \dfrac{RT}{F}$; it corresponds to about 40 mV per decade of concentration. The contribution of unscreenable potential to the total boundary potential increases this coefficient to $2.303 * \dfrac{RT}{F}$, i.e., to about 60 mV per decade [103, 111]. However, the experimentally measured slope is even higher. To explain this, it was suggested that the interaction of ions adsorbed at some depth inside the membrane should be described by discrete distribution of charge of adsorbed ions in the adsorption plane [103]. The general case of interaction of ions in the adsorption plane inside the membrane was considered by Kozlov et al. [111], where both electrostatic and nonelectrostatic interaction of the ions was considered.

3.2.2
Potential Distribution

As was shown, important information on the localization of the adsorption plane for amphiphilic ions can be obtained from the experimental dependence of boundary potential changes on the ionic strength. Another approach consists of the study of the distribution of the electric field created by the amphiphilic ions across the membrane. It can be done by comparison of the results obtained by different techniques that measure potential averaged at different planes outside or inside the membrane. The most informative becomes the comparison of the boundary potential changes measured by the IFC technique with surface potentials measured by the electrophoretic method. Such measurements with different biologically important amphiphilic ions are shown in Fig. 6. It is clearly visible on all graphics that the potential measured by the IFC method is higher than the ζ-potential calculated from the electrophoretic mobility of liposomes in identical conditions. This difference of the potentials is caused by different positions of the planes where these potentials are measured: the ζ-potential is measured in the slip plane in the aqueous phase, while the total change of boundary potential is measured in the plane inside the membrane. In other words, the ex-

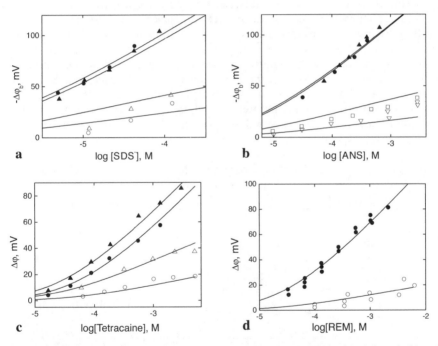

Fig. 6a–d. Changes of boundary potentials measured by the IFC method (*solid symbols*) and zeta potentials calculated from the data of electrophoretic measurements (*open symbols*) initiated by the adsorption of different amphiphilic ions: sodium dodecyl sulfate (**a**), aniline naphthalene sulfonate (**b**), tetracaine (**c**), and remantadine (**d**). BLM were formed either from dioleoyl phosphatidylcholine (**a** and **b**), soybean phosphatidylcholine (**d**), or diphytanoyl phosphatidylcholine (**c**). Water solution contained KCl in concentration 0.1 M (*circles*) or 0.01 M (*triangles*) and low concentrations of buffers which were different in different pictures. *Solid lines* are the theoretical curves plotted from numerical solution of Eqs. 12, 14, and 15. The data were reproduced from [106–109]

ternal part of the boundary potential, measured by the electrophoretic method, is less than the total measured by the IFC method. The internal part of the potential may be attributed to unscreenable potential. Therefore, application of this approach provides direct proof of the existence of unscreenable potential created by adsorption of amphiphilic ions. Remarkably, this approach does not require any theoretical assumptions on the nature of the unscreenable potential. Figure 6 demonstrates that the unscreenable potential is present for all of the studied amphiphilic compounds. Quantitative discrepancies may be caused by different positions of their adsorption planes inside the membrane.

This approach yields some evidence in favor of the model [103] that the unscreenable potential arises due to "immersion" of the adsorption plane of amphiphilic ions into the membrane. According to this model, the depth of immersion, and hence the contribution of unscreenable potential into the total change of boundary potential, depends on the hydrophobicity of the amphiphilic ions. Comparison of the results obtained with hematoporphyrin derivatives differing in their hydrophobic groups demonstrated that more hydrophobic molecules give a higher contribution to the unscreenable potential [110].

Another approach to analyze the influence of adsorption of amphiphilic ions on the distribution of electric field inside the membrane is based on measurements of kinetic parameters of transport of hydrophobic ions. As was shown above, the total change of boundary potential can be divided by this approach into two components: between the bulk aqueous solution and the adsorption plane of hydrophobic ions, and between this plane and the middle of the hydrophobic region inside the membrane. An application of this approach for investigation of adsorption of sodium dodecyl sulfate (SDS) gives a comparison of

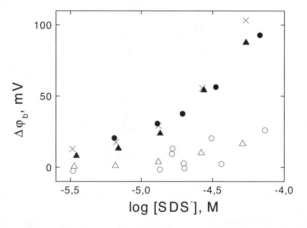

Fig. 7. Changes of boundary potential of BLM from soybean phosphatidylcholine (40 mg/ml) due to adsorption of dodecyl sulfate anions as measured by the IFC method (*crosses*) or from the transport parameters of hydrophobic tetraphenylborate (*circles*) or dipicrylamine (*triangles*) anions. The changes in $\Delta\varphi_b$ (*solid symbols*) and $\Delta\varphi_\tau$ (*open symbols*) were calculated using Eqs. 1 and 4 from the parameters of the exponential decay of current in the presence of hydrophobic anions. The data were reproduced from [70]

the positions of the adsorption planes for these ions and for hydrophobic ions. Figure 7 shows the results obtained by the current relaxation method with tetraphenylborate (TPB) as hydrophobic ion. It follows from Fig. 7 that the adsorption of SDS leads to a decrease of the distribution coefficient of TPB, having only a minor influence on the rate constant of its transport. According to the considerations presented above, one can conclude that the main change of boundary potential due to adsorption of SDS takes place between the aqueous phase and the adsorption plane of TPB. Therefore, SDS ions adsorb inside the membrane at a smaller depth compared to TPB ions. The total change of boundary potential determined from the initial conductivity of BLM for TPB ions coincides with the change of boundary potential measured by the IFC method (Fig. 7).

3.2.3
Transfer Through the Membrane

One of the restrictions of the IFC method discussed above is that it cannot be applied if the changes of boundary potential are caused by adsorption of molecules which can diffuse through the membrane. In this case the difference of the boundary potentials between the two membrane sides measured by the IFC method is less than the true change of the boundary potential on one side. Therefore, a difference between the results obtained by the IFC method for asymmetric additions of amphiphilic molecules and the results obtained by the conductivity method for symmetric additions of these molecules can indicate permeability of these molecules through the membrane. Unfortunately, as it was pointed out above, the conductivity method, being based on strong assumptions, cannot give true results in each case.

In the case of amphiphilic ions, the measurement of boundary potential by the IFC method can be used to study their transfer through the membrane, and a true change of boundary potential on one side of the membrane can be measured even in the complex case when they penetrate through the membrane. The approach will be illustrated with remantadine (REM). This amphiphilic ion carrying positive charge is widely used as medicine against the influenza virus. It was shown that the molecules of REM penetrate through the membrane in neutral form. The transfer is coupled with the release of hydrogen ions in the unstirred layer at the cis side of the membrane (the side at which the REM was added to the water solution), and their absorption on the opposite, trans, side (Fig. 8). The electroneutral transport of weak acids and bases through the BLM was studied earlier by Antonenko and Yaguzhinskii [112], who suggested measurement of the change of local concentrations of hydrogen ions in the unstirred layers by measuring the membrane potential in the presence of a protonophore, making the membrane permeable for hydrogen ions. A theoretical model of electroneutral transport was developed by Markin et al. [113].

It was shown that the distribution of charged forms of amphiphilic ions between two boundaries of the membrane depends on the difference of concentrations of hydrogen ions in the solutions. At some value of ΔpH the limiting case of distribution is achieved, when the charged form, RH^+, is concentrated entirely on one BLM side [113, 114]. If the pH is higher on the trans side, the

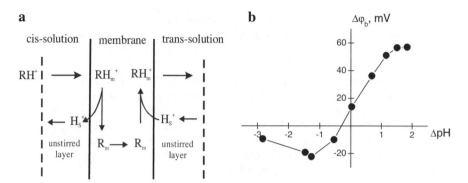

Fig. 8a,b. Transport of the amphiphilic ion remantadine through the membrane (a) decreases the boundary potential difference. The effect depends on the transmembrane pH gradient (b) and can be diminished if the pH gradient is greater than 2

charged form is concentrated at the cis-side boundary. The maximum value of $\Delta\varphi_b$ (Fig. 8) corresponds to this distribution of the charged form. In the case where the pH is higher on the cis side, the concentration of the charged form on the trans side (where the solution is free of REM) is increased. In this case, an opposite sign of the potential is measured (Fig. 8).

This explains the obtained dependence of $\Delta\varphi_b$ on ΔpH. The inner field compensation method measures the difference between the boundary potentials on each BLM side. The boundary potential value at each side is determined by the concentration of the charged REM form, $[RH^+]$, on the BLM surface. So the boundary potential difference is sensitive to the distribution of the charged form between the surfaces. The distribution depends on the pH difference between the two membrane sides.

The proposed model explains the influence of the buffer concentration on the value of the boundary potential difference. Indeed, the transport of REM through the membrane leads to a shift of pH in the unstirred layers. This shift depends on the buffer concentration. The effect is especially pronounced at low buffer concentrations [109, 114]. As a result, a pH gradient is created, and it influences the $\Delta\varphi_b$ values like bulk pH differences. As REM transport leads to a pH decrease on the cis side and to an increase on the trans side, the boundary potential difference increases at low buffer concentration compared with that at high buffer concentration, where the pH gradient is almost equal to zero.

The proposed model of REM transport requires the existence of the neutral form of REM in the membrane. This substance is a weak base. Therefore the amount of its neutral form depends on the pH of the solution. A decrease in the concentration of the neutral REM form in the membrane should lead to a decrease of the REM flux through the membrane. In turn, this should decrease the influence of REM transport on the value of the boundary potential difference. Indeed, when the initial pH of the solution was 4.2, the $\Delta\varphi_b$ became independent of ΔpH [114]. These experiments support the validity of the proposed model.

The results of this work demonstrate the possibility of measuring the changes in the boundary potential due to adsorption of REM at one BLM interface, despite its penetration through the membrane. The IFC method measures the difference between the boundary potentials on each BLM side. It is necessary to exclude the influence of the trans-side boundary. When a substance under investigation cannot pass through the membrane, the measurement can be carried out easily. But in the case of REM we have a quite different situation. REM can pass through the membrane and adsorb on both membrane sides. The results obtained in this work show two ways to exclude the influence of the trans side. Firstly, it is possible to block the transport of REM by decreasing the concentration of the neutral REM form in the membrane. This can be achieved by changing the pH in both cell compartments. The experimental conditions in this case are partly limited. The second way is preferable: it assumes the ΔpH between the compartments is changed. The appropriate transmembrane pH gradient causes this type of distribution of a charged form of REM when it is confined to the cis side of the membrane. This case corresponds, practically, to an equilibrium distribution of REM molecules between the solution and the membrane surface in the absence of REM transport through the membrane. This method of measurement was used for the investigation of adsorption on the BLM of the amphiphilic ions that permeate through it [107, 109, 110].

3.3
Effects of Neutral Molecules

3.3.1
Dipole Potentials

We discussed above the adsorption of ions that can be detected as a change of electrostatic potential on the boundary of the membrane. The adsorption of electroneutral molecules can also be detected by similar methods if these molecules have dipole moments that have a preferred orientation on the boundary of the membrane. The adsorption of such molecules will change the boundary potential that can be measured. Such potential change will be called the dipole potential. The first substance capable of changing the dipole potential of BLM was discovered in 1974. It was found that cholesterol changes the dipole potential in a positive direction, thus increasing the conductivity of BLM to negative ions of iodide and decreasing it for positively charged complexes of nonactin with K^+ [74]. Later it was shown that the effect of cholesterol depends on the lipid content of the membrane, and that perhaps this effect is caused by the change of orientation of the lipid head rather than by the dipole moment of cholesterol [75]. The more significant positive change of the dipole potential in BLM induces the derivative of cholesterol, ketocholestanol [115]. Recently it was found that many styryl dyes are capable of changing the dipole potential to a positive direction [72, 116, 117]. Interestingly, these dyes were designed as probes of electric fields that can be detected from the shift of their absorption or fluorescence spectra by an electrochromic mechanism. The same group of the dye molecule that serves as a sensor of electric field provides large dipole moments of the molecule (it is created by a negative sulfo group localized near the surface of the membrane and the delocalized positive charge of the chromophore immersed in the membrane).

All substances mentioned above change the dipole potential toward the positive direction, i.e., make the potential difference between the lipid and water phases more positive. Many substances were found that change the dipole potential in a negative direction. The most studied substance of such a type is phloretin. The effect of phloretin on the dipole potential was discovered in 1976. It was found that phloretin increases the conductance of BLM to positive complexes carried by nonactin or valinomycin, and decreases the conductance to negatively charged hydrophobic ions [58, 68, 69, 73]. The change of dipole potential due to adsorption of phloretin was determined not only from the conductivity of BLM, but also by other methods: by fast change of membrane potentials in open-circuit conditions [103], by spin labels [118], by fluorescent probes [116, 119], by the IFC method [120, 121] (Fig. 9), and on the lipid monolayers by vibrating electrode [68, 122]. The kinetics of adsorption of phloretin and similar substances on the lipid bilayer were studied by the method of laser temperature jump [123, 124]. Phloretin is a weak acid. It presents in solution in two forms: neutral and negatively charged. It was shown that the dipole potential is created by the neutral form, because it decreased with increasing pH; this correlates with the predicted dependence of the concentration of the neutral form on pH [58, 120]. The derivatives of the phloretin are also capable of changing the dipole potential, but they are less effective [120, 123,125]. A considerable effect on the dipole potential, similar to that of phloretin, was found with the herbicide 2,4-D [71, 126, 127].

The theoretical description of dipole potential is very simple. The equation deriving the dipole potential φ_d as a function of the surface concentration of dipolar molecules n and the projection of their dipole moment to the normal p is similar to the equation of a planar capacitor:

$$\varphi_d = \frac{pn}{\varepsilon \varepsilon_0} \tag{d}$$

If n is proportional to the concentration of dipolar molecules in the solution, the dipole potential should depend on this concentration linearly. The linear dependence was found with styryl dyes [72] (Fig. 9). However, the dependence of dipole potential on the concentration of phloretin is close to linear only at low concentrations and tends to saturation at high concentrations. This can be explained by mutual interaction of the dipoles of the adsorbed molecules of phloretin, as well as by the saturation of the binding sites on the surface of the membrane [68, 128].

The IFC method also allows the study of the transfer of dipolar molecules through the membrane by an approach similar to that discussed above with amphiphilic ions. It was found that the potential measured by the IFC method when phloretin is present in the solution on one side of the membrane is considerably less than the potential determined from the change of conductivity of the membrane after symmetric addition of phloretin to both sides [120]. This observation was explained by the transfer of phloretin through the lipid membrane. The mechanism of the transfer is similar to the one discussed above for amphiphilic ions: phloretin permeates through the BLM in neutral form, and its transport is coupled with protonation/deprotonation reactions on the boundaries of the membrane [121]. Phloridzin, the molecule of which differs from that of phloretin by the presence of an additional glucose group, cannot penetrate through the membrane, and the potentials measured with phloridzin by the two methods coincide.

3.3.2
Potential Distribution and the Depth of Adsorption

The distribution of the electric field across the membrane created by the adsorption of neutral dipolar molecules can contain important information about the localization depth of their adsorption plane inside the membrane. The most useful method for this purpose was based on the measurement of induced conductivity in the presence of either ion carriers or hydrophobic ions. As was shown in previous sections in the present review, the boundary potential can influence either the partition coefficient of charged species between membrane and water, or their rate of transfer across the membrane. The first effect can be observed if the boundary potential changes between water and the adsorption plane of these charged particles, the second, if this potential changes in the region of the membrane deeper than the adsorption plane. Application of such an approach showed that the depth of adsorption of neutral dipolar molecules, in contrast to adsorption of ions, has considerable variation, depending on their structure. Cholesterol or styryl dyes influence only the rate of transport of hydrophobic ions through the membrane, while their partition coefficient remains unaffected [72, 73]. It means that these dipolar molecules adsorb to a considerable depth inside the membrane and change the dipole potential in the region of the membrane which is much deeper than the plane of adsorption of hydrophobic ions. Other dipolar molecules, such as phloretin or 2,4-D, influence both the partition coefficient and the rate of transport of hydrophobic ions; therefore, the depth of the adsorption plane of these molecules and hydrophobic ions is comparable [58, 69, 71, 73, 127].

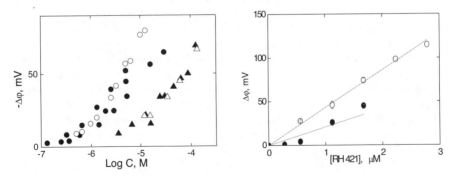

Fig. 9. Changes of dipole potentials on the BLM due to adsorption of phloretin or phloridzin (*left picture*) or styryl dye RH421 (*right picture*) as measured by the IFC method (*solid symbols*) or from the changes of conductivity (*open symbols*) in the presence of nonactin (*left picture*) or hydrophobic TPB- anions (*right picture*). The data were reproduced from [72, 120]

Another way of estimating the adsorption depth of neutral molecules was based on the fluorescence quenching of dyes. It allowed comparison of the depth of localization inside the membrane of the molecules of dye and dipolar molecules – quenchers. The location depth of the dipolar molecules could also be evaluated by the IFC method. It was shown that this method can correctly measure the boundary potential only if the region of its changing is located in the polar layer of the membrane near its surface. In the case of deep adsorption of dipolar molecules inside the hydrophobic region of the membrane, the IFC method yields a potential that is much less than the true change of boundary potential. The deviation depends on the depth of adsorption, which can be used for its evaluation [129].

4
Change of Boundary Potential due to Lipid Hydrolysis

Many biochemical processes on the membrane surface lead to modification of the boundary potential. These effects can be used for analytical applications. For example, variation of boundary potential due to hydrolysis of membrane lipids by phospholipases was used to develop a label-free assay for phospholipase activity [130, 131]. The approach was demonstrated on phospholipases A from different sources (bean, snake venom, pancreas) and on bacterial phospholipase C.

The hydrolysis of phospholipids by phospholipase A_2 leads to the formation of fatty acids and lysolipids. The effects of the phospholipase action as well as additions of the hydrolysis products on the boundary potential were investigated and compared [40, 132] (Fig. 10). Several independent approaches were used: kinetic measurements of boundary potentials were performed on planar lipid bilayers by the IFC technique; ζ-potential measurements of liposomes were made by correlation spectroscopy; and kinetic measurements of the potential changes relative to the adsorption plane of hydrophobic ions [132] (φ_q and φ_r) were evaluated from the transport of TPB- through BLM. This approach gave not only the values of the corresponding components of the boundary potential changes, but also a time dependence of these changes.

The results demonstrated that the phospholipase activity leads to modification of the boundary potential. This phospholipase is a Ca-dependent enzyme, and it was demonstrated that the potential changes are activated by Ca^{2+} and blocked by EDTA. The changes of the boundary potentials due to phospholipase A_2 action measured by the IFC method were much higher than the changes of the ζ-potential. This observation allows us to assume that these changes are caused by the appearance of the products of the reaction, which are either amphiphilic ions or neutral dipolar molecules.

The effects of the hydrolysis products from phospholipase A_2 activity on the boundary potential changes were tested in separate measurements [132]. These experiments performed by different methods also yielded additional information about the influence of these products on the potential distribution across the membrane. Lysolecithin led to a modification of the potential profile between the adsorption plane of tetraphenylborate and the membrane median,

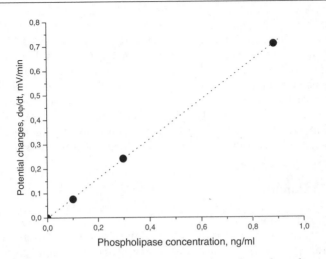

Fig. 10. Relation between the initial rate of changes of membrane boundary potential and the phospholipase A_2 concentration [42]

while fatty acid modified only the potential drop outside of the adsorption plane of hydrophobic ions. Under physiological conditions, both effects were approximately similar. In contrast with the tetraphenylborate technique, the IFC method did not detect any changes of the boundary potential; this is an indication of the high membrane permeability for this hydrolysis product. In biological systems, the symmetric modification of the surface potential induced by symmetrically distributed fatty acids cannot lead to any electrical influence on membrane transport systems, but may provide an autocatalytic effect in the phospholipase action [133, 134].

The electrostatic real-time monitoring of phospholipase A_2 activity allowed the authors to make a simple test of two alternative mechanisms of activity [135] – distinguishing between the scooting and hopping modes [132]. It was demonstrated that the enzyme does not leave the membrane surface after each hydrolysis cycle. The same conclusion was reached on the basis of phospholipase A_2 interaction with lipid monolayers [136].

The boundary potential changes due to the activity of bacterial phospholipase C were investigated on planar lipid membranes and liposomes from soybean phosphatidylcholine [40]. The addition of the enzyme leads at first to some increase of the membrane boundary potential (the sign corresponds to the appearance of positive charges on the membrane surface); then the potential runs in the negative direction. Addition of calcium excludes the first phase of the response. Treatment of the phospholipase with o-phenanthroline diminishes the second phase of the response.

Commercially available soybean phosphatidylcholine is a mixture consisting mainly of phosphatidylcholine, phosphatidylethanolamine, and phosphatidylinositol. At neutral pH, phosphatidylcholine and phosphatidylethanolamine are neutral, and phosphatidylinositol is negatively charged. Hydrolysis of these phospholipids by phospholipase C leads to the formation of neutral diglycerides and water-soluble products. However, no significant changes of ζ-potential, and hence surface potential, were observed during the phospholipase action. Therefore, the boundary potential changes have mainly a dipole nature. The first step in the kinetics may be caused by the formation of the enzyme–substrate complex; the second step is most probably changes of dipole potential due to formation of diglycerides from phospholipids. The latter is supported by the results of monolayer measurements [137], where it was shown that the dipole potential of diglycerides is more negative than that of phospholipids.

5
Analytical Applications

Analytical applications of the described electrostatic effects include quantitative characterization of membrane interaction with different compounds and quantification of substance concentrations in probes. The characterization of the compound–membrane interactions can be useful in secondary screening, as the first two components of early ADME (adsorption, distribution, metabolism, excretion) studies in the drug discovery process, or in the development of new membrane probes. Numerous examples of this approach include investigation of local anesthetics [109], antiviral compounds [114, 138], fluorescent probes for membrane analysis [72, 105–107, 129], antibacterial compounds [139], photosensitizers [110, 140, 141], β-blockers [142], and different biologically effective species [54, 108, 120, 121]. The analysis gives information on binding constants, position of the adsorption plane in the membrane, and in many cases on membrane permeability, even for uncharged species. In combination with more "traditional" techniques of BLM application (measurements of conductivity, capacitance, single-channel characteristics, lifetime and others), which are reviewed elsewhere [143, 144], the electrostatic approach provides comprehensive analysis of mechanisms of membrane activity of chemical compounds.

Application of electrostatic techniques to quantification of concentrations of substances (development of chemical sensors or biosensors) requires incorporation of corresponding receptors. In the case of phospholipase detection, the phospholipids themselves operate as the receptors. It has been demonstrated that the rate of change of boundary potential depends linearly on the phospolipase A_2 concentration; the detection limit is about $2 \cdot 10^{-4}$ units/ml or 0.2 ng/ml [130]. This detection limit is close to that for colorimetric techniques and essentially higher than that for methods based on monolayers. The triggering effect of calcium on the lipid hydrolysis facilitates the determination of the enzyme activity even in the presence of impurities known to cause changes of membrane boundary potentials, such as inductors of lipid peroxidation, Ca^{2+}-independent lipolytic enzymes, etc. For phospholipase C, a quantitative assay has not been evaluated, but the results [40] demonstrated that it also can be developed.

As has been discussed above, measurements of the boundary potential of BLM provide detailed information on modification of the potential profile but demand relatively complicated electronics. A more simple approach is based on measurements of short circuit currents (see the corresponding section above).

This technique has been successfully applied to different analytes. Simply by a variation of lipid content it was possible to get selective transient current responses of supported or nonsupported BLM either to herbicides [145], aflatoxin M1 [146], or sweeteners [147, 148]. A modification of BLM by immobilization of receptors provided additional feasibility. Immobilization of water-soluble hydrolytic enzymes by a technique which is similar to the usually used for incorporation of integral proteins [149, 150] made the membrane sensitive to substrates of these enzymes [151]. Immobolization of antibodies resulted in formation of immunosensors [152]; calix[4]resorcinarene made the membrane selective to dopamine [153] and immobilization of single-stranded DNA made the current response selective to hybridization of complementary DNA strands [154]. Strictly speaking, all these measurements were performed not in true short circuit conditions, but at the applied potential of 25 mV; therefore, the response can include both recharge current and transient changes of membrane conductance. Usually, the authors of these publications explain the generation of transient current responses in these systems as a two-step process: redistribution of charges due to modification of surface charge followed by structural modifications of membranes. Unfortunately, the lack of data with other methods (ζ-potentials, boundary potentials by means of IFC, investigations with hydrophobic ions) does not allow us to reach experimentally proved conclusions on the potential redistribution due to analyte binding. It would be especially interesting to make this analysis for the amazing chromatography-like current kinetics due to addition of a mixture of several compounds [155].

The main advantage of the assay based on the electrostatic technique is that this is label free and uses long-chain natural enzyme substrates. However, the low stability of BLM became a serious limitation in the experiments with several types of real probes. For example, our attempts (in collaboration with V.V. Cherny and M.V. Sukhareva) to apply this approach to measurements in peritoneal exudate or in serum were complicated by the strong influence of these liquids on the membrane stability. Our own attempts to form BLM on different supports (polyacrylamide gel, filters, sBLM were tested) were not successful – we observed either no stability increase or too high leakages. However, these and similar ways became very fruitful in transient current measurements [156–158]. To overcome the stability problems, we (in collaboration with V. Cherny and E. Donath) developed an easy automated technology for the formation of BLM (Fig. 11). The planar lipid bilayers were formed by deposition of a hanging drop of electrolyte on the surface of a thin layer of phospholipid solution swimming on the electrolyte solution. Due to gravity, the drop presses through the lipid solution, and a lipid bilayer is formed between these (the drop and the lower solution) aqueous phases. We used this setup for determination of the phospholipase A_2 activity; the boundary potential was monitored by the second harmonic technique. The facts that the membrane can be formed simply by switching on the motor connected with the syringe, and that both lipid and lower aqueous solutions can be easily modified or replaced (by peristaltic pump or direct additions), provide a simple way for automated membrane preparation. Phospholipid solutions in squalene can be used for the preparation of solvent-free BLM.

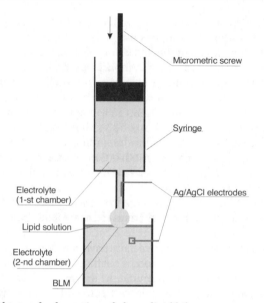

Fig. 11. Automated setup for formation of planar lipid bilayers

6
Conclusion

An investigation of the distribution of the electric fields through planar lipid bilayers can be performed by a number of methods, including compensation of the intramembrane field, measurement of membrane conductivity and transport kinetics in the presence of hydrophobic ions, analysis of electrokinetic effects, or fast measurements of transient transmembrane potential or current. These methods are based on different physical processes, thus providing independent and mutually complementary information on modification of the electrical profile. Possible applications of these methods are summarized in Table 1. Characterization of the membrane–compound interactions can be used in the secondary screening in drug discovery. In addition to the already established techniques, analysis of the electrostatic effects can predict a number of biological effects caused by the modification of the electrostatic potential profile, such as effects on ionic pumps and voltage-dependent channels, binding of charged species with membrane receptors, and electrostatic inhibition or activation of enzyme activity. The second application field is development of biosensors with transducers based on the modification of boundary potentials. In this sense, the technology is physically equivalent to the well-established applications of field-effect transistors, which are also sensitive to the modification of boundary potentials. However, the BLM approaches allow one to detect modifications of different parts of the boundary potential which gives additional advantages. The main limitation of these techniques is the relatively time-consuming membrane preparation and low temporal stability. Automation of membrane preparation,

Table 1. Analytical applications of analysismeasurement of potential profiles through planar lipid bilayers

Analytical information	Application fields	Main limitations	Possible solution
Characterization of membrane/compound interactions: adsorptive properties; localization of the adsorption plane; membrane permeability	Secondary screening in drug discovery (ADME); analysis of localization of membrane probes; prediction of biological effects caused by the potential modification	Time-consuming analysis Time-consuming membrane preparation	Parallelization of the analysis Automation of the membrane formation
Monitoring of ligand-receptor interactions	Development of biosensors	Low stability of BLMs, especially with real biological probes Time-consuming membrane preparation	Supported membranes Automation of the membrane formation

parallelization of the analysis, and development of very small or/and supported planar lipid bilayers could be a perspective solution.

Acknowledgements: The authors thank Dr. K. Pavlov for the critical reading of the manuscript and Prof. O.S. Wolfbeis for fruitful discussions. V.S.S. was supported by INTAS 2001-0224 and RFBR (01-0449246).

References

1. Mueller P, Rudin DO, Tien HT, Wescott WC (1963) J Phys Chem 67:534
2. Chernomordik LV, Sukharev SI, Popov SV, Pastushenko VF, Sokirko AV, Abidor IG, Chizmadzhev Yu (1987) Biochim Biophys Acta 902:360
3. Chizmadzhev YA, Chernomordik LV, Pastushenko VF, Abidor IG (1982) Electrical breakdown of bilayer lipid membranes. In: Progress in science and technique. VINITI, Moscow, p 161
4. Chernomordik LV, Melikyan GB, Chizmadzhev YA (1987) Biochim Biophys Acta 906:309
5. Markin VS, Albanesi JP (2002) Biophys J 82:693
6. Stegmann T, Teissie J, Winterhalter M (2001) In: Katsaras J, Gutberlet T (eds) Lipid bilayers. Springer, Berlin Heidelberg New York, p 265
7. Zimmerberg J, Chernomordik LV (1999) Adv Drug Deliv Rev 38:197
8. Zimmerberg J, Vogel SS, Chernomordik LV (1993) Annu Rev Biophys Biomol Struct 22:433
9. Zimmermann U, Friedrich U, Mussauer H, Gessner P, Hamel K, Sukhorukov V (2000) IEEE Trans Plasma Sci 28:72
10. Neumann E, Kakorin S, Toensing K (2001) Methods Mol Med 37:1
11. Zimmermann U, Vienken J, Halfmann J, Emeis CC (1985) Adv Biotechnol Processes 4:79

12. Mir LM (2001) Bioelectrochemistry 53:1
13. Hofmann GA, Dev SB, Nanda GS, Rabussay D (1999) Crit Rev Ther Drug Carrier Syst 16:523
14. Weaver JC (1993) J Cell Biochem 51:426
15. Miller C (ed) (1986) Ion channel reconstitution. Plenum, New York
16. Montal M (1987) J Membr Biol 98:101
17. Darszon A (1983) J Bioenerg Biomembr 15:321
18. Schindler H (1989) Meth Enzymol 171:225
19. Latorre R, Alvarez O, Cecchi X, Vergara C (1985) Annu Rev Biophys Biophys Chem 14:79
20. Mirsky VM, Sokolov VS, Markin VS, Chekulaeva LN (1984) Biol Membr 1:1143
21. Markin VS, Mirsky VM, Chismadzhev Y (1987) Photoelectrical activity of bacterirhodopsin in planar lipid bilayer. In: Ovchinnikov YA, Hucho F (eds) Receptors and ion channels. Walter de Gruyter, Berlin
22. Mirsky VM, Sokolov VS, Dyukova TV, Mel'nik EI (1983) Bioelectrochem Bioenerg 11:327
23. Portnov VI, Mirsky VM, Markin VS (1990) Bioelectrochem Bioenerg 23:45
24. Bamberg E, Butt HJ, Eisenrauch A, Fendler K (1993) Q Rev Biophys 26:1
25. Christensen B, Gutweiler M, Grell E, Wagner N, Pabst R, Dose K, Bamberg E (1988) Jerus Symp Quantum Chem Biochem 21:527
26. Bamberg E, Dencher NA, Fahr A, Heyn MP (1981) Proc Natl Acad Sci USA 78:7502
27. Fendler K, Hartung K, Nagel G, Bamberg E (1998) Meth Enzymol 291:289
28. Borisova MP, Babakov AV, Kolomytkin OV (1984) Biol Membr 1:187
29. Drachev LA, Jasaitis AA, Kaulen AD, Kondrashin AA, Liberman EA, Nemecek IB, Ostroumov SA, Semenov AY, Skulachev VP (1974) Nature 249:321
30. Fahr A, Lauger P, Bamberg E (1981) J Membr Biol 60:51
31. Tien HT, Ottova-Leitmannova A (eds) (2003) Planar lipid bilayers (BLM) and their applications. Elsevier, Amsterdam
32. Tien HT (1974) Bilayer lipid membranes (BLM): theory and practice. Marcel Dekker, New York
33. Antonov VF (1979) Biofizika membran (Membrane biophysics), vol 1. VINITI, Moscow
34. Antonenko Y, Yaguzhinskii LS (1986) Biochim Biophys Acta 861:337
35. Antonenko YN, Yaguzhinskii LS (1984) Bioelectrochem Bioenerg 13:85
36. Mirsky VM, Stozhkova IN, Szito TV (1991) J Photochem Photobiol B 8:315
37. Stozhkova IN, Mirsky VM, Sokolov VS (1993) Biol Membr 10:44
38. Stozhkova IN, Mirsky VM (1990) Bull Exp Biol Med 110:45
39. Cherny VV, Stozhkova IN, Mirsky VM, Yastrebova TN, Sokolov VS (1992) Biol Membr 9:66
40. Cherny VV, Mirsky VM, Sokolov VS, Markin VS (1989) Bioelectrochem Bioenerg 21:373
41. Cherny VV, Sikharulidze MG, Mirsky VM, Sokolov VS (1993) Biol Membr 6:971
42. Mirsky VM, Cherny VV, Sokolov VS, Markin VS (1990) J Biochem Biophys Meth 21:277
43. Tocanne JF, Teissie J (1990) Biochim Biophys Acta 1031:111
44. Benz R, Frohlich O, Lauger P, Montal M (1975) Biochim Biophys Acta 394:323
45. Lauger P, Neumcke B (1973) In: Eisenman G (ed) Membranes, vol 2. Marcel Dekker, New York, p 1
46. Markin VS, Chizmadzhev YA (1974) Induced ion transport. Nauka, Moscow
47. Haydon DA, Hladky SB (1972) Q Rev Biophys 5:187
48. Benz R, Kolb HA, Lauger P, Stark G (1989) Meth Enzymol 171:274
49. McLaughlin S (1977) Electrostatic potentials at membrane–solution interfaces. Curr Top Membr Trans 9:71
50. Guggenheim EA (1929) J Phys Chem 33:842

51. Guggenheim EA (1930) J Phys Chem 34:1540
52. Hunter RJ (1981) Zeta potentials in colloid science. Principles and applications. Academic, New York
53. Alvarez O, Brodwick M, Latorre R, McLaughlin A, McLaughlin S, Szabo G (1983) Biophys J 44:333
54. Ermakov YA, Cherny VV, Sokolov VS (1992) Biol Membr 6:254
55. Hong FT (1976) Photochem Photobiol 24:155
56. Petrov AG (1998) The liotropic state of matter. Molecular physics and living matter physics. Overseas Publ Assoc, Amsterdam
57. MacDonald RC, Bangham AD (1972) J Membr Biol 7:29
58. Andersen OS, Finkelstein A, Katz I, Cass A (1976) J Gen Physiol 67:749
59. Cros D, Seta P, Gavach C, Benz R (1982) J Electroanal Chem Interfacial Electrochem 134:147
60. Nikolelis DP, Hianik T, Krull UJ (1999) Electroanalysis 11:7
61. Hall JE, Mead CA, Szabo G (1973) J Membr Biol 11:75
62. Lesslauer W, Richter J, Lauger P (1967) Nature 213:1224
63. McLaughlin SG, Szabo G, Eisenman G, Ciani SM (1970) Proc Natl Acad Sci USA 67:1268
64. Laprade R, Ciani S, Eisenman G, Szabo G (1975) Membranes 3:127
65. Hladky SB (1979) J Membr Biol 46:213
66. Lauger P, Benz R, Stark G, Bamberg E, Jordan PC, Fahr A, Brock W (1981) Q Rev Biophys 14:513
67. Benz R, Frohlich O, Lauger P (1977) Biochim Biophys Acta 464:465
68. Cseh R, Benz R (1998) Biophys J 74:1399
69. Melnik E, Latorre R, Hall JE, Tosteson DC (1977) J Gen Physiol 69:243
70. Sokolov VS, Cherny VV, Simonova MV, Markin VS (1990) Biol Membr 7:872
71. Smejtek P, Paulis-Illangasekare M (1979) Biophys J 26:467
72. Malkov DY, Sokolov VS (1996) Biochim Biophys Acta 1278:197
73. Wang CC, Bruner LJ (1978) Nature 272:268
74. Szabo G (1974) Nature 252:47
75. Pickar AD, Benz R (1978) J Membr Biol 44:353
76. Muller RU, Finkelstein A (1972) J Gen Physiol 60:285
77. Hladky SB (1974) Biochim Biophys Acta 352:71
78. Latorre R, Hall JE (1976) Nature 264:361
79. Schoch P, Sargent DF, Schwyzer R (1979) J Membr Biol 46:71
80. Babakov AV, Ermishkin LN, Liberman EA (1966) Nature 210:953
81. Hianik T, Passechnik VI (1995) Bilayer lipid membranes: structure and mechanical properties. Ister, Bratislava
82. Requena J, Haydon DA, Hladky SB (1975) Biophys J 15:77
83. Leikin SL (1985) Biol Membr 2:820
84. Abidor IG, Glazunov IY, Leikin SL, Chizmadzhev YA (1986) Biol Membr 3:627
85. Schoch P, Sargent DF (1976) Experimentia 32:811
86. Alvarez O, Latorre R (1978) Biophys J 21:1
87. Usai C, Robello M, Gambale F, Marchetti C (1984) J Membr Biol 82:15
88. Abidor IG, Aityan SH, Cherny VV, Chernomordik LV, Chizmadzhev YA (1979) Dokl Akad Nauk SSSR 245:977
89. Sokolov VS, Kuz'min SG (1980) Biofizika 25:174
90. Carius W (1976) J Colloid Interface Sci 57:301
91. Passechnik VI, Hianik T (1977) Kolloid Zh 38:1180
92. Benz R, Janko K (1976) Biochim Biophys Acta 455:721
93. Cevc G (1990) Biochim Biophys Acta 1031:311
94. Adamson AW (1967) Physical chemistry of surfaces. Wiley-Interscience, New York
95. Davies JT, Rideal EK (1963) Interfacial phenomena. Academic, New York

96. Cevc G, Marsh D (1987) Phospholipid bilayers. Physical principles and models. Wiley-Interscience, New York
97. Eisenberg M, Gresalfi T, Riccio T, McLaughlin S (1979) Biochemistry 18:5213
98. McLaughlin S, Mulrine N, Gresalfi T, Vaio G, McLaughlin A (1981) J Gen Physiol 77:445
99. McLaughlin S (1989) Annu Rev Biophys Biophys Chem 18:113
100. Ohki S, Sauve R (1978) Biochim Biophys Acta 511:377
101. Ohki S, Kurland R (1981) Biochim Biophys Acta 645:170
102. Ermakov YA, Averbakh AZ, Yusipovich AI, Sukharev SI (2001) Biophys J 80:1851
103. Andersen OS, Feldberg S, Nakadomari H, Levy S, McLaughlin S (1978) Biophys J 21:35
104. Wang CC, Bruner LJ (1978) J Membr Biol 38:311
105. Cherny VV, Kozlov MM, Sokolov VS, Ermakov YA, Markin VS (1982) Biofizika 27:812
106. Ermakov YA, Cherny VV, Tatulian SA, Sokolov VS (1983) Biofizika 28:1010
107. Simonova MV, Cherny VV, Donat E, Sokolov VS, Markin VS (1986) Biol Membr 3:846
108. Sokolov VS, Cherny VV, Simonova MV, Markin VS (1990) Bioelectrochem Bioenerg 23:27
109. Cherny AV, Sokolov VS, Cherny VV (1993) Russ J Electrochem 29:321
110. Stozhkova IN, Cherny VV, Sokolov VS, Ermakov YA (1997) Membr Cell Biol 11:381
111. Kozlov MM, Chernyi VV, Sokolov VS, Ermakov YA, Markin VS (1983) Biofizika 28:61
112. Antonenko YN, Yaguzhinskii LS (1984) Biofizika 29:232
113. Markin VS, Portnov VI, Simonova MV, Sokolov VS, Cherny VV (1987) Biol Membr 4:502
114. Cherny VV, Simonova MV, Sokolov VS, Markin VS (1990) Bioelectrochem Bioenerg 23:17
115. Franklin JC, Cafiso DS (1993) Biophys J 65:289
116. Clarke RJ, Kane DJ (1997) Biochim Biophys Acta 1323:223
117. Clarke RJ (2001) Adv Colloid Interface Sci 89–90:263
118. Franklin JC, Cafiso DS (1993) Biophys J 65:289
119. Verkman AS (1980) Biochim Biophys Acta 599:370
120. Sokolov VS, Cherny VV, Markin VS (1984) Biofizika 29:424
121. Pohl P, Rokitskaya TI, Pohl EE, Saparov SM (1997) Biochim Biophys Acta 1323:163
122. Cseh R, Benz R (1999) Biophys J 77:1477
123. Awiszus R, Stark G (1988) Eur Biophys J 15:321
124. Awiszus R, Stark G (1988) Eur Biophys J 15:299
125. Reyes J, Motais R, Latorre R (1983) J Membr Biol 72:93
126. Smejtek P, Paulis-Illangasekare M (1979) Biophys J 26:441
127. Barth C, Bihler H, Wilhelm M, Stark G (1995) Biophys Chem 54:127
128. deLevie R, Rangarajan SK, Seelig PF, Andersen OS (1979) Biophys J 25:295
129. Passechnik VI, Sokolov VS (2003) Biol Membr 20:343
130. Mirsky VM, Cherny VV, Sokolov VS, Markin VS (1990) J Biochem Biophys Methods 21:277
131. Mirsky VM, Cherny VV, Sokolov VS, Markin VS (1989) Elektrokhimiya 25:1336
132. Cherny VV, Sikharulidze MG, Mirsky VM, Sokolov VS (1992) Biol Membr 9:733
133. Mirsky VM (1995) Biol Membr 12:165
134. Mirsky VM (1994) Chem Phys Lipids 70:75
135. Gelb MH, Min JH, Jain MK (2000) Biochim Biophys Acta 1488:20
136. Ksenzhek OS, Gevod VS, Aianyan AE, Miroshnikov AI (1981) Bioorg Khim 7:1680
137. Pickar AD, Benz R (1978) J Membr Biol 44:353
138. Cherny VV, Paulitschke M, Simonova MV, Hessel E, Ermakov YA, Sokolov VS, Lerche D, Markin VS (1989) Gen Physiol Biophys 8:23
139. Cherny VV, Stozhkova IN, Mirsky VM, Yastrebova TN, Sokolov VS (1992) Biol Membr 9:66
140. Rokitskaya TI, Block M, Antonenko YN, Kotova EA, Pohl P (2000) Biophys J 78:2572
141. Sokolov VS, Block M, Stozhkova IN, Pohl P (2000) Biophys J 79:2121

142. Pohl EE, Krylov AV, Block M, Pohl P (1998) Biochim Biophys Acta 1373:170
143. Tien HT, Barish RH, Gu LQ, Ottova AL (1998) Anal Sci 14:3
144. Tien HT, Ottova AL (1998) Electrochim Acta 43:3587
145. Siontorou CG, Nikolelis DP, Krull UJ, Chiang KL (1997) Anal Chem 69:3109
146. Andreou VG, Nikolelis DP, Tarus B (1997) Anal Chim Acta 350:121
147. Nikolelis DP, Pantoulias S, Krull UJ, Zeng J (2001) Electrochim Acta 46:1025
148. Nikolelis DP, Pantoulias S (2000) Electroanalysis 12:786
149. Montal M, Mueller P (1972) Proc Natl Acad Sci USA 69:3561
150. Montal M, Darszon A, Schindler H (1981) Quart Revs Biophys 14:1
151. Nikolelis DP, Siontorou CG (1995) Anal Chem 67:936
152. Nikolelis DP, Siontorou CG, Andreou VG, Viras KG, Krull U (1995) Electroanalysis 7:1082
153. Nikolelis DP, Petropoulou SS, Pergel E, Toth K (2002) Electroanalysis 14:783
154. Krull U, Nikolelis DP, Jantzi SC, Zeng J (2000) Electroanalysis 12:921
155. Nikolelis DP, Pantoulias S (2001) Anal Chem 73:5945
156. Nikolelis DP, Siontorou CG (1997) J Autom Chem 19:1
157. Novotny I, Rehacek V, Tvarozek V, Nikolelis DP, Andreou VG, Siontorou CG, Ziegler W (1997) Mat Sci Eng C 5:55
158. Siontorou CG, Nikolelis DP, Piunno PAE, Krull UJ (1997) Electroanalysis 9:1067

Electrostriction of Supported Lipid Membranes and Their Application in Biosensing

Tibor Hianik

Abstract. This review reports the significance of bilayer lipid membranes on a solid support (sBLM) for the construction of biosensors. The methods of formation of lipid membranes on different solid supports, including different metals (silver, gold, stainless steel, mercury), agar, and conducting polymers, are presented. Several examples of the application of electrostriction and dielectric relaxation methods for the study of the mechanical properties and dynamics of solid-supported bilayers are shown. We demonstrate that these methods are useful for studying the physical properties of chemically modified supported membranes, the interaction of surfactants and nucleic acids with lipid membranes, the binding of enzymes and antibodies to sBLM, and hybridization of nucleic acids on the membrane surface. A comparison of the mechanical properties of sBLM of various compositions and method of formation is presented.

Keywords: Electrostriction, Dielectric relaxation, Supported lipid membranes, DNA–lipid interactions, Affinity interactions

Abbreviations and Symbols

Ab	Antibody
Ag	Antigen
ACH	Acetylcholinesterase
A-GOX	Avidin-modified glucose oxidase
A-IgG	Avidin-modified immunoglobulin G
B-BLM	Bilayer lipid membrane composed of biotinylated phospholipids
BLM	Bilayer lipid membranes
(dT)15	5′-Deoxypentadecylthymidylate
(dA)15	5′-Deoxypentadecyladenylate
C	Capacitance
C_s	Specific capacitance
CH(dT)15	5′-Cholesterolphosphoryl deoxypentadecyldeoxythymidic acid
C16(dT)15	5′-Palmitylphosphoryl deoxypentadecyldeoxythymidic acid
cmc	Critical micellar concentration
COB	Crude ox brain fraction
CTAB	Cetyltrimethylammonium bromide
d	Thickness
DDA	Dodecylamine
DNA	Deoxyribonucleic acid
DPPC	Dipalmitoylphosphatidylcholine
DTAB	Dodecyltrimethylammonium bromide
e_0	Elementary charge
$E_\perp$	Elasticity modulus
f	Frequency
f_r	Reduction factor
g_s	Specific conductivity
GMO	Glycerol monooleate
GOX	Glucose oxidase
HDA	Hedadecylamine
HDM	Hexadecylmercaptan
I	Current

K	Volume compressibility modulus
K_b	Binding constant
m	Hill coefficient
N	Number of binding sites
p	Pressure
Q_d	Total charge transferred due to dipole reorientation
Q_e	Total charge transferred due to changes of potential
R	Resistance
S	Area
SBPC	Soybean phosphatidylcholine
sBLM	Supported bilayer lipid membranes
tBLM	Tethered bilayer lipid membranes
TBL	Total fraction of phospholipids from bovine brain
V	Voltage
12T	Dodecanethiol
α	Electrostriction coefficient
$\Delta\Phi_m$	Intrinsic membrane potential
ε	Relative dielectric permittivity
ε_0	Permittivity of free space
η	Coefficient of dynamic viscosity
τ	Relaxation time

1
Introduction

The interest in solid-supported bilayer lipid membranes (sBLM) as a tool for the construction of highly sensitive and selective biosensors is considerable. sBLM are composed of a solid support that serves as an electrode, and a lipid bilayer on its surface. Bilayer lipid membranes (BLM) represent a relatively biocompatible structure for the development of new types of electrochemical sensors and biosensors with fast response times (on the order of a few seconds) and high sensitivity (i.e., nanomolar detection limits). The use of biomembranes as recognition elements started with the pioneering work by Mueller et al. [1], using BLM in a circular hole of small diameter (~0.5 mm) in the wall of a Teflon cell immersed in electrolyte. Only a short period after the development of a method of formation of BLM, work directed at their use as immunosensors was reported [2]. Considerable progress has been achieved due to the development of stable supported bilayer lipid membranes (sBLM). In the early design the lipid bilayer floated on an ultrathin layer covering the hydrophilic substrate [3]. Almost simultaneously Tien and Salamon [4] developed a simple method of forming sBLM on a freshly cut tip of stainless steel, silver or platinum wire coated with Teflon.

sBLM prepared from lipids that were modified to contain enzymes proved to be very sensitive biosensors for detecting glucose, and had a functional lifetime and mechanical stability of over 48 h [5–8]. Other work has investigated the electrical properties of lipid monolayers formed on thin gold layers using hydrocarbon-hexadecylmercaptan [9], bilayers containing alkylthiols [10], and bilayers on thin interdigitated platinum layers [11]. Stable lipid bilayers were also successfully created on agar-agar surfaces [12–14] and on the surface of a conducting polymer [15]. Stable supported membranes have also been obtained on the surface of a hanging mercury drop electrode [16].

The sBLM are suitable for application in biosensing when functional units are immobilized mostly on the surface of the lipid bilayer. The interaction of the lipid bilayer with a solid support, however, caused certain steric problems for incorporation of integral proteins. The problem with this limitation has recently been solved by involving an additional, tethering layer. This layer links the BLM to the solid support in a mechanically and chemically stable way by establishing covalent bonds between some of the lipid molecules in the BLM monolayer and the tethering units. The important structural peculiarity of tethered BLM (tBLM) consists in separation of BLM from the solid support and in creation of certain space. In the case of hydrophilic tethers, this space can be filled by water. Thus, tBLM allow incorporation into BLM of even bulky proteins (see Knoll et al. [17] for review) and ionic carriers or ionic channels [18].

Supported hybrid cell membranes that mimic the biomembrane represent new structures with a number of potential applications in biosensing [19]. The advantage of this approach is the high biocompatibility of these systems that utilize a number of advantages of living organisms. It has been shown that the presence of lipid membranes preserves the functional protein unit against degradation and provides the necessary conformational mobility for performance of the receptor or catalytic function [20]. Supported lipid membranes can be stabilized by saccharides [21], polymers [22] or bacterial surface proteins (S-layers) [23]. In the latter case, the S-layers form a regular crystalline net on the membrane surface that is convenient for immobilization of protein macromolecules [24]. Besides glucose oxidase, other proteins such as bacteriorhodopsin [16, 25], urease [15], immunoglobulins [26], H^+-ATPase [27], cytochrome oxidase [28], and nucleic acids [29–31] were successfully incorporated into sBLM.

The study of the physical properties of supported membranes requires the application of relevant physical methods. In the present contribution, we would like to show the usefulness of some special methods – electrostriction and dielectric relaxation – to study various properties of sBLM. Such properties include membrane stability and dynamics, as well as membrane binding properties during immobilization of enzymes to the sBLM surface and interaction with sBLM immunoglobulins, surfacatants, and nucleic acids. Furthermore, the electrostriction method is useful for the detection of interaction of uncharged compounds with sBLM, which might be difficult or impossible to study by traditional methods such as amperometry.

2
Methods of Preparation of Supported Lipid Membranes and Their Modification by Functional Macromolecules

Freely suspended BLM have been used for the last four decades for investigation of transduction of electrochemical signals through lipid membranes, and the preparation of these systems has been extensively described elsewhere [32].

These are mechanically fragile systems that are unsuitable for most practical applications. The preparation of stabilized lipid membranes on solid supports depends on the applications, and the solid support may consist of a metallic surface, agar or ultrafiltration membranes. The former two types are mainly used for the construction of disposable biosensors, whereas the latter is used for continuous monitoring of an analyte.

2.1
Supported Lipid Membranes

Metal-supported lipid bilayers formed on the tip of freshly cut Teflon-coated stainless steel, silver or platinum wire represent one of the simplest supported systems which can be prepared (Teflon-coated wires are commercially available; in addition, coating of the wire by electrochemical polymerization can be used [33]). This system was reported by Tien and Salamon [4]. The technique of formation of sBLM is based on the interaction of an amphiphilic lipid molecule with a nascent metallic surface. The procedure consists of two steps. First, one end of a Teflon-coated metal wire (the diameter of the wire is usually 0.1–0.5 mm) is immersed in a lipid solution (this can be, for example, 2% phospholipid in n-hexadecane or other organic solvent) and then, while still immersed, the tip is cut off with a miniature guillotine (good reproducibility of sBLM requires reproducible cutting of the wire). Second, the fresh tip of the wire, having become coated with lipid, is placed in electrolyte (usually 0.1–1 M KCl buffered with, e.g., Tris-HCl), whereupon the lipid film spontaneously thins, forming a self-assembled lipid bilayer. Solvent-free sBLM can be prepared using the Langmuir–Blodgett technique. A similar approach can be used to prepare filter-supported lipid membranes. Lipid at an air/water interface can be deposited on a vertical filter paper by adjusting the level of the solution to drop below the filter and immediately rise above the filter again [34–36].

Smooth platinum layers sputtered on glass or silicon can be used as supports [11, 37]. Puu et al. [38] reported formation of a lipid bilayer on smooth platinum layers by use of the Langmuir–Blodgett technique and liposome fusion.

The application of alkylthiols permits the preparation of a chemisorbed coating of monolayer on metals such as gold, and thus results in higher stability of the biosensor [10]. A second monolayer can be added to complete a BLM by various techniques, e.g., by immersion into lipid solution (in n-hexane) or by vesicle fusion [39]. When erythrocyte ghosts instead of vesicles are used for preparation of a second monolayer, then hybrid bilayer membranes are formed [19, 40]. The advantage of this approach consists in the fact that the lipid monolayer already contains functionalized molecules, like acetylcholinesterase (ACH). Thus there is no necessity to use any additional chemical modification of these membranes for fabrication of biosensors, e.g., based on detecting ACH activity. Also, disulfide-functionalized amphiphilic copolymers (hydromethacrylates) [41, 42] and thiolipids with hydrophilic oxyethylene spacers of defined length [43] form lipid layers on gold substrates. In addition to alkylthiols, mercaptoundecanoic acid with a mixture of alkylthiols can be used for formation of monomolecular layers for direct chemical attachment of functionalized molecules [44, 45].

The procedure for the formation of sBLM on a conducting polymer support is similar to that on a metal support. A polypyrrole surface was recently tested and has been used for the preparation of sensors for ammonia and urea [15].

Agar-supported sBLM have been reported in the literature [12–14]. In the simplest configuration, the sBLM can be formed on the tip of an agar (2% agar in 1 M KCl)-filled Teflon tube (inner diameter 0.5 mm). The procedure is similar to that used for formation of solvent-free metal-supported sBLM (see [14, 46]). A detailed review of supported membranes is provided in an article by Sackmann [22], and on tethered membranes in the review by Knoll et al. [17]. This review also reports a number of biophysical studies of tBLM. Application of thin organic films can be found in the book by Ulman [47].

2.2
Immobilization of Functional Macromolecules on the Surface of BLM and sBLM

Incorporation of proteins into lipid layers or immobilization on sBLM or liposomes is a crucial stage in the preparation of biosensors. The method of immobilization should fulfill certain requirements: (1) stability of lipid–protein complex for sufficient length of time; (2) optimal conformational lability of macromolecules; and (3) access to reactive sites of enzymes, antibodies or receptors.

For membrane-integral proteins or receptors which contain hydrophobic constituents, the incorporation into membranes can be done either by mixing the proteins with lipid in the membrane-forming solution, or by fusion of proteoliposomes with lipid layers. Mixtures of proteins with lipids in solution have been successfully used for incorporation of bacteriorhodopsin [19, 48], insulin receptors [49], and antibodies [26]. Vesicle fusion for incorporation of proteins has also been extensively reported (e.g., bovine serum albumin [39], acetylcholinesterase [38, 50], cholera toxin [50], bacteriorhodopsin [19, 51], cytochrome oxidase, nicotinic acetylcholine receptor [38], and H^+-ATPase [27]).

Another immobilization method which is irreversible in a practical context has been developed by Bayer, Rivnay, and Wilchek [52, 53]. This method consists of the use of the high affinity of streptavidin and/or avidin for biotin. Thus, if a streptavidin- or avidin-modified macromolecule is added to a BLM, sBLM or liposome that is prepared from biotinylated phospholipids, a stable complex between protein and phospholipid forms. sBLM biosensor applications based on streptavidin or avidin have been reported [5, 6].

The avidin–biotin technology has been reported in a number of publications about immobilization of protein on liposome surfaces [24, 54]. A further novel approach to immobilization consists of using a bacterial S-layer matrix for immobilization of protein macromolecules [24, 55]. For immobilization of oligonucleotides to sBLM the effective approach consists of modification of a short sequence of single-stranded DNA by a hydrophobic chain, e.g., palmitic acid [29, 30] or cholesterol [31].

3
Basic Principles of the Electrostriction and Dielectric Relaxation

3.1
Electrostriction Method

The membrane can be represented by an equivalent electrical circuit consisting of a capacitor and a resistor in parallel. It is well known that unmodified free-standing membranes are characterized by low specific conductance, g_s, (typically 10^{-4} Ω^{-1} m^{-2}), while the specific conductance of sBLM formed on the tip of metal wire have a g_s value that is one order of magnitude higher [4]. The specific capacitance, C_s, of BLM that contain solvent sBLM are similar and reach 0.4×10^{-3} F·m^{-2}, while twofold higher values of C_s are typical for solvent-free membranes. Application of dc or ac voltage with amplitude V_0 to the membrane will compress the bilayer with pressure $p=C_s V_0^2/2d$, where d is the thickness of the hydrophobic part of the membrane (typically d=5 nm for BLM that contain solvent), and consequently the thickness of the membrane decreases. This phenomenon can be used for determination of the mechanical parameters of a membrane in the direction perpendicular to its plane: Young's modulus of elasticity, $E_\perp$, and coefficient of dynamic viscosity, η. According to definition: $E_\perp = -p/(\Delta d/d)$. This value determines the ability of the membrane to deform upon application of external pressure. The value of η corresponds to the internal friction between hydrocarbon chains of phospholipids during the compression of the membrane.

The membrane capacitance is directly associated with the thickness by: $C=\varepsilon\varepsilon_0 S/d$ (S is the membrane area, ε is relative dielectric permittivity of the hydrophobic part of the membrane, typically $\varepsilon \sim 2$, and $\varepsilon_0 = 8.85 \times 10^{-12}$ F/m is the permittivity of free space). Therefore, the changes in membrane thickness result in changes of its capacitance. For certain conditions, when a membrane is considered as a volume incompressible body (i.e., K>>$E_\perp$, where K is the volume compressibility modulus, which really holds for the lipid bilayer [56]), $\Delta d/d=\Delta C/2C$. Thus, the elasticity modulus is $E_\perp = 2p/(\Delta C/C)$. The membrane capacitance depends nonlinearly on the amplitude of the voltage, $C=C_m(1+\alpha V^2)$, where $C_m=C$ (V=0) and α is the electrostriction coefficient. Then according to the electrostriction method developed by Passechnik and Hianik [57], when ac voltage $V=V_0 \sin 2\pi ft$ with frequency, f, and amplitude, V_0, is applied to the BLM, this results in the generation of a component of the membrane current with frequency 3f and amplitude I_3, in addition to the first harmonic (with frequency f and amplitude I_1). As shown by Carius [58], upon simultaneous application of direct and alternating voltage, the membrane current contains additionally the second current harmonic with frequency 2f and amplitude I_2.

An additional parameter is the phase shift φ, defined as the difference of phase between the measured third harmonic component and that expected for an ideally elastic body. The phase shift can vary from 0° for an ideal loss-free membrane to 90° for a purely viscous body. The values of phase shift and elasticity modulus determine the coefficient of dynamic viscosity, η. I_1, I_2, I_3 and φ can be used to calculate the electrical capacitance of a membrane, the intrinsic membrane potential, $\Delta\Phi_m$, the modulus of elasticity, and the coefficient of dynamic viscosity. In terms of the measured quantities

$$C = I_1 / \left(2\pi f V_0\right) \tag{1}$$

$$\Delta\Phi_m = I_2 V_0 / \left(4 I_3\right) \tag{2}$$

$$E_\perp = 3 C_s V_0^2 I_1 / \left(4 d I_3\right) \tag{3}$$

$$\eta = E_\perp \sin\varphi / (2\pi f) \tag{4}$$

Thus, to obtain C, $\Delta\Phi_m$, $E_\perp$, and η it is sufficient to measure the amplitudes I_1, I_2, and I_3 and phase shift φ. The values of the above parameters can be obtained using standard electronic equipment including resonance amplifiers [56].

3.2
Dielectric Relaxation

The dielectric relaxation method is based on analyses of the time course of changes in the capacitance following sudden changes in the voltage applied across the bilayer [59]. Using this method, we can obtain information about, e.g., reorientation of molecular dipoles and cluster formation. A symmetric voltage jump (–V to +V) across a bilayer can orient naturally occurring dipoles. The magnitude and time course of these effects will depend on the structure of the bilayer and bulk phase. Such effects will also be reflected by changes in membrane capacitance. The capacitance of the bilayer depends on the dielectric constant ε of the bilayer material, the membrane area S, and the thickness of the membrane d (see Sect. 3.1.). An electric field could affect all these parameters, so each must be considered separately. This has been done by Sargent [59] and Hianik and Passechnik [56].

3.2.1
Dielectric Constant

The dielectric constant of the membrane reflects both the dielectric properties of the individual molecules and their organization in the bilayer. In molecular reorientation, both intrinsic dipoles and the axis of highest polarizability of anisotropic molecules will tend to align themselves in the field [60]. The amplitude and time course of such motions are reflected in the displacement current induced in the external circuit. For small field strength the charge displaced is proportional to the scalar product of the dipole moment and the electric field, i.e., it depends on the field direction as well as magnitude. In the case of application of a symmetric voltage jump across the membrane, the total charge Q_d transferred due to dipole reorientation will be the same for a voltage change of V volts, independent of the initial voltage. Specifically, apart from saturation effects,

$$Q_d(-V \div +V) = Q_d(0 \div 2V) \tag{a}$$

In liquids, saturation often starts above 10^7 V m^{-1}. Such field strengths are readily obtained in BLM, so that saturation effects could be expected at the higher voltages.

3.2.2
Membrane Thickness

The dependence of membrane thickness on applied voltage – electrostriction – has been described in Sect. 3.1. (see [56] for more detail). The associated changes in capacitance have been found to depend approximately linearly on the square of applied voltage V, so that $\Delta C = Q_e/V = \alpha C_m V^2$, where $C_m = C$ (V=0) and α is the electrostriction coefficient. Note that ΔC does not depend on the sign of the applied voltage, so that suddenly changing the membrane voltage from -V to +V should not result in electrostriction relaxation. Both dipole reorientation and membrane thinning affect the specific capacity of the bilayer, i.e., the corresponding Q values will be proportional to the membrane area.

3.2.3
Membrane Area

The formation of a ring (torus) of bilayer outside the original bilayer region due to electrostrictive forces has been described in earlier works by Babakov et al. [61] and White and Thompson [62] and analyzed in detail by Hianik and Passechnik [56]. As the area change is also proportional to the square of the applied voltage, no distinction between an area or thickness change is possible on the basis of voltage dependence. However, the method of electrostriction based on measurements of higher current harmonics (see Sect. 3.1.) does not depend on membrane area or on effects connected with changes of area of the border region of BLM (see [56]). This method can therefore be preferably applied to studying BLM electrostriction. As has been shown by Sargent [59], no correlation was found between the normalized dielectric relaxation parameters and membrane area, showing that the effects were indeed related to the bilayer rather than the border region.

3.2.4
Basic Principles of the Measurement of Dielectric Relaxation

The dielectric relaxation was determined by measuring the time course of the displacement current following a step change in potential. A detailed description of the construction and operation of the apparatus is given elsewhere [59], but the principle of the technique is presented in Fig. 1. A positive voltage is applied to the electrodes at time t=0 (Fig. 1a). This causes a large charging current (I_0) to flow, which decays with a time constant of $R_s C_m$, (R_s=solution+electrode resistance, C_m=membrane capacitance). In addition, there may be relaxation currents caused by voltage- or time-dependent changes in C_m, and a dc component through R_m (Fig. 1b). At time t=0 a negative voltage is applied to an R-C analog circuit which models the parameters of the experimental system (R'_s and

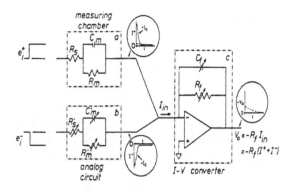

Fig. 1. Outline of the technique used for measurement of current relaxation in BLM following a sudden change in voltage (for explanation see the text) [59]

C'_m are adjusted to match the peak current and the initial decay of the charging peak, respectively; R'_m can be found by measuring the dc resistance, R_m, of the bilayer). In this way a current is generated that is equal in magnitude but opposite in sign to that generated by the membrane charging current and the dc component. The currents from circuits a and b are combined, resulting in a canceling of the charging peaks and dc currents. The net current is fed into the current-voltage converter (Fig. 1c), which produces a voltage (V_0) proportional to the input current (I_{in}). The gain of the current–voltage converter is controlled by the feedback resistor R_f and the time is given by $R_f C_f$. For unmodified BLM, which are characterized by high resistance, R_m was always greater than $10^9\ \Omega$, so that the dc current was less than 10^{-10} A (as this is below the resolution of the apparatus, the dc component could be neglected).

Any capacitive relaxation is assumed to have an exponential time course. After adjusting the analog circuit parameter to balance the charging peak ($R'_m = R_m$, $C'_m = C_m$ ($t \cong 0$)), and for $t > 4\tau_c$ ($\tau_c = R_s C_m$ = charging time constant of BLM),

$$I(t) \cong \sum_i I_0^i e^{-t/\tau_i} \tag{5}$$

The amplitudes are only meaningful when normalized in some manner; as an initial trial, the values were expressed per unit area, of which the simplest measure is the membrane capacity at zero voltage $C(0) = C'_m$. Thus it is convenient to present the relaxation amplitude (I_r^i) as a fractional or percentage change in capacitance, for which the complete expression is:

$$I_r^i \equiv \Delta C_i / C'_m = I_0^i \tau_i \left(V_0 C'_m \right) \tag{6}$$

This is related to the "dielectric increment" through the dielectric constant. The latter is, however, not known for all the conditions met. Therefore the phenomenological "relaxation amplitude" was used for analysis [59]. The resolution of the apparatus allows the detection of ΔC_i of about 1 pF with time constant $\tau > 1\ \mu s$. All relaxation phenomena reported here are considerably slower than this, so that no inaccuracy is introduced from this source. Figure 2 shows a typical exam-

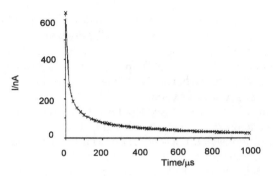

Fig. 2. Plot of the amplitude of displacement current for sBLM prepared from SBPC+1-do-decanethiol on a thin gold layer (*solid line*) and the best fit (*crosses*)

ple of time dependence for relaxation of the amplitude of the displacement current for sBLM prepared from soybean phosphatidylcholine and 1-dodecanethiol on a thin gold layer (solid line), and the best fit (crosses) of the experimental curve by the sum of three exponential functions (see Eq. 5) with relaxation times $\tau_1=40$ µs, $\tau_2=290$ µs, and $\tau_3=1.9$ ms (see Sect. 4.6). We can see good coincidence of the results of calculations with the results of measurements.

4
Electrical and Mechanical Properties of sBLM

As a class of membranes, sBLM maintain the structural and thermodynamic properties of freely suspended BLM and liposomes. In addition, sBLM allow the avoidance of certain artifacts that are common for lipid dispersions (e.g., large curvature of small liposomes). The study of sBLM as model membranes by various techniques has been reviewed recently by Sackmann [22] and that of tethered membranes (tBLM) by Knoll et al. [17]. In this section we will focus on those properties of sBLM which are closely connected with their use in practical biosensing applications as enzymatic and affinity biosensors, as well as application of sBLM to studying the influence of surfactants and nucleic acids on lipid membranes.

4.1
Specific Conductivity, Capacitance, and Stability

The electrical parameters of supported membranes depend on various factors, such as lipid composition, hydrocarbon solvent content, ionic strength of supporting electrolyte, and frequency of ac voltage. The conductivity and capacitance of sBLM formed from a solution based on hydrocarbon solvent on the tip of freshly cut Teflon-coated metal wire have been studied by Tien and Salamon [4]. They showed that in general the specific conductivity, g_s, of sBLM depends on lipid composition, and is 5–50-fold higher than that of freely suspended BLM (10–100 nS·cm^{-2}). Only small differences are observed in the specific capacitance, C_s, of these membranes (~0.46 µF·cm^{-2}). Detailed studies of the electrical properties of sBLM formed on the tip of Teflon-coated stainless steel wires showed that the specific capacitance of an sBLM depends on the frequency of ac voltage and on the ionic strength of the electrolyte [63]. The value of C_s decreased with ionic strength and with frequency (in the measured range of 1 Hz–30 kHz). Interestingly, for higher ionic strength (0.1 M KCl) and at higher frequencies (f>1 kHz), the specific capacitance of sBLM becomes similar to that of the electrolyte/metal interface. Moreover, the C_s increased with time. The behavior of specific capacitance suggests that the structure of sBLM is likely a mixture of a continuous monolayer, three layers, and in certain cases also a bilayer interrupted by bare "islands" of metal [63]. When a negative potential of amplitude –350 mV was applied to the sBLM relative to the reference Ag/AgCl electrode, the covering of the tip of the metal wire by lipid film was improved. The lipid film had, however, a multilayer structure [64].

For freely suspended BLM, Coster and Simons [65] observed anomalous dispersion of BLM capacitance at low ionic strength. They explained this observation as due to different properties of water within and adjacent to the membrane. The dispersion of the capacitance of BLM with frequency started at f>1 kHz for lower ionic strength (5×10^{-3} M NaCl) and at f>10 kHz for higher ionic strength (0.1 M NaCl) [66]. The dispersion of the capacitance of lipid films with frequency is related to the complex nature of capacitance $C=C'-iC''$, where C'' is frequency dependent (see [67]). The values of g_s, (5–10 nS cm^{-2}) and C_s (0.36–0.61 µF cm^{-2}) of agar-supported solvent-containing BLM [14] are similar to those of freely suspended membranes.

sBLM with solvent and various lipid compositions formed on smooth gold supports exhibited conductivity in the range 9–600 nS cm^{-2} and specific electrical capacitance in the range 0.57–0.69 µF cm^{-2}. These films were characterized by higher conductivity in the gel state rather than in the liquid-crystalline state, which can be explained by an increased amount of structural defect of the gel state of sBLM [6]. In our recent work [68], we observed a frequency dependence of the specific capacitance of sBLM formed on gold layers (the first monolayer adjacent to the thin smooth gold layer was 1-dodecanethiol, while the second monolayer was from a crude ox brain fraction (COB)). However, in contrast to sBLM formed on the rather rough surface of the tip of stainless steel wire, the C_s of sBLM on gold layers were characterized by the dispersion with frequency only at f>1 kHz for ionic strengths lower than 0.3 mM and at f>10 kHz for an ionic strength of 0.1 M. The specific capacitance for a frequency range of 100 Hz–1 kHz was ~0.5 µF cm^{-2} and that for the alkylthiol layer (without phospholipid monolayer) was around 1 µF.cm^{-2}, which is in good agreement with the results obtained by Florin and Gaub [9]. For a clean gold support the value of C_s considerably decreased with frequency and with decreasing ionic strength of electrolyte. However, at frequencies f>10 kHz and ionic strength 0.1 M the C_s of the gold layer/electrolyte interface was similar to that of the C_s of sBLM at frequencies in the range 100 Hz–10 kHz. At lower ionic strength, this equality began at even lower frequencies (f~300 Hz). These experiments suggest that when the analytical detection parameter is chosen to be membrane capacitance, frequencies of ac voltage below 1 kHz and ionic strength of 0.1 M are more appropriate for biosensor development.

The results of the determination of specific capacitance, conductivity, and thickness of supported films are summarized in Table 1. Optical thicknesses of sBLM were determined by the reflectance method [9]. The higher value of the thickness of sBLM obtained by optical methods in comparison with that calculated from specific capacitance of the membrane (the value of 2.1 for relative dielectric permittivity of the hydrocarbon part of the membrane, ε, and 8.85×10^{-12} F/m for dielectric permittivity of free space, ε_0, were used to calculate the membrane thickness, d_{EL}, using the equation $d_{EL}=\varepsilon\varepsilon_0/C_s$) is due to the fact that the electric field "sees" only the hydrocarbon chain region. This is caused by penetration of ions into the region of the polar head groups (see [9]).

The mechanism of sBLM conductance is not clear yet. While for sBLM formed on the rather rough surfaces of the cut tip of Teflon-coated wire, the higher specific conductivity (in comparison with that for freely suspended and gold-supported membranes) can be due to pinholes and uncovered parts of the wire directly in contact with electrolyte, for smooth membranes formed on gold supports the contribution of pinholes to conductivity is very small [9].

Table 1. Specific capacitance, C_s, specific conductivity, g_s, and electrical, d_{EL}, and optical, d_{OPT}, thicknesses of BLM and sBLM formed by various methods and made from different lipids. GMO – glycerol monooleate, SBPC – soya bean phosphatidylcholine, COB – crude ox brain fraction, TBL – total fraction of phospholipids from bovine brain, HDM – hexadecylmercaptane, DPPC – dipalmitoylphosphatidylcholine, 12T – dodecanethiol

Support	Lipid	C_s, (μF cm^{-2})	g_s, (nS cm^{-2})	d_{EL} (nm)	d_{OPT} (nm)	Reference
Freely suspended	Lecithin	0.35–0.70	10	2.6–5.3	–	[4]
Freely suspended	GMO	0.50–0.78	100	2.4–3.7	–	[4]
Agar-agar	SBPC	0.36–0.61	5–10	3.0–5.1	–	
Stainless steel wire	Lecithin 5% n-decane	0.5	500		–	[4]
Stainless steel wire	COB 2% n-decane	0.5	–	3.7	–	[63], f=1 kHz 1 mM KCl
Thin gold layer	HDM 1% n-hexane	1.0	–	1.86	1.86	[9]
Thin gold layer	DPPC n-decane	0.57	400	3.6	4.06	[9]
Thin gold layer	12T 2% n-hexane	1.0	–	1.86	–	f=0.1–10 kHz 0.1 M KCl
Thin gold layer	TBL 2% n-decane	0.5	–	0.37	–	f=0.1–10 kHz 0.1 M KCl

The stability of supported sBLM to electrical breakdown voltage is considerably higher than that for freely suspended BLM. The amplitude of the breakdown voltage of BLM is in the range 100–150 mV. For agar-supported sBLM the breakdown voltage reaches 450 mV [14], and for sBLM formed on metal or polymer supports the value surpasses 1 V [69]. The greater stability of supported membranes is due to physical binding of the monolayer adjacent to the support, and the fact that the appearance of pores in the outer monolayer will cause exposure of hydrophobic parts of the membrane to water, which is thermodynamically unfavorable (see [9]).

The measurement of the electrical properties of sBLM allowed the structure of the lipid membrane on a metal support to be established. While there were no special problems in understanding the structure of a lipid membrane formed on a gold support on a chemisorbed alkylthiol layer, the view of the structure of the lipid film formed on a tip of metal wire changed as soon as new results from investigation of these films appeared. The first schematic representation of the structure of these sBLM was presented by Tien and Salamon [4]. They suggested that this structure is analogous to the free-standing BLM and is based on a lipid bilayer in which one monolayer is closely adjacent to the wire support (Fig. 3a). Later, based on the work by Passechnik et al. [63], it was suggested that these layers do not have an exact bilayer structure and that, in addition to the lipid bilayer regions, monolayers, multilayers, and uncovered parts of the metal could also appear (Fig. 3b). This suggestion was confirmed in the paper by Haas et al. [64]. These authors showed that the structural defects in the sBLM could be considerably reduced by application of a negative voltage to the working electrode during formation of the lipid film. In this case, however, multilayers are formed (Fig. 3c).

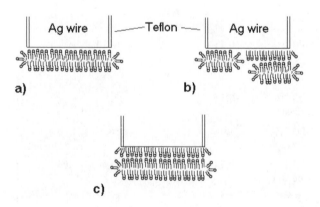

Fig. 3a–c. Schematic representation of the structure of sBLM formed on a tip of metal wire: **a** according to [4] and **b** modified structure according to [63]. **c** Possible sBLM structure obtained after continuous application of negative voltage on the membrane during its formation

4.2
Mechanical Properties of BLM and sBLM

The elasticity modulus, $E_\perp$, and the coefficient of dynamic viscosity, η, characterize the structural state of the hydrophobic part of the membrane, and changes of the ordering of this hydrophobic region should affect these parameters. Therefore the changes of membrane mechanical parameters upon bilayer formation and/or application of an external ac or dc voltage, as well as steady-state values of $E_\perp$ and η, can serve as a sensitive indicator of the stability and structural state of a membrane. These parameters also reflect how the type of support (metal, conducting polymer, agar-agar), as well as the degree of roughness of the support, influence membrane physical properties. In this part we therefore show some examples of the behavior of the parameters of membranes formed on different solid supports. The formation of BLM and sBLM is accompanied by a thinning process as a consequence of various driving forces. By measuring $E_\perp$ and C, we can observe this thinning process.

As we already mentioned above, Haas et al. [64] showed that formation of sBLM under application of a negative potential is accompanied by a decrease of electrical current and membrane capacitance. This has been explained by a decrease of the structure defect in the lipid film due to formation of a multilayer lipid structure at the hydrophobic surface of the silver. A similar effect was also observed in our work [70]. In addition to the membrane capacitance we measured changes in elasticity modulus and phase shift. The values of elasticity modulus and phase shift have been used for calculation of the coefficient of dynamic viscosity according to Eq. 4. These values are plotted in Fig. 4 as a function of time. We can see that all values decrease with time, except for the phase shift. An analogous decrease of membrane capacitance was also observed in the paper by Haas et al. [64]. In this work the final steady-state value of capacitance was in the range 170–210 pF. The average capacitance obtained in our experiments was 194±36.5 pF. The larger average values obtained in our experiments are due to the larger diameter of the wire: 0.5 mm, instead of 0.38 mm used in [64]. The average capacitance values obtained in Ref. [70] correspond to the thickness of the hydrophobic part of the lipid film: d=18.5±3.5 nm. Considering that the hydrophobic part of a lipid bilayer containing n-decane has a thickness of approx. 5 nm [56], it is possible to conclude that the lipid film at the metal support has a multilayer structure. This conclusion is in agreement with that reported by Haas et al. [64].

The estimation of film thickness, however, did not take into account the thick inhomogeneities and edge part of sBLM. Therefore the real value of the film area should be less than that calculated on the basis of wire diameter, and hence also the film thickness should be lower than presented above. From Fig. 4 it is also seen that both the elasticity modulus and coefficient of dynamic viscosity decrease and the phase shift increases with time (i.e., under continuous application of –350 mV at the sBLM). In the calculation of elasticity modulus and coefficient of dynamic viscosity according to Eqs. 3 and 4 it has been assumed that the sBLM has a bilayer structure. However, the decrease of membrane capacitance indicates that corrections on the "true" value of capacitance and thickness

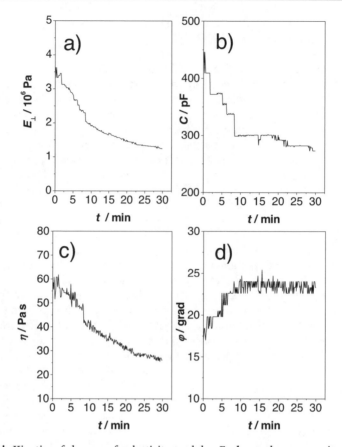

Fig. 4a–d. Kinetics of changes of a elasticity modulus, $E_\perp$, **b** membrane capacitance, C, **c** coefficient of dynamic viscosity, η, and **d** phase shift, φ, of sBLM formed at the tip of freshly cut Ag wire following application of negative potential (–350 mV) vs. Ag/AgCl electrode (negative terminal is on the Ag wire). sBLM were from soybean phosphatidylcholine dissolved in n-decane (50 mg mL^{-1})

are necessary in order to obtain the real value of the elasticity modulus $E'_\perp$ and coefficient of dynamic viscosity η′:

$$E'_\perp = E_\perp \left(I_{10}/I_0\right)^2 \tag{7}$$

$$\eta' = \eta \left(I_{10}/I_0\right)^2 \tag{8}$$

where I_{10} is the amplitude of first current harmonics at t=0. The corrected values of both $E'_\perp$ and η′ are presented in Fig. 5 together with uncorrected values, i.e., $E_\perp$ and η. We can see that the corrected values revealed considerably fewer changes with time and after approx. 30 min of application of negative potential of sBLM, steady-state values were obtained.

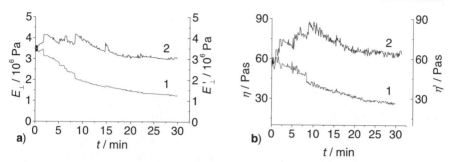

Fig. 5a,b. Kinetics of changes of a elasticity modulus $E_\perp$ (1) and that corrected on increased thickness of the membrane, $E'_\perp$ (2) according to Eq. 7. **b** The same for coefficient of dynamic viscosity, η (1) and that corrected on increased thickness of membrane, η' (2) according to Eq. 8

It is interesting to compare the behavior of sBLM formed under application of negative potential with that of sBLM formed at open circuit potential. In the latter case, the formation of sBLM was accompanied by increase of membrane capacitance and decrease of elasticity modulus [69]. For sBLM formed on conducting polymers, we observed similar behavior of C and $E_\perp$ with time as that for metal-supported sBLM [15]. It has been shown by Passechnik et al. [63] that these sBLM (i.e., formed without application of negative potential) are unstable with time. The changes of their electrical and mechanical parameters are due to the formation of islands of uncovered metal parts (which are characterized by considerably higher capacitance due to formation of electric double layers of cations a close distance from the metal surface) and inhomogeneities caused by solvent redistribution. The inhomogeneities which contained the solvent are more compressible than ordered parts of lipid bilayers. The elasticity moduli of these membranes were $(1.6\pm1.2)\times10^6$ Pa, while for sBLM of the same composition, but formed under application of negative potential, 2.5 times larger values of $(4.12\pm2.34)\times10^6$ Pa were observed. Thus, application of a negative voltage during formation of sBLM resulted in formation of a compact lipid layer at the metal surface, which is more ordered in comparison with sBLM formed at open circuit potential.

Different behaviors of capacitance and elasticity modulus were observed for solvent-free sBLM formed from SBPC on thin gold layers covered by 1-dodecanethiol. The capacitance practically did not change, while $E_\perp$ slowly increased with time, reaching the steady-state value after about 20 min. (Fig. 6). The behavior of C and $E_\perp$ for agar-supported BLM was the same as that for free-standing BLM – both parameters slightly increased with time, reaching a plateau value at approximately 20 min. [37]. Typical steady-state parameters for various membrane systems are shown in Table 2.

sBLM show a considerable dispersion of $E_\perp$, but we can see that the Young's moduli of the membranes of SBPC made from two different hydrocarbon solvents do not differ significantly from each other in contrast to the conventional BLM. Moreover, the values of $E_\perp$ for sBLM formed on a rough surface at the tip of Ag or stainless steel wire (including that covered by a layer of conducting

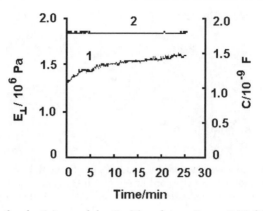

Fig. 6. Changes in the elasticity modulus $E_\perp$ (1) and capacitance C (2) following the application of ac voltage to solvent-free sBLM formed on a thin gold layer covered by a monolayer of 1-dodecanethiol. Phospholipid: SBPC. Electrolyte: 0.1 M KCl. V_0=40 mV, f=1 kHz, T=20 °C

Table 2. The mechanical parameters of free- standing BLM and sBLMs formed on different solid supports (silver wire (Ag- wire), stainless steel wire (ss- wire), ss- wire with polypyrrole (PPY), agar, thin gold film) from soya bean phosphatidylcholine (SBPC) and/or crude ox brain fraction (COB), with and without monolayer from 1-dodecanethiol (12T), with and without solvent (n-hexane (C_6), n-decane (C_{10}), n-hexadecane(C_{16}), solvent free (sf)). S.D. obtained from the measurement of at least five membranes in each series

Support	Lipid/Solvent	$E_\perp$, Pa	η, Pa s
No support (free-standing BLM)	SBPC/C_{10}	$(1.3\pm0.3)\times10^7$	550±150
	SBPC/C_{16}	$(6.2\pm1.2)\times10^7$	1950±500
Agar	SBPC/C_6+C_{16}	$(2.0\pm0.9)\times10^7$	–
Ag-wire	SBPC/C_{10}	$(1.6\pm1.2)\times10^6$	60±35
Ag-wire	SBPC/C_{16}	$(1.5\pm1.3)\times10^6$	65±40
Ag-wire (-350 V applied to sBLM during its formation)	SBPC/C_{10}	$(4.1\pm2.3)\times10^6$	65±35
ss-wire	SBPC/sf	2×10^6	–
ss-wire	COB/C_{10}	$(2.4\pm1.3)\times10^6$	–
ss-wire+PPY	COB/C_{10}	$(2.7\pm2.9)\times10^6$	–
Ag-wire	SBPC+12T/sf	$(4.2\pm3.0)\times10^7$	–
Thin gold layer	SBPC+12T/sf	$(1.0\pm0.5)\times10^8$	–
Thin gold layer	SBPC+12T/C_6	$(1.6\pm1.1)x10^7$	–

polypyrrole) are one order of magnitude smaller than those of BLM. The measured moduli $E_\perp$ did not depend on membrane capacitance, i.e., on the bilayer/torus area ratio. Thus, it shows that the formation of a new bilayer region from the edges does not determine the values of the elasticity modulus [56]. In the case of membranes that contain solvent, the value of $E_\perp$ depends on the support roughness [68]. sBLM formed on a rough surface at the tip of a wire are more compressible than those formed on smooth gold layers covered by alkylthiol. Tight coverage of a solid support by a monolayer has an important effect on $E_\perp$. We can see from Table 1 that the $E_\perp$ values of an sBLM formed from SBPC without and with alkylthiols differ about tenfold. One of the possible reasons is that the metal surface of an sBLM is not fully covered by lipids without alkylthiols.

4.3
Changes of Electrostriction of sBLM Following Interaction with Surfactants, Surfactant–DNA Complexes, and DNA

The study of the mechanisms of DNA–lipid interactions has an important role for understanding the processes connected with gene expression [71] and electroporation of DNA through plasmatic membranes of the cell [72]. Because DNA is negatively charged, it does not induce substantial changes of the physical properties of the membranes itself, but only in the presence of cations, like Ca^{2+} or Mg^{2+}. In the presence of these cations so-called triple complexes are formed [71]. Also the transfer of DNA into the cell is more effective when nucleic acid is in complexes with cationic lipids [73], cationic antibiotics [74], or cationic surfactants [75].

It has been shown that DNA in the presence of bivalent cations induces a shift of the phase transition of saturated phosphatidylcholines toward higher temperatures. This effect has been explained by partial denaturation of DNA at the lipid bilayer surface, which makes possible interaction of bases with the hydrophobic part of the membrane [73]. A recently performed ultrasound velocimetry study of DNA–DPPC interactions in the presence of calcium ions showed induction of a biphasic stage in the melting curves, which was explained by the coexistence of free and strongly bound lipids [76]. DNA interacts with cationic surfactants cooperatively, as has been shown recently by Petrov et al. [75]. These authors found different effects of the cationic surfactants dodecylamine (DDA) and dodecyltrimethylammonium bromide (DTAB) on the secondary structure of DNA. While DTAB stabilized the DNA structure, DDA resulted in its destabilization. The destabilizing effect of DDA has been explained by the displacement of intramolecular hydrogen bonds in complementary base pairs with intermolecular H bonds between unsubstituted DDA amino groups and proton-accepting sites of nucleic bases. The existence of DNA–lipid interaction was also revealed by increased surface pressure of lipid monolayers following addition of plasmid DNA [72]. DNA can also interact with positively charged peptides [77, 78].

Despite extensive study of DNA–lipid interactions, only a little is known about the mechanisms of these processes. The sBLM and the electrostriction method could be very helpful in improving our understanding of the mechanisms of surfactant–lipid and DNA–lipid interactions [70]. The next sections are concerned with the study of the effect of the cationic surfactant hexadecylamine (HDA), DNA–HDA complexes, and DNA in the presence of Mg^{2+} ions on the electrostriction of sBLM. The sBLM were formed under the application of a negative potential of –350 mV against the reference Ag/AgCl electrode as shown in a previous section.

4.3.1
Interaction of HDA with sBLM

Hexadecylamine (HDA) is a molecule composed of an amino group and a hydrophobic chain. The amino group is positively charged at neutral pH. At low concentration the HDA is in the form of the monomer, while above the critical micellar concentration (2.75 mM) it forms micelles. In this section we will present results of studies of the interaction of HDA with stabilized sBLM, i.e., formed under application of a negative potential. The sBLM were kept at this potential approx. 30 min. Then only an alternating potential of relatively small amplitude (50 mV) with a frequency f=1 kHz was applied to the membrane in order to measure changes of the physical properties of sBLM.

In most cases the HDA at lower concentration (up to 0.3 mM) did not cause changes of membrane capacitance and elasticity modulus. However, at higher concentration addition of this surfactant resulted in increase of elasticity modulus and decrease of membrane capacitance. Relative changes of these values as a function of HDA concentration are presented in Fig. 7. We can see that the shape of these changes is sigmoidal, which may indicate a certain cooperation of interaction of HDA with sBLM. Probably the accumulation of HDA monomers at the membrane surface at higher concentration resulted in the formation of aggregates, which are more efficient at changing the properties of sBLM in comparison with monomers [79]. We should, however, note that the changes of the physical properties of sBLM following addition of HDA below the critical micellar concentration were not very reproducible. In certain cases we observed an increase of elasticity modulus and decrease of membrane capacitance at lower concentrations (0.1 mM). In other cases, addition of HDA at a concentration of

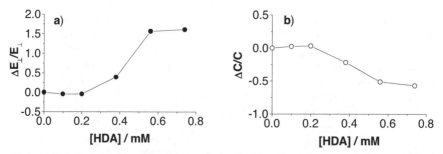

Fig. 7a,b. Relative changes of elasticity modulus, $E_\perp$ (a) and membrane capacitance, C (b) of sBLM as a function of concentration of HDA

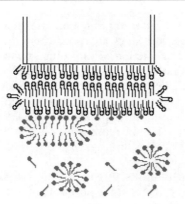

Fig. 8. Schematic representation of HDA monomers and micelles at the surface of sBLM

0.1 mM resulted in an increase of elasticity modulus without substantial changes of membrane capacitance. Then for c>0.3 mM HDA the elasticity modulus and capacitance decreased.

Different changes of the values of $E_\perp$ and C following addition of HDA can be particularly explained by the different aggregation state of the surfactant at the surface of the lipid film. The surfactant can be in both the monomeric and micellar state (Fig. 8). Incorporation of monomers into the bilayer could result in a decrease of the free volume in the lipid layer and, due to electrostatic attractive forces, also an increase of the ordering of the polar part of the sBLM. These two effects should explain the increase of the elasticity modulus. The formation of micelles, on the other hand, could cause appearance of additional layers at the sBLM surface and thus a decrease of membrane capacitance. The layers of HDA could also contribute to the overall elasticity modulus of the sBLM. Considering that the elasticity modulus $E_\perp$ of BLM composed of diacylphosphatidylcholines (e.g., SBPC) is higher than that for α-monoglycerides (i.e., molecules composed of one hydrophobic chain), we should expect that the decrease of elasticity modulus could be due to both the lower elasticity modulus of HDA layers and the increased number of layers at the surface of sBLM (see the effect of increased thickness of sBLM on $E_\perp$ value discussed in Sect. 4.2.). We should note that a different direction of changes of elasticity modulus in the presence of the cationic surfactant cetyltrimethylammonium bromide (CTAB) or neutral surfactant Triton X-100 have been reported in our earlier work. We showed that the changes of physical properties of BLM depended also on the initial state of the elasticity modulus of the membrane (see [56]).

4.3.2
Interaction of HDA–DNA Complexes with sBLM

Due to the electrostatic interactions between positively charged amino groups of HDA and negatively charged phosphate groups of DNA, HDA–DNA complexes are formed [75, 80, 81]. The interaction of the complexes (HDA:DNA=1:1 mol/mol) with sBLM was studied by measurement of the elasticity modulus, membrane capacitance, and intrinsic membrane potential, $\Delta\Phi_m$, following addition of complexes to the buffer from stock solution, where the complexes were at a concentration of 3.3 mM. Interaction of the complexes with sBLM resulted in different changes of the values $E_\perp$, C, and $\Delta\Phi_m$. In Fig. 9 we can see the typical kinetic changes observed following addition of the complexes: increase of elasticity modulus, membrane capacitance, and intrinsic membrane potential, $\Delta\Phi_m$. The changes of relative values of these parameters are shown in Fig. 10. However in certain cases, after an initial increase of elasticity modulus and membrane capacitance a decrease of these values was also observed at higher concentration of complexes (not shown).

Considering the obtained results, three types of processes could be suggested to explain the observed changes of the physical properties of sBLM following adsorption of HDA–DNA complexes:

1. The complex could be incorporated into the sBLM (see Fig. 11), which could be accompanied by an increase of relative dielectric permittivity of the hy-

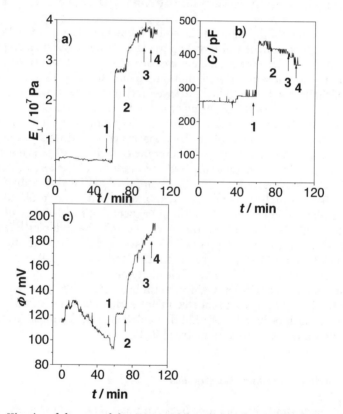

Fig. 9a–c. Kinetics of changes of elasticity modulus, $E_\perp$ (**a**), electrical capacitance, C (**b**), and intrinsic membrane potential, $\Delta\Phi_m$ (**c**) of sBLM following addition of HDA–DNA complexes at concentrations 1–300; 2–400; 3–500; and 4–550 μM. Moment of addition is shown by arrows

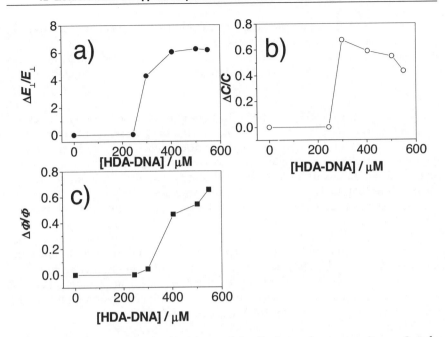

Fig. 10a–c. Relative changes of **a** elasticity modulus, $E_\perp$, **b** membrane capacitance, C, and **c** intrinsic membrane potential, $\Delta\Phi_m$, of sBLM as a function of concentration of HDA–DNA complexes

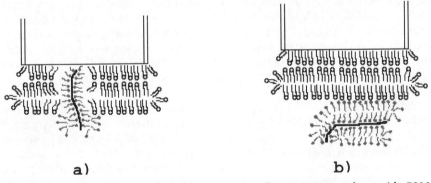

Fig. 11a,b. Schematic representation of the interaction of HDA–DNA complexes with sBLM. **a** HDA-DNA complex is incorporated into the lipid film; **b** complex is adsorbed at the surface of the lipid film

drophobic part of the lipid layer and consequently by an increase of membrane capacitance. The increase of membrane capacitance can, however, also be caused by structural defects in the lipid layers that can be induced by HDA–DNA interactions.

2. In certain cases an increase of elasticity modulus was observed, but the capacitance did not change. This effect can be explained by incorporation of

monomers of HDA into the lipid layer. These monomers are not involved in the formation of HDA–DNA complexes.

3. Adsorption of the complexes could result in an increase of sBLM thickness (Fig. 11), which was observed in certain cases as a decrease of membrane capacitance.

Thus we can see that HDA and HDA–DNA complexes caused different effects on the sBLM physical properties. While interaction of HDA with sBLM was in all cases accompanied by a decrease of membrane capacitance, the interaction of complexes resulted in a capacitance increase or did not induce changes of this value with simultaneous increase of the elasticity modulus. The obtained results clearly show that the HDA–DNA complexes are not only adsorbed on the membrane surface but could also be incorporated into the lipid interior and/or could induce changes in the hydrophobic part of the lipid membrane. We should, however, note that the observed differences could be caused by changes in the process of micelle formation in the presence of DNA, as has been recently shown by Sukhorukov et al. [82]. They found that two different DNA–surfactant complexes are formed in the presence of DNA and that the cmc for surfactants was 15–40 times lower than that without DNA. It has been assumed that amphiphile molecules act as linkers which favor DNA aggregation.

4.3.3
Interaction of the DNA Complexes with sBLM: Influence of Mg²⁺

In a first series of experiments we studied the interaction of DNA with sBLM without Mg^{2+} ions. In this case, addition of DNA to the buffer in the concentration range 21–170 µg mL^{-1} did not result in any substantial changes of elasticity modulus, membrane capacitance or intrinsic membrane potential. This means that negatively charged DNA cannot adsorb on the negatively charged surface of sBLM at sufficiently close distance in order to induce changes of the membrane physical properties.

Substantial changes of sBLM properties were observed when Mg^{2+} was added to the electrolyte where DNA was present at a concentration of 170 µg mL^{-1} (0.52 mM). This is seen in Fig. 12. Addition of Mg^{2+} at relatively low concentration (up to 7.5 mM) resulted in a substantial increase of elasticity modulus, an increase of intrinsic membrane potential, and modest changes of electrical capacitance. It is interesting that we observed changes in the physical properties of sBLM at a lower concentration of Mg^{2+} than that characteristic for triple complexes (DNA–Mg^{2+}–phospholipid) reported earlier for liposomes [71]. We should note that Mg^{2+} itself (without DNA) at concentrations up to 7.5 mM did not induce changes of sBLM physical properties. This is also in agreement with the result obtained by Haas et al. [64]. In this work the authors did not observe changes of membrane capacitance in the presence of a 1 mM concentration of Mg^{2+}.

The increase of elasticity modulus and membrane potential give clear evidence about the formation of DNA–Mg^{2+}–lipid complex head groups at the surface of sBLM. This binding by means of the Mg^{2+} cations probably results in re-

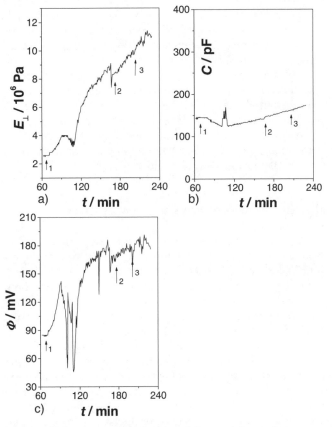

Fig. 12a–c. Kinetics of changes of elasticity modulus, $E_\perp$ (**a**), membrane capacitance, C (**b**), and intrinsic membrane potential, $\Delta\Phi_m$ (**c**) of sBLM following addition of Mg^{2+} to the electrolyte containing 10 mM NaCl and 0.52 mM DNA. Mg^{2+} was added in final concentrations of 1–3.8; 2–5.6; and 3–7.4 mM. The moment of addition of Mg^{2+} is shown by arrows

striction of the mobility of phospholipids and consequently could result in an increase of the ordering of the lipid layer. A small increase of elasticity modulus following addition of bivalent cations could be caused, with competitive binding of Na^+ and Mg^{2+} ions to DNA. Initially the DNA molecules are surrounded by Na^+ ions. These ions are, however, not able to mediate interaction of DNA with the membrane. Addition of Mg^{2+} results in adsorption of these ions on the negatively charged surface of sBLM and they also start to compete with Na^+ to bind with DNA. This competitive binding could be one of the possible reasons for the slow process of binding DNA to the membrane. The speed of change of the physical properties of sBLM induced by DNA–Mg^{2+} complexes is really much slower than that induced by HDA–DNA complexes. While in the former case the changes last several tens of minutes, in the latter case they last less than 1 min.

Thus, the interaction of the cationic surfactant hexadecylamine (HDA) with sBLM resulted in a decrease of membrane capacitance and induced two-directional changes of elasticity modulus (increase or decrease), which can be caused by a different aggregation state of surfactant at the surface of sBLM. In contrast, the HDA–DNA complexes resulted in most cases in an increase of elasticity modulus and increase of membrane capacitance, which can be caused by incorporation of these complexes into the hydrophobic interior of sBLM. Certain parts of these complexes can, however, remain adsorbed on the sBLM surface. DNA itself does not cause substantial changes of the physical properties of sBLM; however, addition of bivalent cations of Mg^{2+} to the electrolyte caused substantial increase of elasticity modulus and membrane potential. These changes are, however, much slower than that for HDA–DNA complexes, which can be caused by slow competitive exchange between Na^+ and Mg^{2+} ions.

4.4
sBLM Modified by Short Oligonucleotides as a Possible Genosensor

In the previous section we showed that interaction of DNA with sBLM in the presence of bivalent cations resulted in considerable changes of the membrane electrostriction. The effect of the sensitivity of the membrane electrostriction can therefore also be used to study the peculiarities of the interaction of short oligonucleotides with sBLM. This task has crucial importance for fabrication of genosensors based on amphiphilic films, and for direct detection of DNA hybridization at the surface of these films [30, 83, 84].

Both BLM and sBLM allow DNA to be immobilized on a lipid surface in such a way that it provides sufficient conformational freedom of immobilized oligonucleotides and thus good access for a complementary DNA chain. For this purpose the oligonucleotides are chemically modified by a hydrophobic anchor, allowing immobilization of DNA at the surface of the lipid film. It has been shown that various membrane properties, such as conductance [30, 84, 85], capacitance relaxation [30] or properties of lipid monolayers [86, 87], change following hybridization. We also showed that detection of hybridization depends on the orientation of oligonucleotides relative to the thin film surface [88].

The interpretation of the results of the detection of hybridization using the lipid films requires knowledge of the mechanisms of interaction of oligonucleotides with BLM or sBLM. In this work we therefore present the peculiarities of interaction of short oligonucleotides with free-standing BLM and with lipid films formed on a different solid support [31]. In the experiments we used 5′-deoxypentadecylthymidylate (($dT)_{15}$) chemically modified by cholesterol and/or by palmitic acid. The obtained 5′-cholesterolphosphoryl ($CH(dT)_{15}$) and/or 5′-palmitylphosphoryl ($C16(dT)_{15}$) compounds were then used as a probe for anchoring single-stranded DNA to sBLM by cholesterol or palmitic acid residues. We showed that interaction of $CH(dT)_{15}$ with BLM and/or with sBLM formed on agar or metal supports resulted in considerable increase of the elasticity modulus in a direction perpendicular to the membrane plane, $E_\perp$. An increase of membrane potential of sBLM was also observed. The interaction of complementary 5′-deoxypentadecyladenylate ($dA)_{15}$ with sBLM modified by $CH(dT)_{15}$ resulted in a further increase of membrane potential, while the $E_\perp$ only slightly decreased. In contrast with $CH(dT)_{15}$, ($dT)_{15}$ modified by palmitic acid did not result in considerable changes of elasticity modulus; however, it induced changes of membrane potential of lipid membranes comparable with that caused by $CH(dT)_{15}$.

An example of the changes of elasticity modulus $E_\perp$, membrane capacitance C, and intrinsic membrane potential $\Delta\Phi_m$ of BLM and sBLM formed on an agar support following addition of $CH(dT)_{15}$ and complementary oligonucleotide $(dA)_{15}$ are shown in Figs. 13 and 14, respectively. Addition of $CH(dT)_{15}$ at a final concentration of 100 nM to BLM composed of a mixture of egg phosphatidyl-choline (eggPC) and hexadecylamine (HDA) (eggPC:HDA=2:1 w/w) resulted in a sharp increase of elastic modulus (by 1.6 times; trace 1), while the capacitance slightly decreased (trace 2) (Fig. 13a). Incorporation of oligonucleotide was accompanied by an increase of the membrane potential (Fig. 13b). Addition of the complementary chain $(dA)_{15}$ resulted in a further increase of membrane poten-

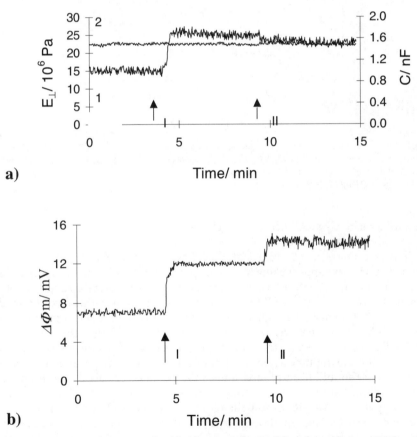

Fig. 13a,b. Kinetics of changes of **a** elasticity modulus, $E_\perp$ (1) and capacitance, C (2), and **b** intrinsic membrane potential, $\Delta\Phi_m$, of BLM following addition of $CH(dT)_{15}$ (I) and $(dA)_{15}$ (II) at final concentration of 100 nM. Electrolyte: 1 M KCl+10 mM HEPES, pH 7

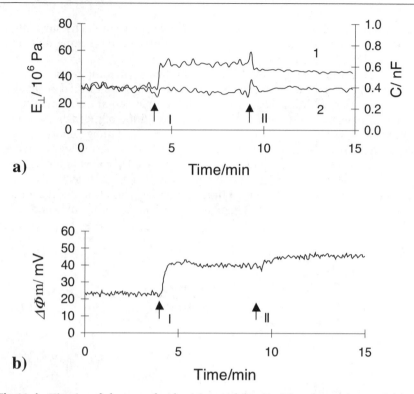

Fig. 14a,b. Kinetics of changes of **a** elasticity modulus, $E_\perp$ (1) and capacitance, C (2), and **b** intrinsic membrane potential, $\Delta\Phi_m$, of sBLM formed on the agar support following addition of $CH(dT)_{15}$ (I) and $(dA)_{15}$ (II) at final concentration of 100 nM. Electrolyte: 1 M KCl+10 mM HEPES, pH 7

tial and a relatively small decrease of elasticity modulus, while the capacitance practically did not change.

Similar changes of the values studied have also been observed for sBLM formed on an agar support using the same lipid composition as in the case of BLM (Fig. 14). In contrast with BLM, these membranes were characterized by higher values of elasticity modulus $(3.2\pm0.1)\times10^7$ Pa and higher initial surface potential difference 23 ± 0.8 mV. The higher value of the elasticity modulus for agar-supported BLM was particularly caused by the restricted mobility of the phospholipid head groups adjacent to the agar support. The larger potential difference can be due to both membrane asymmetry connected with slightly changed orientation of phospholipid head groups, and to the certain electrolyte asymmetry at both membrane sides. Similar results have been obtained on sBLM formed on a gold support [31]. No significant changes of elasticity modulus, membrane capacitance or surface potential were observed when noncomplementary oligonucleotide was added to the electrolyte for both BLM and sBLM.

The increase of elastic modulus might be due to at least two effects: (1) incorporation of cholesterol residues in the membrane and a condensation effect of cholesterol; and (2) an increase of ordering of the polar part of the membrane due to electrostatic interaction of opposite charges (positive surface charge of the membrane surface and negative charge of oligonucleotide). The condensing effect of cholesterol [89], particularly causing an increase of elasticity modulus [56, 90], is well known. Another effect resulting in an increase of elasticity modulus, i.e., decrease of compressibility of sBLM in a direction perpendicular to the membrane plane, might be a decrease of repulsive forces between positively charged HDA molecules at the polar part of the membrane due to incorporation of negatively charged oligonucleotides. For both BLM and sBLM we observed a slight decrease of elasticity modulus following addition of a complementary chain (i.e., $(dA)_{15}$). This peculiarity can be explained as follows. The flexibility of single-stranded DNA is higher than that of double-stranded DNA [91]. The single-stranded DNA is randomly oriented relative to the membrane plane, with a preference for parallel orientation due to attractive forces between the negative charges of DNA and positive charges of HDA. Formation of a double helix of DNA following addition of the complementary chain resulted in changes of orientation of the nucleic acid. Due to a rather high persistent length (approx. 50 nm, see Ref. [91]), the double helix should be oriented mostly perpendicular to the membrane plane. Therefore, the attractive forces between the negative charges of DNA and positive charges of HDA should be lower than for parallel orientation of single-stranded DNA. Thus, weaker attractive forces will cause more freedom in the mobility of phospholipid head groups and consequently result in higher compressibility of the membrane (i.e., lower elasticity modulus).

In order to analyze the interaction of single-stranded DNA with the lipid membrane and the hybridization of DNA at the membrane surface, we analyzed the behavior of the surface potential of sBLM [31]. We constructed a Scatchard plot of $\Delta\Phi/\Phi_0 c$ vs. $\Delta\Phi/\Phi_0$ for changes of surface potential following addition of $CH(dT)_{15}$ and $(dA)_{15}$ ($\Delta\Phi=\Phi-\Phi_0$, where Φ is the surface potential of BLM at corresponding concentration of $CH(dT)_{15}$ or $(dA)_{15}$ and Φ_0 is the initial surface potential of BLM, i.e., prior to addition of $CH(dT)_{15}$ or $(dA)_{15}$). This approach allowed us to analyze the possible cooperativity pattern. When the above relationship holds:

$$\Delta\Phi/\Phi_0 c = c^{m-1}K_b\left(N-\Delta\Phi/\Phi_0\right), \tag{9}$$

where K_b is the binding constant and N indicates the availability of binding sites at the gold support, and represents the distribution coefficient for the concentration of free and bound sites. Three different cooperativity patterns can be obtained depending on the Hill coefficient m: for m>1 the processes are positively cooperative, for m=1 they are noncooperative, and for m<1 they show negative cooperativity [92]. The function $\Delta\Phi/\Phi_0 c$ linearly depends on the changes of surface potential induced by addition of $CH(dT)_{15}$ (correlation coefficient R=0.7±0.22), i.e., m≈1 (see [31]). This means that the process of incorporation of cholesterol-modified oligonucleotides in the sensor surface is noncooperative. In the case of hybridization of DNA, the value of $\Delta\Phi/\Phi_0 c$ practically does not depend on the $\Delta\Phi/\Phi_0$ value. This fact reflects a weak effect of DNA hybridization on the changes of surface potential of BLM. Recently we obtained similar results in a study of hybridization of DNA on a gold surface [88]. The 18-mer probe was immobilized on a gold surface via a thiol group. Addition of the complementary chain resulted in an increase of conductivity of the DNA layer. The process of change of the conductivity was not cooperative.

In order to obtain more information about the adsorption of oligonucleotides on sBLM, we determined the surface charge density of sBLM in a shielding experiment. For this purpose the sBLM (on agar support) were formed at low electrolyte concentration (10 mM KCl+10 mM HEPES, pH 7). The electrolyte concentration in the Teflon tube was 0.11 M KCl. The increase of ionic strength of the electrolyte in which the membrane was formed allowed us to determine the surface charge density of the sBLM. An increase of ionic strength should result in a decrease of Gouy–Chapman potential, due to shielding of surface charges by ions [93]. Therefore for sBLM formed initially in asymmetrical electrolyte conditions (i.e., 0.01 M/0.11 M), we should expect a decrease of the overall surface potential following achievement of symmetrical conditions (i.e., 0.11 M/0.11 M). The degree of this potential decrease will depend on the surface charge density. Therefore having this value, we can determine the surface charge density (see [56, 93] for detailed procedure). We showed that an unmodified sBLM of eggPC+HDA was characterized by a surface charge density of 0.096 ± 0.01 e/nm^2, i.e., one elementary charge per 10.4 nm^2. Addition of CH(dT)$_{15}$ in a final concentration of 100 nM resulted in an increase of surface potential by approx. 20 mV. This increase of surface potential corresponded to the appearance of a negative surface charge on the membrane, of a surface charge density 0.83 ± 0.06 e/nm^2. The charge Q carried by one single-stranded DNA molecule with n anionic phosphate groups is given by $Q=2ne_0f_r$, where $e_0=1.6\times10^{-19}$ C is the elementary charge, factor 2 refers to the contribution of both cation and the polyanion, and $f_r\approx0.4$ is the reduction factor due to counterion screening in the relative motion of the cations and the polyanion [91]. Hence for a 15-mer oligonucleotide, $Q=12e_0$. Considering that the area of an sBLM is $\sim2\times10^{-3}$ cm^2, there are approximately 1.4×10^{10} molecules of CH(dT)$_{15}$ on the membrane surface. This number of molecules is about 40 times lower than in the case of chemisorption of thiol-modified DNA onto a gold support (Hianik et al. [88]). We should, however, note that the estimation of the number of DNA molecules was performed assuming that the cholesterol-modified oligonucleotides are lying flat with their full length on the BLM surface. Thus this estimation represents the lower limit of the number of DNA molecules immobilized on a BLM.

In contrast to interaction of CH(dT)$_{15}$ with BLM and sBLM, palmitoyl-modified oligonucleotide (C16(dT)$_{15}$) did not induce substantial changes of compressibility of both BLM and sBLM; however, both CH(dT)$_{15}$ and C16(dT)$_{15}$ induced similar changes of the surface potential. At the same time we should note that modification of oligonucleotide by a hydrophobic chain allowing its incorporation into the lipid layer is crucial for detection of DNA hybridization at the membrane surface by the electrostriction method. Unmodified oligonucleotides (dT)$_{15}$, (dA)$_{15}$ or dA$_1$T$_8$C$_6$G$_7$ at concentrations up to 100 nM have no significant effect on the physical properties of BLM and sBLM.

4.5
Changes of Physical Properties of sBLM During Binding of Enzymes and Immunoglobulins

The preparation of biosensors based on sBLM requires the immobilization of enzymes, antibodies or receptors on the bilayer surface. One convenient way to bind macromolecules to sBLM is the so-called avidin–biotin technology [52]. In this case the macromolecule (e.g., enzyme, antibody) is modified by avidin. In parallel, the phospholipid head groups are modified by biotin. Because the interaction of avidin with biotin results in stable avidin–biotin complexes, the addition of avidin-modified enzyme into the electrolyte with biotinylated sBLM leads to immobilization of functional molecules on the membrane surface. We can expect that immobilization of avidin-modified compounds on the membrane surface will result in changes of the structural state of the membrane. The degree of such binding could therefore be observed by the measurement of membrane compressibility, capacitance, and other macroscopic parameters. As an example of results from the electrostriction method, we show how immobilization of avidin-modified glucose oxidase and avidin-modified immunoglobulin influence the membrane capacitance and the value of $E_\perp$.

To measure the degree of binding of the avidin–glucose oxidase complex (A-GOX) to sBLM we measured the time course of changes of elasticity modulus, membrane capacitance, and intrinsic potential of biotinylated sBLM following the addition of A-GOX to a preformed membrane [6]. The changes of parameters $E_\perp$ and C of biotin-modified sBLM following the addition of A-GOX to a final concentration of 30 and 60 nM, respectively, are shown in Fig. 15a. The first addition of A-GOX (final concentration 30 nM) induced an approximately 20% increase of elasticity modulus, a twofold decrease of membrane capaci-

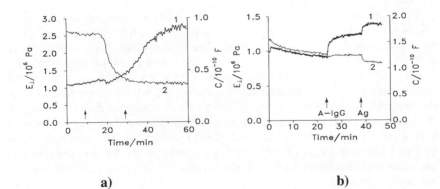

a)   b)

Fig. 15a,b Time course of changes of elasticity modulus $E_\perp$ (1) and electrical capacitance C (2) for sBLM prepared from crude ox brain extract, modified by biotin, following the addition of **a** A-GOX at a final concentration of 30 and 60 nM (the moment of the addition of A-GOX is indicated by arrows) and **b** A-IgG at a final concentration of 43 μg mL^{-1} and Ab at 31 μg mL^{-1}. The moment of the addition of A-IgG and Ag is indicated by arrows. T=20 °C

tance, and an increase of $\Delta\Phi_m$ by about 100 mV (not shown). The increase of the elasticity modulus indicates an increase of the ordering of the hydrophobic core of the lipid bilayer. This could be due to the specific binding of A-GOX to the biotin sites in the polar region of sBLM. This chelation-like interaction probably leads to a decrease of the specific area per phospholipid molecule and thus to an increase of ordering of the hydrophobic part of the membrane. The subsequent addition of A-GOX (final concentration 60 nM – second arrow in Fig. 15a) resulted in an additional increase of the elasticity modulus, while the capacitance and intrinsic membrane potential changed only slightly. This may suggest that at a concentration of 30 nM, the A-GOX complex practically covers the whole surface of the membrane.

Unmodified sBLM were characterized by a value of the intrinsic membrane potential of about 40–100 mV. Modification of lipids by biotin resulted in a slight decrease of $\Delta\Phi_m$ by 10 mV. These potentials can be due to the asymmetry of sBLM caused by the difference between the lipid/water and lipid/metal interfaces. At present, it is difficult to determine the real cause of changes of $\Delta\Phi_m$ values, but the change on adsorption of A-GOX shows that this complex bears a significant negative charge, which is consistent with [94]. An example of a binding study of the interaction of avidin-modified swine immunoglobulin (A-IgG) with sBLM is shown in Fig. 15b. Addition of the A-IgG to the electrolyte at a final concentration of 43 µg mL^{-1} resulted in an increase of elasticity modulus by 30%, and a decrease of membrane capacitance by 1%. When the membrane parameters stabilized, the antigen (Ag) – the mouse monoclonal antibody IVA 44 – was added to the electrolyte at a final concentration of 0.31 µg mL^{-1}. Addition of Ag resulted in a further increase of $E_\perp$ and a decrease of C. The nature of the changes of $E_\perp$ following the addition of A-IgG is similar to that of A-GOX. Further changes of $E_\perp$ probably reflect the increase of structural changes of the membrane caused by the A-IgG–Ag complex. This example also shows the usefulness of the measurement of membrane mechanical parameters to check an antigen–antibody reaction.

The important role of the influence of the structural changes of macromolecules on the mechanical properties of supported lipid membranes can also be demonstrated when small molecules, like the herbicide 2,4-D, interact with monoclonal antibody immobilized on lipid bilayers on a gold support [95]. Figure 16 shows changes of elasticity modulus $E_\perp$ as a function of 2,4-D concentration. We can see that $E_\perp$ increases with increasing 2,4-D concentration. The detection limit of this biosensor was relatively small. However, the present experiment gives certain information about the peculiarities of the Ag–Ab interaction, when Ab is immobilized on a membrane surface. Dramatic changes of the mechanical parameters of the membrane proved that the Ag–Ab binding process had a considerable influence on the membrane ordering. Probably, the Ag–Ab binding resulted in substantial conformation changes of the macromolecular complex. These changes can be subsequently widespread over a large membrane area. As a result, a wide membrane region will be characterized by a new structural state, different from that of an untreated membrane.

Let us suppose that the lipid film consists of two kinds of regions – disturbed (due to binding of avidin-modified antibody to the membrane) and undis-

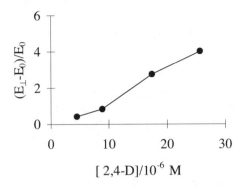

Fig. 16. Changes of elasticity modulus $E_\perp$ as a function of 2,4-D concentration for biosensors based on gold-supported lipid film with avidin-modified monoclonal antibody against 2,4-D immobilized by addition of a small amount of antibody on the film prior its immersion in the electrolyte

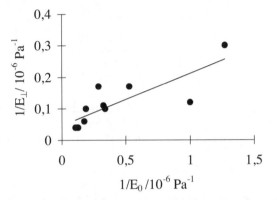

Fig. 17. Plot of $1/E_\perp$ versus $1/E_0$ for 11 independently prepared membranes modified by monoclonal antibody. The value of $1/E_\perp$ corresponds to the established value of elasticity modulus at 30 µM concentration of 2,4-D in electrolyte. E_0 is the elasticity modulus of sBLM prior to addition of 2,4-D

turbed. Then, as we showed earlier, the overall elasticity modulus of the membrane will be characterized by the following equation [56]:

$$1/E_\perp = s/E_m + (1-s)/E_0 \tag{10}$$

where s is the relative area of changed membrane structure, E_m is its elasticity modulus, and E_0 is the elastic modulus of undisturbed membrane (prior to addition of Ag). If our assumption is right, then the dependence of $1/E_\perp$ on $1/E_0$ should be a straight line. The slope of this line and its intersection with the $1/E_\perp$ axis then make it possible to determine both s and E_m. For quantitative analysis of the peculiarities of changes in the mechanical properties of sBLM, we therefore plotted the final value of compressibility ($1/E_\perp$) (established after addition of 2,4-D) as a function of the initial value of lipid film compressibility ($1/E_0$) (i.e., prior to addition of 2,4-D. The value of $1/E_0$ corresponded to the compressibility of the lipid film with immobilized avidin-modified Ab). This plot for 11 membranes is presented in Fig. 17, and can be fitted by straight line (correlation coefficient r=0.83). Using Eq. 10, from the parameters of the straight line we can conclude that due to Ag–Ab interaction, the relative area of the altered structure of the membrane is s=0.83, i.e., approximately 83% of the membrane structure is changed. The elasticity modulus of the distorted region is $E_m=1.84\times10^7$ Pa.

4.6
Dielectric Relaxation of Lipid Bilayers upon Binding Processes

In this section, we will show how the method of dielectric relaxation allows one to study the dynamic properties of supported bilayers, and selective binding of the macromolecules on the membrane surface. The method of dielectric relaxation allows us to determine the characteristic time of the reorientation of dipole moments of phospholipid head groups. Due to the domain structure of lipid bilayers and the different size of the clusters, we can expect different collective movement of the reoriented dipole moments following the application of symmetrical voltage jumps to the membrane.

The curves for each voltage setting were analyzed independently to eliminate bias in "choosing" the time constants. We confirmed the conclusion obtained by Sargent [59], i.e., no correlation was found between τ_i and the voltage jumps. The results at a series of different voltages and with different membranes were then combined to calculate the corresponding relaxation amplitudes (Eq. 6). The voltage dependence was checked graphically by plotting the relaxation amplitude $I_0^i \tau_i C'_m$ versus voltage V_0 in a double log plot. All relaxations followed a linear relationship (slope=1). This means that relaxation times do not depend on the amplitude of the applied voltage in the range of voltage that was studied.

The results of determining these relaxation times are presented in Table 3. We can see considerable differences of relaxation times between different membrane systems. While free-standing BLM are characterized by two relaxation times, the dynamics of sBLM formed on the tip of stainless steel wire can be

Table 3. Relaxation times of the reorientation of dipole moments of membranes prepared from SBPC, or SBPC, and alkylthiol, formed by different methods

Membrane system	τ_1, µs	τ_2, µs	τ_3, µs	τ_4, µs	τ_5, µs	τ_6, µs
BLM in n-decane	6.9±1.9	18±6				
sBLM in n-decane, (stainless steel wire)	5.9±0.4	13.6±2.0	27.6±3.6	51.7±4.2	85±9	120±12
sBLM with alkylthiol (thin gold layer)	40±4.0	290±100	1903±760			

characterized by up to six relaxation times. The reason for the increase in the number of relaxation components could be due to physical and chemical adsorption of phospholipids on the metal support. This adsorption could result in a different degree of immobilization of lipid molecules and thus lead to the appearance of different and more relaxation times. A similar result is also seen for membranes formed on thin gold layers with alkylthiols. In this case, the number of relaxation components is lower. However, a further increase of the duration of relaxation times takes place. This provides evidence about the strong restriction of movements of dipole moments. This result is in agreement with the increased values of elasticity modulus of the latter membrane system in comparison with that of sBLM formed on wires.

The physical origins of different relaxation times for BLM have been analyzed by Sargent [59]. It was suggested that the fastest relaxation times (several microseconds to tens of microseconds) could correspond to small amplitude reorientations of individual dipoles about an axis lying in the plane of the membrane, while the times of about 100 µs reflect a rotational reorientation of individual molecules. Slow relaxation components (several hundred microseconds to milliseconds) probably indicate the reorientation movements of domains or clusters of dipoles in the membrane plane. For comparison, an NMR study by Davis [96] gave dipole correlation times of 1–5 µs for lecithin vesicles, which is consistent with the results obtained from conventional BLM as used in our experiments. Relaxation times in the micro- and millisecond range obtained by various techniques on phospholipid bilayers were recently reported by Laggner and Kriechbaum [97].

Dielectric relaxation experiments allowed us to study the binding of enzymatic complexes on the membrane surface and confirmed the strong binding of the avidin-GOX complex to the biotinylated membranes. These experiments were performed on a free-standing BLM. We checked in a stepwise manner how dipole relaxation times of phospholipid head groups changed upon the modification of lipids and membranes. The results obtained are summarized in Table 4. In this experiment current relaxation curves were averaged and the standard deviation taken as the experimental uncertainty. Native BLM formed from crude ox brain extract exhibited one relaxation time of 5 ± 1 µs. Additional relaxation components (115 ± 27 and 26 ± 1 µs) appeared in BLM modified by biotin.

Table 4. Capacitance relaxation components of BLM of different compositions and following the absorption of avidin–GOX complex from both sides of the membrane at thea final concentration of 30 mM:. 1. BLM from crude ox brain extract (COB);. 2. B-BLM: BLM from biotinylated COB. 3. B-BLM+A-GOX: BLM from biotinylated COB modified by avidin–GOX complex

System	τ_1, µs	τ_2, µs	τ_3, µs
BLM	5.3±1.0	–	–
B-BLM	5.0±0.3	26.1±5.0	115±27
B-BLM+A-GOX	16.5±1.0	26.5±1.0	505±16

Addition of the A-GOX complex to the electrolyte (final concentration 30 nM) on both sides of the biotinylated BLM resulted in the appearance of a slow component (505±16 µs). The appearance of this slow component presumably represents a collective motion of coupled dipole moments and reflects clustering in the membrane, induced by binding of the A-GOX complex.

Considerable changes of relaxation times have also been observed when short oligonucleotides modified by palmitic acid were incorporated into the BLM and sBLM, as well as during hybridization with complementary oligonucleotide chains at the membrane surface [30].

Thus, the electrostriction and dielectric relaxation methods can be very useful for the study of the physical properties of solid-supported lipid bilayers, and to observe and study the binding of various macromolecules on the membrane surface. In addition, these methods can be used for the direct detection of various substances by sBLM-based biosensors.

Acknowledgements: This work was financially supported by the NATO SfP Program (Project No. SfP–978003), by INTAS (Project 01-0224), and by the Slovak Grant Agency (Project No. 1/1015/04).

References

1. Mueller P, Rudin DO, Tien HT, Wescott WC (1962) Nature 194:979
2. Del Castillo J, Rodriguez A, Romero CA, Sanchez V (1966) Science 153:185
3. McConnell HM, Watts TH, Weis RM, Brian AA (1988) Biochim Biophys Acta 864:95
4. Tien HT, Salamon Z (1989) Bioelectrochem Bioenerg 22:211
5. Snejdarkova M, Rehak M, Otto M (1993) Anal Chem 65:665
6. Hianik T, Snejdarkova M, Passechnik VI, Rehak M, Babincova M (1996) Bioelectrochem Bionerg 41:221
7. Nikolelis DP, Hianik T, Krull UJ (1999) Electroanalysis 11:7
8. Hianik T (2000) Rev Mol Biotechnol 74:189
9. Florin EL, Gaub HE (1993) Biophys J 64:375
10. Mirsky VM, Mass M, Krause C, Wolfbeis OS (1998) Anal Chem 70:3674
11. Hianik T, Dlugopolsky J, Passechnik VI, Sargent DF, Ivanov SA (1996) Colloids Surf A 106:109
12. Uto M, Araki M, Taniguchi T, Hoschi S, Inoue S (1994) Anal Sci 10:943
13. Lu XD, Ottova-Leitmannova A, Tien HT, Bioelectrochem Bioenerg (1996) 39:285
14. Ziegler W, Gaburjakova J, Gaburjakova M, Sivak B, Rehacek V, Tvarozek V, Hianik T (1998) Colloids Surf A 140:357
15. Hianik T, Cervenanska Z, Krawczynsky vel Krawczyk T, Snejdarkova M (1998) Mater Sci Eng C 5:301
16. Guidelli R, Aloisi G, Becucci L, Dolfi A, Moncelli MR, Buoninsegni FT (2001) J Electroanal Chem 504:1
17. Knoll W, Frank CW, Heibel C, Naumann R, Offenhäuser A, Rühe J, Schmidt EK, Shen WW, Sinner A (2000) Rev Mol Biotechnol 74:137
18. Cornell BA, Braach-Maksvytis VLB, King L, Raguse BDJ, Wieczorek I, Pace RJ (1997) Nature 387:580
19. Rao NM, Plant AL, Silin V, Wight S, Hui SW (1997) Biophys J 73:3066
20. Hianik T, Snejdarkova M, Passechnik VI, Rehak M, Babincova M (1996) Bioelectrochem Bioenerg 41:221

21. Hianik T, Dlugopolsky J, Gyepessová M, Sivák B, Tien HT, Ottová-Leitmannová A (1996): Bioelectrochem Bioenerg 39:299
22. Sackmann E (1996) Science 271:43
23. Sleytr UB, Sara M (1997) Trends Biotechnol 15:20
24. Breitwieser A, Küpcü S, Howorka S, Weigert S, Langer C, Hoffmann-Sommergruber K, Scheiner O, Sleytr UB, Sara M (1996) BioTechniques 21:918
25. Seifert K, Fendler K, Bamberg E (1993) Biophys J 64:384
26. Hianik T, Krivanek R, Masar E, Dujsik J, Snejdarkova M, Rehak M, Stepanek I, Nikolelis DP (1998) Gen Physiol Biophys 17:239
27. Naumann R, Jonczyk A, Kopp R, Van Esch J, Ringsdorf H, Knoll W, Graber P (1995) Angew Chem Int Ed Engl 34:2056
28. Pracker O, Naumann R, Rühe J, Knoll W, Frank CW (1999) J Am Chem Soc 121:8766
29. Siontorou CG, Nikolelis DP, Piunno PAE, Krull UJ (1997) Electroanalysis 9:1067
30. Hianik T, Fajkus M, Sivak B, Rosenberg I, Kois P, Wang J (2000) Electroanalysis 12:11
31. Fajkus M, Hianik T (2002) Talanta 56:895
32. Nikolelis DP, Krull UJ (1993) Electroanalysis 5:539
33. Rehak M, Snejdarkova M, Otto M (1994) Biosens Bioelectron 9:337
34. Nikolelis DP, Siontorou CG (1995) Anal Chem 67:936
35. Nikolelis DP, Siontorou CG, Andreou VG, Krull UJ (1995) Electroanalysis 7:531
36. Nikolelis DP, Krull UJ (1992) Talanta 39:1045
37. Tvarozek V, Tien HT, Novotny I, Hianik T, Dlugopolsky J, Ziegler W, Leitmannová-Ottova A, Jakabovic J, Rehacek V, Uhlar M (1994) Sens Actuators B 19:597
38. Puu G, Gustafson I, Artursson E, Ohlsson PA (1995) Biosens Bioelectron 10:463
39. Lang H, Duschl C, Grätzel M, Vogel H (1992) Thin Solid Films 210–211:818
40. Hubbard JB, Silin V, Plant AL (1998) Biophys Chem 75:163
41. Spinke J, Yang J, Wolf H, Liley M, Ringsdorf H, Knoll W (1992) Biophys J 63:1667
42. Erdelen C, Haussling L, Naumann R, Ringsdorf H, Wolf H, Yang J, Liley M, Spinke J, Knoll W (1994) Langmuir 10:1246
43. Lang H, Duschl C, Vogel H (1994) Langmuir 10:197
44. Mirsky VM, Riepl M, Wolfbeis O (1997) Biosens Bioelectron 9–10:977
45. Snejdarkova M, Csaderova L, Rehak M, Hianik T (2000) Eletroanalysis 12:940
46. Novotny I, Rehacek V, Tvarozek V, Nikolelis DP, Andreou VG, Siontorou CG, Ziegler W (1997) Mater Sci Eng C 5:55
47. Ulman A (1991) An introduction to ultrathin organic films. From Langmuir–Blodgett to self assembly, 1st edn. Academic, San Diego
48. Hianik T, Vozár L (1985) Gen Physiol Biophys 4:331
49. Hianik T, Kavecansky J, Zorad S, Macho L (1988) Gen Physiol Biophys 7:191
50. Ohlsson PA, Tjarnhage T, Herbai E, Lofas S, Puu G (1995) Bioelectrochem Bioenerg 38:137
51. Dencher NA (1989) Methods Enzymol 171:265
52. Rivnay B, Bayer EA, Wilchek M (1987) Meth Enzymol 149:121
53. Wilchek M, Bayer EA (1990) Meth Enzymol 184:746
54. Noppl-Simson AA, Needham D (1996) Biophys J 70:1391
55. Pum D, Sleytr UB (1999) Trends Biotechnol 17:8
56. Hianik T, Passechnik VI (1995) Bilayer lipid membranes: structure and mechanical properties, 1st edn. Kluwer, Dordrecht
57. Passechnik VI, Hianik T (1997) Kolloid Zh 38:1180
58. Carius W (1976) J Colloid Interface Sci 57:301
59. Sargent DF (1975) J Membrane Biol 23:227
60. Böttcher CJF (1952) Theory of electric polarization, 1st edn. Elsevier, Amsterdam
61. Babakov AV, Ermishkin LM, Liberman EA (1966) Nature 210:953
62. White SH, Thompson TE (1973) Biochim Biophys Acta 323:7
63. Passechnik VI, Hianik T, Ivanov SA, Sivak B (1998) Electroanalysis 10:295
64. Haas H, Lamura G, Gliozzi A (2001) Bioelectrochemistry 54:1

65. Coster HGL, Simons R (1970) Biochim Biophys Acta 203:17
66. Hanai T, Haydon DA, Taylor JL (1964) Proc R Soc Lond A 281:377
67. Kruglyakov PM, Rovin YuG (1978) Physicochemistry of black hydrocarbon films, 1st edn. Nauka, Moscow
68. Hianik T, Passechnik VI, Sokolikova L, Snejdarkova M, Sivak B, Fajkus M, Ivanov SA, Franek M (1998) Bioelectrochem Bioenerg 47:47
69. Hianik T, Dlugopolsk J, Gyepessová M (1993) Bioelectrochem Bioenerg 31:99
70. Hianik T, Labajova A (2002) Bioelectrochemistry 58:97
71. Kuvichkin VV, Kuznetsova SM, Emeljanenko VI, Zhdanov RI, Petrov AI (1999) Biophys USSR 44:430
72. Spassova M, Tsoneva I, Petrov AG, Petkova JI, Neumann E (1994) Biophys Chem 52:267
73. Zuidam NJ, Barenholz Y (1998) Biochim Biophys Acta 1368:115
74. Blanc I, Chazalet MSP (2000) Biochim Biophys Acta 1464:309
75. Petrov AI, Khalil DN, Kazaryan RI, Sukhorukov BI (2002) Bioelectrochemistry 58:77
76. Kharakoz DP, Khusainova RS, Gorelov AV, Dawson KA (1999) FEBS Lett 446:27
77. Hui SW, Langner M, Zhao Y-L, Ross P, Hurley E, Chan K (1996) Biophys J 71:590
78. Zhdanov RI, Podobed OV, Vlasov VV (2002) Bioelectrochemistry 58:53
79. Schreirer S, Malheiros SVP, De Paula E (2000) Biochim Biophys Acta 1508:210
80. Sukhorukov BI, Montrel MM, Sukhorukov GB, Shabarchina LI (1994) Biophys USSR 39:273
81. Montrel MM, Sukhorukov GB, Petrov AI, Shabarchina LI, Sukhorukov BI (1997) Sens Actuators B 42:225
82. Sukhorukov BI, Petrov AI, Kazarian RL, Kuvichkin VV (2000) Biophys USSR 45:245
83. Siontorou CG, Brett AMO, Nikolelis DP (1996) Talanta 43:1137
84. Siontorou CG, Nikolelis DP, Piunno PA, Krull UJ (1997) Electroanalysis 9:14
85. Oretskaya TS, Romanova EA, Andreev SYu, Antsypovich SI, Toth C, Gajdos V, Hianik T (2002) Bioelectrochemistry 56:47
86. Hianik T, Fajkus M, Tomcik P, Rosenberg I, Kois P, Cirak J, Wang J (2001) Chem Monthly 132:141
87. Hianik T, Vitovic P, Humenik D, Andreev S-Yu, Oretskaya TS, Hall EAH, Vadgama P (2003) Bioelectrochemistry 59:35
88. Hianik T, Gajdos V, Krivanek R, Oretskaya T, Metelev V, Volkov E, Vadgama P (2001) Bioelectrochemistry 53:199
89. Yeagle PL (1985) Biochim Biophys Acta 822:267
90. Hianik T, Haburcak M (1993) Gen Physiol Biophys 12:283
91. May S, Harries D, Ben-Shaul A (2000) Biophys J 78:1681
92. Scatchard G (1949) Ann N Y Acad Sci 51:660
93. Sargent DF, Hianik T (1994) Bioelectrochem Bioenerg 33:11
94. Degani Y, Heller A (1989) J Am Chem Soc 111:2357
95. Hianik T, Snejdarkova M, Sokolikova L, Meszar E, Krivanek R, Tvarozek V, Novotny I, Wang J (1999) Sens Actuators B 57:201
96. Davis JH (1983) Biochim Biophys Acta 737:117
97. Laggner P. Kriechbaum M (1991) Chem Phys Lipids 57:121

Transport Proteins on Solid-Supported Membranes: From Basic Research to Drug Discovery

Klaus Fendler, Martin Klingenberg, Gerard Leblanc, Jan Joep H. H. M. DePont,
Bela Kelety, Wolfgang Dörner, Ernst Bamberg

Abstract. Solid-supported membranes (SSM) can be used as capacitive electrodes for the investigation of electrogenic transport proteins (ion pumps). Membrane fragments or liposomes that contain the transport protein are adsorbed on the SSM. The proteins are activated by supplying a substrate via a rapid solution exchange at the SSM. The charge translocations during the reaction cycle of the transport protein can be measured via the capacitance of the SSM. The SSM and the adsorbed membrane fragments or proteoliposomes represent a rugged structure that can be used in basic research and drug discovery. On the basis of a few examples we demonstrate how this technique can be applied to the investigation of the transport mechanism of membrane proteins. In addition, we present the application of the system in drug discovery. A sensor based on the SSM technology represents a promising system for the rapid screening of pharmaceutically relevant compounds for transport proteins.

Keywords: Solid-supported membrane, Electrogenic, Membrane protein, Drug discovery, Drug screening,

Abbreviations

SSM   Solid-supported membrane
AAC   ATP/ADP exchanger
CAT   Carboxyatractyloside
BKA   Bongkrekic acid
DTT   Dithiothreitole
NEM   N-ethyl maleimide

1
Introduction

Cells communicate with the surrounding medium using an apparatus of highly selective and effective proteins located in the cellular membrane. Apart from receptors and ion channels, transporters are among the most important membrane proteins. They transport molecules of all imaginable sizes and properties across the cell membrane, ranging from hydrogen atoms to macromolecules like peptides and even proteins. Actually, 20% of the energy generated in our body is devoted to these processes. Transporters are involved in the uptake of nutrients, the buildup of membrane potential, the generation of muscle tension and so on.

For many years scientists have been interested in the molecular mechanism which renders the transport processes so effective and specific. However, only in recent years has the interplay of structural information and functional investigation allowed light to be shed on these questions. It turns out that these proteins have developed quite different strategies to accomplish their transport functions.

Among the many techniques used for the investigation of transport processes electrical techniques play a central role. They are sensitive and allow a high time resolution. The latter is important because partial reactions in these proteins occur on the timescale of microseconds up to several hundred milliseconds. At the same time, the charge translocation in a single protein is very small. In contrast to ion channels, where up to 10^6 ions per second cross the pore, transport in transporters occurs typically with a rate of ca. 10 to 100 s^{-1}. Therefore, the collective charge movement of more than 10^5 molecules has to be observed to obtain a detectable electrical current. This leads to the problem of synchronization of all these molecules.

Special techniques had to be developed to observe the tiny charge movements of the synchronized transporters. One of them takes advantage of the high specific capacitance of planar lipid bilayers [1]. Rugged planar lipid bilayers can be prepared on solid supports [2]. We call this a solid-supported membrane or SSM. Membrane fragments containing the transporter or liposomes in which the transporter has been incorporated (proteoliposomes) are adsorbed on the SSM and the currents generated by the transporter activity are recorded via capacitive coupling. The compound membrane formed by the adsorbed membrane fragments or proteoliposomes and the SSM can withstand high flow velocities, allowing for a fast solution exchange at the surface. By rapidly changing from a solution containing no substrate for the transporter to one that contains a substrate, the transporter can be activated in a synchronized manner. A transient current can be recorded that contains information about the temporal development of the transport process [3].

While this technique is interesting for the investigation of basic questions pertaining to the transport mechanism of transport proteins it also contains potential for industrial applications. The SSM sensor is rugged and inexpensive and can be used for many solution-exchange processes. The measurement is rapid and automation and computer control are straightforward. The SSM has already been used in basic research for the investigation of several transport proteins belonging to different families. Therefore it promises to become a platform technology for drug development as described below.

2
Rapid Solution-Exchange Technique for the Investigation of Electrogenic Ion Pumps

2.1
Sensor Preparation

The SSM was prepared by linking an alkanethiol (octadecyl mercaptan) monolayer to a gold electrode deposited on a glass support and covering it with a lipid (diphytanoyl phosphatidylcholine) monolayer (Fig. 1). The planar membrane formed has an area of ~1 mm^2. Membrane fragments or proteoliposomes containing the protein of interest adsorb on the SSM as shown in Fig. 1. The compound membrane composed of the SSM and the adsorbed proteoliposomes/membrane fragments form a capacitively coupled system that allows measurement of charge movements in the proteoliposomes/membrane fragments [1] via the gold electrode and a reference Ag/AgCl electrode in the solution.

In the setup the SSM is mounted in a flow-through cuvette with an inner volume of 17 μl (Fig. 2). The gold electrode is connected to an amplifier, and the

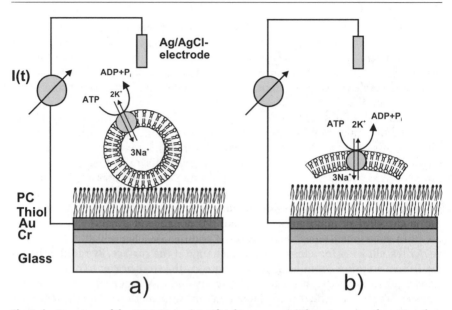

Fig. 1a,b. Structure of the SSM. It consists of a glass support (Glass, 1 mm), a chromium layer (Cr, 5 nm), a gold layer (Au, 150 nm), an octadecyl mercaptan monolayer (Thiol), a diphytanoyl phosphatidylcholine monolayer (PC), and the liposomes (a) or membrane fragments (b) containing the protein (Na⁺/K⁺-ATPase)

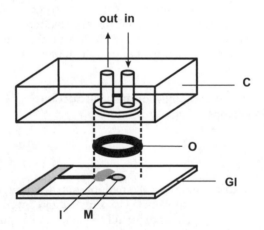

Fig. 2. Cuvette for rapid solution exchange (not to scale). The letters in the figure denote the SSM glass plate (Gl), the SSM membrane area (M), the electrical insulation (I), the o-ring (O, inner diam. 4.6 mm, height 0.9 mm), and the cuvette (C). The diameter of the solution inlet and outlet (in, out) is 1 mm

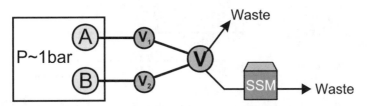

Fig. 3. Experimental setup. Activating (A) and nonactivating (B) solutions are in pressurized containers. Primary valves (V1 and V2) and the main valve (V) are switched on/off according to the protocol given in Fig. 4

reference Ag/AgCl electrode is connected to the solution in the outward channel by a salt bridge. The solution is driven through the cuvette by applying pressure (0.6–1 bar) to the solution containers connected to the cuvette (Fig. 3). Three valves control the solution flow. Opening of the primary valves V_1 and V_2 starts the experiment. The main valve (V) directs the activating solution (A) into the cuvette and the nonactivating solution (B) into the waste container (on state) or vice versa (off state), ensuring a permanent flow in both pathways. A permanent solution flow reduces artifacts due to the solution exchange. A typical flow protocol is given in Fig. 4. The setup described here is an improved version of the one presented in [3] where technical details may be found.

2.2
Measuring Procedure

The experiments were carried out at room temperature (22 °C). After the formation of the SSM its capacitance and conductance were measured until they became constant after a waiting time of ~90 min. Typical values were 300–500 nF/cm^2 for the capacitance and 50–100 nS/cm^2 for the conductance. The proteoliposomes or membrane fragments were thawed and sonicated for ~30 s. Then, 40 μl of the suspension containing 0.37–1.4 mg/ml protein were injected into the cuvette. They were allowed to adsorb on the SSM for 30–50 min. A typical solution-exchange protocol consists of three phases: (1) nonactivating solution (2 s), (2) activating solution (2 s), and (3) nonactivating solution (2 s). The data recording and the solution exchange were controlled via computer as described in [3]. Electrical signals were observed at the concentration jumps taking place at the beginning (on-signal; see, e.g., Fig. 4) and, depending on the properties of the enzyme, also at the end of phase 2 (off-signal; see, e.g., Fig. 7).

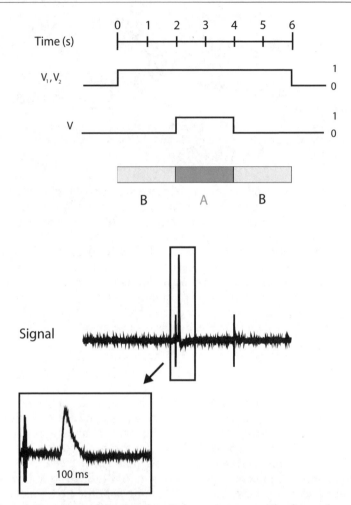

Fig. 4. Typical time protocol for the electrical measurements. The figure shows the status of the valves (0=off, 1=on) and the solution present in the cuvette (A=activating, B=nonactivating) for an ATP concentration jump for activation of Na+/K+-ATPase. In the lower panel the corresponding current signal is displayed. In the time trace the switching of the valve creates artifactual currents at t=2.0 and 4.0 s, respectively. The current generated by the Na+/K+-ATPase starts ca. 100 ms after the main valve (V) has been switched on

3
ATPases

3.1
Na$^+$/K$^+$-ATPase

The Na$^+$/K$^+$-ATPase has important physiological functions in almost all membranes of animal cells. In the physiological pump mode, hydrolysis of one molecule of ATP results in the transport of three Na$^+$ ions out of the cell, and of two K$^+$ ions into the cytoplasm (for a review see [4]). During this cycle, one positive elementary charge is translocated from the cytoplasmic to the extracellular side of the membrane. It has been known for many years that during its transport cycle the enzyme assumes two different conformational states, E$_1$ and E$_2$. In the E$_1$ state the enzyme has a high affinity for Na$^+$ at the cytoplasmic surface so that Na$^+$ can be bound at the cytoplasmic cation binding site. After binding and hydrolysis of ATP the Na$^+$/K$^+$-ATPase is converted to the phosphorylated E$_2$P state, where the cation binding region is accessible to the extracellular side. In the E$_2$P conformational state Na$^+$ is released, the pump is dephosphorylated, and, due to the high affinity for K$^+$ ions on the extracellular surface, K$^+$ can be bound and translocated to the cytoplasm. Then the enzyme is converted into its E$_1$ form again. In the absence of K$^+$ dephosphorylation is slow and the enzyme can be characterized by the simplified reaction scheme:

$$E_1 + Na_c \longrightarrow E_1Na \xrightarrow{\text{ATP}} E_1PNa \longrightarrow E_2P + Na_e \qquad \text{(a)}$$

Here Na$_c$ and Na$_e$ are Na$^+$ ions present in the cytoplasmic and the extracellular medium, respectively. To understand the transport mechanism of the enzyme it is important to know which of the reaction steps are electrogenic. This information can be obtained by electrical measurements at the SSM.

Figure 5 shows the electrical current obtained after an ATP concentration jump in the presence of Na$^+$. At t=0 the valve is switched from the nonactivating to the activating solution. At t~130 ms the activating solution reaches the surface of the SSM. This delay is caused by the mechanical delay of the valve and the transit time of the solution from the valve to the SSM. The signal rises with a time constant of ~10 ms and decays with a time constant of ~19 ms. A component with a negative amplitude (time constant ~113 ms) is also observed. This component is due to the capacitive coupling of the liposomes to the electrode via the SSM [5] and is not related to the protein activity. Under the conditions of the experiment Na$^+$ is already bound to the protein when ATP is added. Therefore, the transient current represents the Na$^+$ translocation reaction E$_1$PNa$\rightarrow$ E$_2$P+Na$_e$.

An important aspect is the electrogenicity of cation binding. This problem could be studied using a Na$^+$ concentration jump. Experiments were done in the absence and presence of ATP. Electrical currents were measured after rapid addition of Na$^+$ in the presence and absence of ATP (Fig. 6). An electrogenic reaction of the protein was observed only with Na$^+$, but not with other monovalent cations (K$^+$, Li$^+$). These currents can be explained only by a cation binding proc-

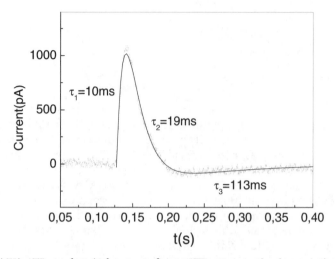

Fig. 5. Na⁺/K⁺-ATPase: electrical current after an ATP concentration jump. Activating solution: 100 µM ATP, 200 mM NaCl, 25 mM imidazole, pH 7.0 (HCl), 3 mM MgCl₂, and 0.2 mM DTT. Nonactivating solution: same composition but no ATP

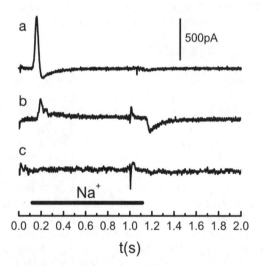

Fig. 6a–c. Na⁺/K⁺-ATPase: Na⁺ concentration jump (10 mM, addition at t=0 s and removal at t=1 s) under different conditions. **a** In the presence of 100 µM ATP electrogenic Na⁺ transport by the NaK-ATPase is activated. **b** The same procedure as in **a**, but in the absence of ATP. After addition of Na⁺ charge translocation takes place. **c** The same experiment as in **b** was done after incubation with the specific inhibitor of P-type ATPases, orthovanadate. This procedure results in complete inhibition of the electrogenic process. In addition all solutions contained 25 mM imidazole, pH 7.0 (HCl), 3 mM MgCl₂, and 0.2 mM DTT

ess on the cytoplasmic side. In fact, this is a direct demonstration of the electrogenicity of the cytoplasmic cation binding step of the Na^+/K^+-ATPase. The data shown in the figure imply that ~30% of the charge transported during the $E_1PNa_c{\rightarrow}E_2PNa_e$ reaction is translocated during cytoplasmic Na^+ binding.

As demonstrated for the Na^+/K^+-ATPase from pig kidney, the SSM represents a convenient tool to study the transport mechanism of ATPases. Electrogenic partial reactions that have not been accessible to other techniques have been investigated and valuable information about the transport mechanism may be obtained [3, 5]. Other ATPases have also been investigated using a fast solution exchange at the SSM, such as the Na^+ transporting F_0F_1-ATPase of *Ilyobacter tartaricus* and the K^+ transporting Kdp-ATPase of *Escherichia coli* [unpublished data].

4
Carriers

4.1
ATP/ADP Exchanger

The ADP/ATP exchanger (also termed ADP/ATP carrier, AAC) is the most abundantly occurring transporter of the inner mitochondrial membrane and is responsible for the membrane-potential-driven exchange of ATP versus ADP between the matrix space and the cytosol [6]. This process is the last step of oxidative phosphorylation. The purified carrier reconstituted into liposomes represents an excellent model system for the investigation of the kinetic properties of the carrier. In particular, rapid charge movements during ATP and ADP transport could be detected and assigned to distinct partial reactions of the reaction cycle [7].

The currents induced by concentration jumps of 100 µM ADP and 100 µM ATP are shown in Fig. 7. The fluid flow with the nucleotide-containing solution was started at t=6 s and stopped at t=8 s causing a short artifact due to the switching of the electrical valve. After a delay time of approximately 100 ms, which is the transit time of the fluid from the electrical valve to the cuvette, an on- as well as an off-signal could be detected. In the case of ATP the positive on-signal corresponds to the transport of negative charge into the liposomes and the negative off-signal to the transport of negative charge out of the liposomes. In the case of ADP the currents are reversed. The on-signal corresponds to the transport of negative charge out of the liposomes and the off-signal to the transport of negative charge into the liposomes. Figure 7b shows the inhibition of the ADP-induced current by 1 µM carboxyatractyloside (CAT) and 1 µM bongkrekic acid (BKA). CAT and BKA act as specific inhibitors and bind to the cytosolic and matrix side of the AAC, respectively [8]. Similar results were obtained for ATP-induced currents. Mg^{2+} forms a complex with ATP and ADP that cannot be transported [6]. This is demonstrated for ATP in Fig. 7d. After addition of 4 mM $MgCl_2$ to the activating solution no electrical current could be detected.

We observed not only charge translocation induced by an ADP and ATP concentration jump but also a charge movement in the opposite direction on removal of the nucleotides. The transported charges for the on- and off-signals

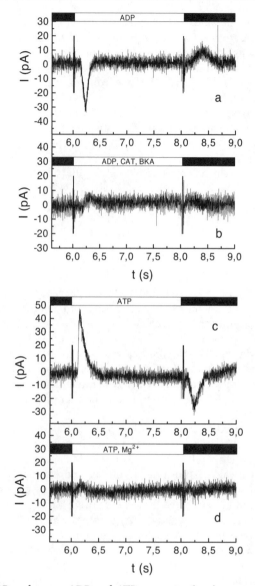

Fig. 7a–d. ATP/ADP exchanger: ADP and ATP concentration jump and inhibition of the AAC transport activity. **a** Activating solution: 100 µM ADP. **b** Activating solution: 100 µM ADP, 1 µM CAT, 1 µM BKA. **c** Activating solution: 100 µM ATP. **d** Activating solution: 100 µM ATP, 4 mM MgCl$_2$; nonactivating solution: 4 mM MgCl$_2$. In addition all solutions contained 100 mM NaCl, 20 mM MES, pH 6.2

were calculated by integrating the signals and are approximately the same. At least 80–90% of the charge transported into the liposomes on activation with the nucleotides is transported back out of the lipsosomes on removal of the nucleotides. Comparison of the transported charge in the on-signal of ATP transport and ADP transport in the same experiment yielded a ratio of approximately 2:1. This implies that by translocation of one ATP molecule twice as much charge is moved compared to the translocation of one ADP molecule. Taking into account the charge of the substrates this finding suggests that an equivalent of 3.3 countercharges resides in the nucleotide binding site and is cotransported with the nucleotides.

4.2
Melibiose Permease

The melibiose permease (MelB) of *Escherichia coli* is a membrane-bound ion-coupled sugar cotransporter or symporter that uses the favorable Na^+, Li^+, or H^+ electrochemical potential gradient to drive cell accumulation of α-galactosides (melibiose, raffinose) or β-galactosides (TMG) [9, 10].

A recombinant transporter harboring a 6-His tag (Mel-6His permease) could be purified in large amounts and it exhibits cation-dependent sugar-binding and transport properties comparable to those of the native permease in its natural environment when reconstituted in liposomes [11, 12]. Information at the molecular level has been obtained using molecular biology, and biochemical and spectroscopic approaches. Biochemical and spectroscopic evidence supports the earlier suggestion that some discrete steps of MelB cycling involve conformational transitions of the transporter [13].

Electrophysiological techniques have proved to be extremely useful tools to investigate the mechanism of ion transfer across the membrane by ion-coupled transporters in eukaryotic cells [14–16]. Partial reactions that are associated with the translocation of electrical charge can be identified, which can then be interpreted in terms of, e.g., an electrogenic binding process, the translocation of ions, or a conformational transition. Although it is admitted that the bacterial counterparts of eukaryotic transporters are electrogenic, the limited size of bacterial cells or derived membrane vesicles has hampered their analysis by the electrical approach. However, the progress in purification and reconstitution of MelB in liposomes opens new perspectives. Indeed, proteoliposomes adsorbed on a SSM can be used for the investigation of these bacterial proteins.

Transient currents were selectively recorded by applying concentration jumps of Na^+ ions (or Li^+) and/or of different sugar substrates of MelB (melibiose, thiomethyl galactoside, raffinose) [17]. Characteristically, the transient current response was fast including a single decay exponential component ($\tau\sim15$ ms) on applying a Na^+ (or Li^+) concentration jump in the absence of sugar. However, in the presence of sugar the electrical transients were biphasic and comprised both the fast and an additional slow ($\tau\sim350$ ms) decay component. Selective inactivation of the cosubstrate translocation step by acylation of MelB cysteines with N-ethyl maleimide (NEM) suppressed the slow response components and had no effect on the fast transient one (Fig. 8). We suggest that the fast tran-

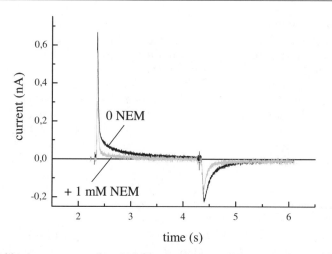

Fig. 8. Melibiose permease: electrical signals after a simultaneous 20 mM melibiose and 10 mM NaCl concentration jump before (0 NEM) and after (1 mM NEM) addition of 1 mM NEM. In addition to the sugar and/or cation the activating and the nonactivating solutions contained 100 mM KP_i at pH 7.0. All solutions were DTT-free

sient response reflects charge transfer within MelB during cosubstrate binding, while the slow component is associated with charge transfer across the proteoliposome membrane. From the time course of the transient currents we estimate a rate constant for Na^+ binding in the absence and presence of melibiose of $k > 50$ s^{-1} and for melibiose binding in the absence of Na^+ of $k \sim 10$ s^{-1}.

Cooperativity of the MelB carrier is demonstrated in Fig. 9. As a first approximation, the peak current is proportional to the activity of the carrier. We have, therefore, determined the peak current as a function of the magnitude of the concentration jump. The figure depicts the variation of the peak current recorded (I_p) on imposing melibiose concentration jumps of varying amplitude. Both in the presence or the absence of Na^+, the concentration dependence shown in the figure was fitted using a hyperbolic function ($I_p = I_p^{max} c / (c + K_{0.5}^{mel})$; c=melibiose concentration). The fit of the data obtained in the presence of 10 mM Na^+ yielded a half-saturation concentration for melibiose of $K_{0.5}^{mel} = 3$ mM. This $K_{0.5}^{mel}$ value estimated from the electrical measurement satisfactorily compares with the apparent affinity of MelB for melibiose determined in binding experiments (0.5–0.9 mM [18, 19]) or with the concentration of sugar inducing half maximal intrinsic fluorescence variation of MelB (1.2 mM [12]). In Na^+-free media the data are best fitted by a hyperbolic function using a half-saturation concentration for melibiose of $K_{0.5}^{mel} = 22$ mM. This higher $K_{0.5}^{mel}$ value parallels the increased K_m for transport of 10 mM [18], or the $K_{0.5}^{mel}$ of 11 mM in fluorescence studies in corresponding Na^+-free media [12].

In conclusion, a fast solution exchange at the SSM was shown to be a useful tool for the investigation of carriers of eukaryotic (AAC) as well as of prokaryotic (MelB) origin [20]. Especially, the investigation of the electrical properties of bacterial membrane proteins represents a challenge for the investigator be-

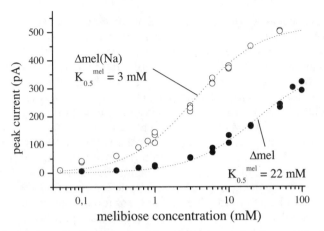

Fig. 9. Melibiose permease: dependence of the peak currents on the sugar concentration. In addition to the sugar and/or cation given below, the activating and the nonactivating solutions contained 100 mM KP_i and 0.1 mM DTT at pH 7.0. Δmel (*filled circles*): activating solution: melibiose at different concentrations; nonactivating solution: glucose at the same concentration as melibiose. The *dashed line* represents a fit with a hyperbolic function; half-saturation concentration $K_{0.5}^{mel}$=22 mM. Δmel(Na) (*open circles*): activating solution: melibiose at different concentrations, 10 mM NaCl; nonactivating solution: glucose at the same concentration as melibiose; 10 mM NaCl. Half-saturation concentration $K^{0.5}_{mel}$=3 mM

cause most of them are not accessible by patch clamp and voltage clamp techniques. Here the SSM provides a unique possibility to obtain information about the transport process.

5
Application: The SSM as a Tool for Drug Discovery

Once the usefulness and applicability of the SSM had been shown in a number of scientific investigations, its value as a tool for drug discovery with transporters and ion pumps was recognized. Starting from the laboratory setup discussed in the previous section, a screening platform which meets the demands of modern pharmaceutical drug research can be established. In the following paragraphs we will first outline the relevance of the targets at which this technology is aimed and discuss methods currently used for their screening. Second, we will give a brief report of the current status of development and will finally turn to the future perspectives.

5.1
Electrogenic Proteins: Pharmaceutical Relevance and Screening

Transporters, ion channels, and ion pumps constitute a group of electrogenic membrane proteins which play important roles in signal transduction, information processing, and metabolism. Ion channels exhibit high transport rates. In consequence, they are accessible by various screening techniques. They have been successfully subjected to pharmaceutical drug research for many years, whereas transport proteins with their significantly lower transport rates are much more difficult to handle. Yet many transporters are possible targets for new drug developments. Some of the most profitable drugs are in fact inhibitors of transport proteins. For example, the gastric proton pump (H^+/K^+-ATPase) from stomach mucosa is a drug target for the treatment of ulcers and gastric reflux. Many of the so-called proton pump inhibitors belong to the group of substituted pyridinemethylbenzimidazoles. Omeprazole (Astra), lansoprazole (Takeda), pantoprazole (Byk Gulden), and pariprazole (Eisai) are the corresponding drugs available on the market today. Ouabain and digoxin exert their effect on the cardiac muscle cell Na^+/K^+-ATPase in patients with congestive heart failure. Fluoxetine (Prozac) and imipramine, inhibitors of the Na^+-coupled serotonin transporter (SERT), are used as antidepressants, as is desipramine which inhibits the reuptake of the neurotransmitter by the Na^+-coupled norepinephrine transporter (NET). It can be anticipated that the progress made in deciphering the human genome will bring about even more new targets in the near future.

5.2
Assays for the Screening of Electrogenic Membrane Proteins

Membrane proteins and transporters in particular are difficult targets with regard to the appropriate screening technology. There exist, however, a number of techniques which have been used in the past for this purpose. We will discuss in the following the existing methods and their advantages and drawbacks. From this it will become clear that the existing technologies are not satisfactory, in particular with respect to high-throughput screening of transporters.

5.2.1
Conventional Assay Technologies

For survival assays organisms are manipulated such that they die either in case a certain transport process occurs or in case the respective transport process does not occur. Test substances are then applied during growth and the growth itself is monitored. These tests are very time-consuming and they are prone to interferences with other biochemical effects of substances on growth, leading to a high number of false positive or false negative hits.

The use of radioactively labeled substrates or ligands enables the measurement of substrate transport with an extremely high sensitivity. Target proteins are usually expressed in cultured cells which can either be used directly for this assay type or from which the protein can be purified for functional reconstitution in liposomes. Cells or proteoliposomes are then incubated with the labeled substance and the uptake or secretion is monitored. Despite its indisputably high sensitivity, matters of environmental safety and cost render this type of assay most difficult, leaving it as the ultimate option for cases in which other methods are not available or unsatisfactory.

A number of dyes are known for their ability to alter their spectral properties in response to changes in ionic strength, pH or membrane potential. These dyes are often used for fluorescence assays. Fluorescence readers can deal with huge amounts of samples, and they are well suited for high-throughput screening. However, many fluorescence-based assays lack sensitivity or suffer from interfering fluorescence of the test substances.

5.2.2
Cell-Based Electrophysiological Methods

A huge number of transporter substrates are electrically charged. They cause an electrical current when crossing a lipid membrane. Time-resolved current measurements can thus be used to analyze the underlying protein activity. The patch-clamp technique has set the standards in this field. However, patch-clamp measurements are very time-consuming and can only be carried out by skilled and well-trained personnel. Their information content is very high, the possible throughput very low. Several companies are currently developing automated patch clamp systems to enable their use in high-throughput screening.

Operated in its whole cell mode a patch-clamp setup can be used to investigate all protein molecules of a single cell at once. Using this technique, the investigation of ion channels is straightforward due to their high turnover. However, the investigation of transporters with their inherent low turnover represents a much more difficult task. A widespread use of automated patch-clamp systems for the screening of transporters is at present rather unlikely.

5.2.3
Sensor-Based Methods

Several new approaches in preclinical drug research are based on the use of functionalized sensor surfaces. These techniques can be divided into three categories:
1. Sensors which specifically bind molecules from the surrounding solution at their surface, resulting in a change in the sensor's optical [21] or electrical [22] properties. Some of these systems have been commercially available for some years. Proteins or their substrates either have to be immobilized or labeled for their measurement.
2. Sensors which enable the detection of enzymatic activity without the need for labels or labeled substrates. They are composed of a surface carrying a lipid membrane which can either incorporate proteins ("tethered bilayers", [23–26]) or offer membrane fragments a surface to adsorb to (see previous section).
3. Sensors suitable for growing cells on [27, 28]. With a few exceptions that are not suited for the investigation of transporters, these methods are still under development.

5.3
An Integrated System for Drug Discovery

One step on the way to a SSM-based screening system for routine use in pharmaceutical research is the integration of all necessary components into a user-friendly and functionally optimized housing. The first integrated system that was built is shown in Fig. 10. The microfluidic flow cell (see Fig. 11) forms the core of a single channel system. It resides in a Faraday cage which shields the

Fig. 10. Integrated system for drug discovery consisting of the workstation (*right side*) controlled by a computer. The lid of the Faraday cage that contains valves, cuvette, and sensor is open

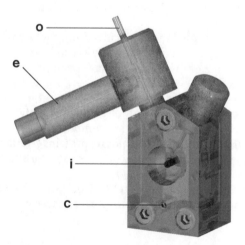

Fig. 11. 3D CAD model of the flow cell. It has been prototyped in PMMA. Inlet (i), outlet (o), salt bridge and reference electrode (e), contact to sensor (c)

sensor as well as the buffer reservoirs, valves, and tubes from electrical interference. The Faraday cage is integrated in the outer housing with the current-to-voltage converter fixed next to it. The latter is needed to translate very small currents of 0.01 to 10 nA into voltage signals in the millivolt range. All low-voltage circuit boards are mounted on the inside walls of the housing, whereas the power supplies and the pressure controller are placed in a separate housing to reduce line voltage interference. A software package has been developed concurrently. It allows to alter the flow rates by controlling the operating pressure and to switch all valves in a user-defined manner. Furthermore, it contains functions for data acquisition, data manipulation, and file handling.

5.3.1
The Gastric Proton Pump Revisited

As mentioned above, the H^+/K^+-ATPase is a validated target for acid-related diseases. Much of the early work, which ultimately led to the development of omeprazole and other extremely specific drugs, was done on animal models. These experiments are expensive and time-consuming. Progress was made when isolated vesicles were used and assayed for phosphate release in the absence and presence of inhibitors. However, real-time monitoring was not possible at this stage since phosphate determination requires the addition of further reagents which are not compatible with the enzyme activity.

The catalytic cycle of the H^+/K^+-ATPase is electroneutral. However, proton transport in the absence of K^+ is electrogenic [29], giving rise to a detectable current. To demonstrate the potential of the SSM-based screening system, the current response to an ATP concentration jump was recorded using membrane fragments with porcine H^+/K^+-ATPase (upper trace in Fig. 12). The preparation of the sensor and the measurement procedure are similar to the ones described in the previous section. Incubation with 10 µM of activated omeprazole leads to

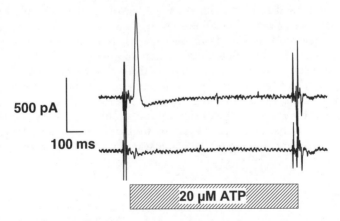

Fig. 12. Current response of a H^+/K^+-ATPase preparation after an ATP concentration jump (20 µM) before (*upper trace*) and after incubation with 10 µM omeprazole (*lower trace*). In addition all solutions contained 25 mM imidazole, pH 6.1 (HCl), and 0.2 mM MgCl$_2$

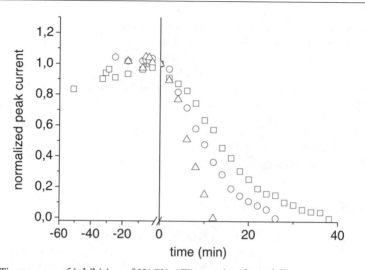

Fig. 13. Time course of inhibition of H^+/K^+-ATPase using three different omeprazole concentrations (*square*=1 µM, *circle*=5 µM, *triangle*=10 µM). The inhibitor was present in both the activating and the nonactivating solution. All other conditions as mentioned in Fig. 12

a complete inhibition of the response (lower trace in Fig. 12). The time dependence of the inhibition was analyzed for three omeprazole concentrations. The results are shown in Fig. 13. This experiment demonstrates monitoring of inhibition with the sensor.

5.4
Outlook

Once an assay has been established on the SSM-based screening system, this technology in its present state allows to carry out several hundred rapid solution-exchange experiments a day in the case where no irreversibly binding inhibitors are used. Yet the manual replacement of the sensor chip after irreversible inhibition is the bottleneck limiting the throughput. New sensor layouts are under development for the single channel system. They will facilitate the work with pharmaceutical test substances by reducing the number of manual operations. The next version of the device will be a multichannel system with disposable sensor units. This state-of-the-art tool will considerably speed up the screening procedure, making preclinical research with transport proteins much more efficient.

References

1. Bamberg E, Butt HJ, Eisenrauch A, Fendler K (1993) Q Rev Biophys 26:1
2. Seifert K, Fendler K, Bamberg E (1993) Biophys J 64:384
3. Pintschovius J, Fendler K (1999) Biophys J 76:814

4. Läuger P (1991) Electrogenic ion pumps. Sinauer, New York
5. Pintschovius J, Fendler K, Bamberg E (1999) Biophys J 76:827
6. Klingenberg M (1980) J Membr Biol 56:97
7. Gropp T, Brustovetsky N, Klingenberg M, Muller V, Fendler K, Bamberg E (1999) Biophys J 77:714
8. Brustovetzky N, Becker A, Klingenberg M, Bamberg E (1996) Proc Natl Acad Sci USA 93:664
9. Leblanc G, Bassilana M, Pourcher T (1988) In: Palmieri F, Quagliariello E (eds) Molecular basis of biomembrane transport. Elsevier, Amsterdam, p 53
10. Tsuchiya T, Raven J, Wilson TH (1977) Biochem Biophys Res Commun 76:26
11. Pourcher T, Leclercq S, Brandolin G, Leblanc G (1995) Biochemistry 34:4412
12. Mus-Veteau I, Pourcher T, Leblanc G (1995) Biochemistry 34:6775
13. Leblanc G, Pourcher T, Bassilana M (1989) Biochimie 71:969
14. Kappl M, Hartung K (1996) Biophys J 71:2473
15. Hazama A, Loo DDF, Wright EM (1997) J Membr Biol 155:175
16. Mager S, Naeve J, Quick M, Labarca C, Davidson N, Lester HA (1993) Neuron 10:177
17. Ganea C, Pourcher T, Leblanc G, Fendler K (2001) Biochemistry 40:13744
18. Pourcher T, Bassilana M, Sarkar HK, Kaback HR, Leblanc G (1990) Philos Trans R Soc Lond B Biol Sci 326:411
19. Zani ML, Pourcher T, Leblanc G (1994) J Biol Chem 269:24883
20. Jung H (2001) Biochim Biophys Acta 1505:131
21. Asanov AN, Wilson WW, Oldham PB (1998) Anal Chem 70:1156
22. Cornell BA, Braach-Maksvytis VLB, King LG, Osman PDJ, Raguse B, Wieczorek L, Pace RJ (1997) Nature 387:580
23. Lindholm-Sethson B, Gonzales JC, Puu G (1998) Langmuir 14:6705
24. Heyse S, Stora T, Schmid E, Lakey JH, Vogel H (1998) Biochim Biophys Acta 1376:319
25. Michalke A, Galla HJ, Steinem C (2001) Eur Biophys J 30:421
26. Naumann R, Baumgart T, Gräber P, Jonczyk A, Offenhäuser A, Knoll W (2002) Biosens Bioelectron 17:25
27. Fromherz P, Offenhäuser A, Vetter T, Weis J (1991) Science 252:1290
28. Offenhäuser A, Knoll W (2001) Trends Biotechnol 19:62
29. van der Hijden HT, Grell E, de Pont JJ, Bamberg E (1990) J Membr Biol 114:245

Subject Index